S0-BVO-546

Parsimony, Phylogeny, and Genomics

Parsimony, Phylogeny, and Genomics

EDITED BY

Victor A. Albert
University of Oslo, Norway

OXFORD
UNIVERSITY PRESS

OXFORD
UNIVERSITY PRESS

Great Clarendon Street, Oxford OX2 6DP

Oxford University Press is a department of the University of Oxford.
It furthers the University's objective of excellence in research, scholarship,
and education by publishing worldwide in

Oxford New York

Auckland Cape Town Dar es Salaam Hong Kong Karachi
Kuala Lumpur Madrid Melbourne Mexico City Nairobi
New Delhi Shanghai Taipei Toronto

With offices in

Argentina Austria Brazil Chile Czech Republic France Greece
Guatemala Hungary Italy Japan Poland Portugal Singapore
South Korea Switzerland Thailand Turkey Ukraine Vietnam

Oxford is a registered trade mark of Oxford University Press
in the UK and in certain other countries

Published in the United States
by Oxford University Press Inc., New York

© Oxford University Press 2005

The moral rights of the authors have been asserted
Database right Oxford University Press (maker)

First published 2005

All rights reserved. No part of this publication may be reproduced,
stored in a retrieval system, or transmitted, in any form or by any means,
without the prior permission in writing of Oxford University Press,
or as expressly permitted by law, or under terms agreed with the appropriate
reprographics rights organization. Enquiries concerning reproduction
outside the scope of the above should be sent to the Rights Department,
Oxford University Press, at the address above

You must not circulate this book in any other binding or cover
and you must impose this same condition on any acquirer

British Library Cataloging in Publication Data
(Data available)

Library of Congress Cataloging-in-Publication Data

Parsimony, phylogeny, and genomics / edited by Victor A. Albert.
    p. cm.
  Includes bibliographical references and index.
  ISBN 0-19-856493-7 (alk. paper)
  1. Cladistic analysis. 2. Genetics. 3. Phylogeny. I. Albert, Victor A. (Victor
Anthony), 1964-
  QH441. P37 2005
  576. 8′8—dc22   2004026054

1 0 9 8 7 6 5 4 3 2 1

Typeset by Newgen Imaging Systems (P) Ltd., Chennai, India
Printed in Great Britain
on acid-free paper by
Antony Rowe, Chippenham

ISBN 0-19-856493 7 (Hbk)   978 0 19 856493 5

Dedication

This book is dedicated to James S. Farris, one of the foremost scholars of phylogenetic biology in the twentieth century. It is now thirty-five years since the two-paper introduction of Wagner parsimony[1]. Wagner parsimony paved the way for all modern parsimony approaches, including the more general algorithms of Fitch and Sankoff. These landmark publications by Dr. Farris are among the best known and highest-cited papers in systematic biology. At the same time, and little known to either the biological or mathematical communities, was Dr. Farris's 1970 development[2] of what has come to be known as the *Farris transform*[3]. This transformation[4] was rediscovered in the context of phylogenetics twice, as well as in Wolf Prize winner Mikhail Gromov's work on hyperbolic groups, for which it was later dubbed the *Gromov product*. The Farris transform also appears, in 'disguise', in distance geometry, where it is known as the *covariance mapping*.

[1] Kluge, A. G. and Farris, J. S. (1969). Quantitative phyletics and the evolution of Anurans. *Syst. Zool.* **18**: 1–32; Farris, J. S. (1970). Methods for computing Wagner trees. *Syst. Zool.* **19**: 83–92.

[2] Farris, J. S., Kluge, A. G., and Eckhart, M. J. (1970). A numerical approach to phylogenetic systematics. *Syst. Zool.* **19**: 172–189.

[3] Dress, A., Holland, B., Huber, K. T., Koolen, J. H., Moulton, V. and Weyer-Menkoff, J. (2005). Δ Additive and Δ ultra-additive maps, Gromov's trees, and the Farris transform. *Discrete Appl. Math.* **146**: 51–73.

[4] The similarity measure $S_a = 1 - D_a$ that one gets from a dissimilarity measure D defined on a set of tree leaves (terminal taxa) X containing leaf a by placing $D_a = D(x,y) - D(a,x) - D(a,y)$ is the *Farris transform* of D relative to a. Here, a can be interpreted as an outgroup root and the values of $D(a,x)$ and $D(a,y)$ as the distances of leaves x and y from that root, in which case $-D_a = D(a,x) + D(a,y) - D(x,y)$ would be twice the distance of the last common ancestor of x and y from a.

Dr. Farris has received the honor Doctor of Philosophy *honoris causa* from the University of Helsinki, Finland.

Personal Dedication

I first met Steve Farris in 1990, at the International Congress of Systematic and Evolutionary Biology meetings held in College Park, Maryland. I was anxious to meet the man whose work had so affected me already, despite having just started my PhD in 1989. Axel Meyer, who was then at Stony Brook, introduced us. Steve had taught statistics at Stony Brook for many years (although few know that he was trained as a systematic ichthyologist). I showed Steve some calculations and graphics I had made with Brent Mishler that we thought had bearing on the issue of consistency of parsimony (see Chapter 1, for example). Seeing an opportunity, Steve spontaneously proceeded to persuade organizers of a Hennig Society symposium at the Congress to let me fill an empty slot. I was lucky enough to have brought overheads. I gave the talk, and suffice it to say that I made some friends, and estranged some others. But this was the true start of my career. Controversy has never been slight in the field of phylogenetics, and Steve has almost never been slight (except when he was young, or in that photo in David Hull's book), and certainly never anything but controversial. Like it or not, debate, duel, divide, and conquer is one approach to science. Regardless, we can thank Steve's relentless pursuit of what he thought was correct for many thousands of papers in the literature using parsimony approaches, as well as for many others that argue against it.

Victor A. Albert

Preface

Parsimony analysis (cladistics) has long been one of the most widely used methods of phylogenetic inference in the fields of systematic and evolutionary biology. Moreover, it has mathematical attributes that lend itself for use with complex, genomic-scale data sets. This book reviews philosophical, statistical, methodological, and mathematical aspects of parsimony analysis, and demonstrates the potential that this powerful hierarchical data-summarization method has for both structural and functional genomic research.

The book is aimed primarily at graduate-level students as well as professional researchers in the fields of phylogenetics and phylogenomics (within both the evolutionary and molecular biology communities). However, mathematicians, statisticians, and philosophers of science will also find the contents of relevance and use.

Readers will discover among the chapters that parsimony analysis does not represent a single research view, but rather a variety of perspectives all based upon a theme. I viewed it of great importance to display this diversity in light of the multiplicity of other phylogenetic methods that have been developed over the years.

My aim with this volume has been to provide parsimony analysis with a benchmark for its current place in science and for judgment of its progress into the future. Previous works focusing on parsimony analysis are surprisingly few given the extremely widespread use of parsimony methods in the academic journal literature. Those books that have been written are mainly introductory treatises, i.e. geared for mid-upper level university courses. A noteworthy exception is Elliott Sober's *Reconstructing the Past: Parsimony, Evolution, and Inference* (1988, MIT Press), which was written for a specialist audience. While there exist advanced texts devoted to phylogenetic analysis of morphological data, concepts of species, cladistic methods in biogeography, and mathematical aspects of phylogenetic inference, there has been no book that specifically incorporates advanced material spanning philosophical, methodological, and mathematical perspectives on the relevance of *parsimony analysis*, particularly as applied to the burgeoning field of *genomic biology*.

My work on this book began at the 21st annual meeting of the Willi Hennig Society, held at the Hanasaari Cultural Centre, Helsinki, Finland. I am grateful to the various chapter authors for their enthusiasm for the project. Mike Steel and David Penny are acknowledged for winning the "First Draft In Prize." I thank Cécile Ané, Joe Felsenstein, Mike Sanderson, Mark Simmons, and several chapter authors for their thoughtful reviews of one or more chapters. Other referees are listed among chapter acknowledgments. Andreas Dress, Katharina Huber, and Vincent Moulton kindly contributed information on the Farris transform. Finally, I thank Oxford University Press Commissioning Editor Ian Sherman for immediate interest in the project, and Editorial Assistants Abbie Headon, Kerstin Demata and Production Editor Anita Petrie for their assistance. Heartfelt thanks also to Charlotte, Torben, and Siri for putting up with me.

Victor A. Albert

Contents

Contributors

Victor A. Albert, Natural History Museum, University of Oslo, P.O. Box 1172 Blindern, NO-0318 Oslo, Norway. e-mail: victor.albert@nhm.uio.no

Vladimir N. Babenko, National Center for Biotechnology Information, National Library of Medicine, National Institutes of Health, 8600 Rockville Pike, Bldg. 38A, Bethesda, MD 20894, USA. e-mail: babenko@ncbi.nlm.nih.gov

Jerrold I. Davis, L.H. Bailey Hortorium and Department of Plant Biology, Cornell University, Ithaca, NY 14850, USA. e-mail: jid1@cornell.edu

Jan E. De Laet, Royal Belgian Institute of Natural Sciences, Vautierstraat 29, Brussels, Belgium. e-mail: jdelaet@natuurwetenschappen.be, jan.delaet@lid.kviv.be

Pablo A. Goloboff, CONICET, INSUE, Instituto Miguel Lillo, Miguel Lillo 205, 4000 San Miguel de Tucumán, Argentina. e-mail: pablogolo@csnat.unt.edu.ar

Arnold G. Kluge, Cladistics Institute, Ann Arbor, MI 48103, USA. e-mail: akluge@umich.edu

Eugene V. Koonin, National Center for Biotechnology Information, National Library of Medicine, National Institutes of Health, 8600 Rockville Pike, Bldg. 38A, Bethesda, MD 20894, USA. e-mail: koonin@ncbi.nlm.nih.gov

David A. Liberles, Computational Biology Unit, Bergen Centre for Computational Science, University of Bergen, NO-5020 Bergen, Norway. e-mail: liberles@cbu.uib.no

Damon P. Little, L.H. Bailey Hortorium and Department of Plant Biology, Cornell University, Ithaca, NY 14850, USA. e-mail: dpl10@cornell.edu, dlittle@nybg.org

Brent D. Mishler, University Herbarium, Jepson Herbarium, and Department of Integrative Biology, University of California, Berkeley, CA 94720, USA. e-mail: bmishler@socrates.berkeley.edu

Kevin C. Nixon, L.H. Bailey Hortorium and Department of Plant Biology, Cornell University, Ithaca, NY 14850, USA. e-mail: kcn2@cornell.edu

David Penny, Allan Wilson Centre for Molecular Ecology and Evolution, Massey University, Palmerston North, New Zealand. e-mail: d.penny@massey.ac.nz

Diego Pol, Division of Paleontology, American Museum of Natural History, Central Park West at 79th Street, New York, NY 10024, USA. e-mail: dpol@amnh.org

Igor B. Rogozin, National Center for Biotechnology Information, National Library of Medicine, National Institutes of Health, 8600 Rockville Pike, Bldg. 38A, Bethesda, MD 20894, USA. e-mail: rogozin@ncbi.nlm.nih.gov

Elliott Sober, Department of Philosophy, University of Wisconsin, Madison, WI 53706, USA. e-mail: ersober@wisc.edu

Mike Steel, Allan Wilson Centre for Molecular Ecology and Evolution, University of Canterbury, Christchurch, New Zealand. e-mail: m.steel@math.canterbury.ac.nz

Ward C. Wheeler, Division of Invertebrate Zoology, American Museum of Natural History, Central Park West at 79th St, New York, NY 10024–5192, USA. e-mail: wheeler@amnh.org

Yuri I. Wolf, National Center for Biotechnology Information, National Library of Medicine, National Institutes of Health, 8600 Rockville Pike, Bldg. 38A, Bethesda, MD 20894, USA. e-mail: wolf@ncbi.nlm.nih.gov

CHAPTER 1

Parsimony and phylogenetics in the genomic age

Victor A. Albert

1.1 Parsimony inference

Parsimony (Ockham's razor) as a method of inference has a long history. Based upon a parsimony argument, Copernicus maintained that his heliocentric solar system theory was superior to the geocentric one of Ptolemy because of its greater simplicity. His reasoning was that Ptolemy's theory required what amounted to independent models for each planet's movement (extra parameters), whereas his own included the simplifying factor of Earth–Sun movement for each planet.[1]

According to Copernicus, his theory "follow[s] Nature, who producing nothing vain or superfluous often prefers to endow one cause with many effects...." An important point about Copernicus's argument is that it represented an appeal to a universal law in nature, in other words, God. Modern considerations of parsimony methodology, especially those following Lamarck and Darwin, have by necessity been occupied with other, non-deist justifications (Sober 2003).

Parsimony today stands as a method of inference from observations. For example, if one has a coin with heads and tails, in the absence of any prior information about the coin other than this observation, the most parsimonious assumption for the result of a coin toss is one or the other, i.e. 50/50 chance. If the toss were to be repeated 1000 times, one could establish a frequency-based probability (with margin of error) that this were so.

If one were a Bayesian, and Joe had already tossed the coin 1000 times and gotten heads for 500 tosses, this prior probability could be used to assess the posterior probability.

In this simple example, parsimony, maximum likelihood (the mean of a normal distribution, as with 1000 coin tosses), and posterior probability all give the same answer. However, this was a *very* simple example, involving a single object with only two alternatives. The relationships between parsimony, likelihood, and Bayesian inference become much less obvious with more objects (characters) and alternatives (states). A biological example that Sober and Steel (2000) and Sober (2003) have examined is Crick's (1968) parsimony-based claim that all life has a common ancestor. Crick's argument was that since many different versions of the genetic code could have been possible, the common use of one (albeit with slight modifications) by all extant organisms strongly suggests their common ancestry. The idea is that selection would operate against code changes in descendants of a given code. In other words, one beginning of life with this attribute is more parsimonious than many (say, X). But Crick's is also a likelihood argument (Sober 2003):

> Pr(code universal in extant organisms | one ancestor) > Pr(code universal in extant organisms | X separate ancestors)

which takes the standard form $Pr(O|M_1) > Pr(O|M_2)$, comparing likelihoods of O observations given models M. The formulation above follows the Law of Likelihood (Royall 1997), which states that a hypothesis with higher likelihood is preferable over one with lesser.

---

[1] Sober (e.g., Sober 1989, 2003; Chapter 3) has been an active student of this history, and I acknowledge his work for this example and several others I present below.

As pointed out by Sober (2003), parsimony and likelihood therefore provide identical evaluations of Crick's common ancestry hypothesis. However, further equivalence postulates between parsimony and likelihood, which I explore later, show the issue to be much more complex.

1.2 Examples of modern uses of parsimony

1.2.1 Curve fitting

Parsimony often plays a role in choosing among models fit to a set of points on an x,y plane. For example, a set of points might be regressed by a line a bit sloppily, or by a parabola far better. The question is, which model to accept? The latter requires an extra adjustable parameter, so it could be considered less parsimonious based on *economy of assumptions* (P_{EA} parsimony), i.e. Ockham's razor. On the other hand, a better *fit* to a parabola, which takes the form of minimization of residual variance between observations and model, is of course the likelihood, $Pr(O|M)$.

There are different criteria to choose among models. In one example, an excellent parabolic fit (or even a higher-order one) might have no logical relationship to the data at hand (e.g. length of a naked mRNA strand vs. number of free bases after chemical degradation *in vitro*), and so a sloppy line would be the better model (through P_{EA}), albeit representing data that may have been collected in an error-prone manner.

Other data with better parabolic fit might have a realistic basis—this would then suggest a defiance of P_{EA} in terms of the number of adjustable parameters. But how to decide between the models? Two well-known criteria offer likelihood-based methods for this choice, the Bayesian and Akaike information criteria (BIC and AIC, respectively; see Sober 2003; Felsenstein 2004). In these criteria, parsimony takes the role of a penalty for complexity (in terms of number of adjustable parameters, p, referring to P_{EA}) among models, M, with different log-likelihoods. For example, with the BIC (the ratio of the average likelihoods for two models), for the most likely parabola to be preferred, it must fit the observations, n, better

enough than the most likely linear model to avoid the complexity factor, which is dependent on sample size:

$$BIC_M = -2\log L_M + p_M \log n$$

Roughly equal likelihoods, L, will likely mean that the line will win.

1.2.2 Trees of species or genes

Data points based on characters (e.g. nucleotides) sampled from species or genes can be analyzed under a *hierarchic model* in order to reconstruct most parsimonious trees. This operation is, of course, the central subject of this book. Most parsimonious trees are hierarchies or partially collapsed hierarchies with changes minimized across all characters that could show evidence for grouping. Here, groups are defined as two or more species or genes partitioned from two or more other species or genes. Not all character-state distributions can show evidence for grouping, and the specifics of information use is a major difference between parsimony, likelihood, and distance matrix methods. An illustration of this, as well as what trees demonstrate among the methods, will be useful.

For four species or genes, A, B, C, and D, there are 2^{n-1} different ways (in this case $n = 4$) for binary characters to partition species or genes:

{AB} {CD}	{ABC} {D}	{ABCD} {}
{AC} {BD}	{ABD} {C}	
{AD} {BC}	{ACD} {B}	
	{BCD} {A}	

Parsimony can use only $2^{n-1} - (n+1)$ partitions, i.e. three—the two-item splits shown to the left. A character (in isolation from other characters) that argues for such a partition incurs one state change between such splits, yielding two groups (Fig. 1.1). None of the other partitions produce groups, although the middle four 3:1 splits incur state changes (these merely show a *difference*, between, say D vs. A, B, and C; Fig. 1.1). No changes are implied in the 4:0 split. However, likelihood methods use all of the eight partitions (see below). Distance matrix methods use information at rate $(n^2 - n)/2$. As species/gene number increases, it

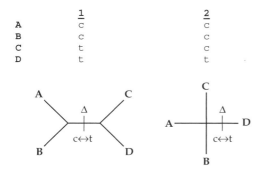

Figure 1.1 Examples of two characters and their states that (1) can show evidence for hierarchy vs. (2) evidence for difference. Trees implied by 1 vs. 2 alone are also shown. Δ indicates where a character-state change could occur. Note that only character 1 could support two groups, a group defined as comprising two or more items; character 2 can only support a fully collapsed tree in which one branch (not a group) is different from the others. A–D, species or genes. c/t, different pyrimidine bases.

can be seen that $2^{n-1} - (n+1)$ approaches 2^{n-1}, but that $(n^2 - n)/2$ lags far behind. Thus, for a given large n (such as with genomic-scale estimation of gene family phylogenies), parsimony utilizes the majority of all available evidence while only incorporating characters that could show evidence for grouping.

Likelihood trees demonstrate relationships among species or genes that maximize $Pr(O|M)$, where M is an evolutionary model. As such, likelihood methods need *all* of the observations O, including the non-grouping partitions, to maximize the likelihood of the data; anything less would compromise the calculation. Branches of likelihood trees have lengths in terms of character-state change probabilities. Parsimony trees display relationships in terms of character-state changes along branches. Distance matrix methods of building trees show raw or model-adjusted differences between species or genes.

These illustrations are not meant as justifications for one method over the other in phylogeny reconstruction; rather, my goal has been to draw attention to differences in information use among the different methods, and to what trees derived from them demonstrate. This has often been confused in the theoretical and biological literature.

1.2.3 Phylogenetic models for which parsimony and likelihood are equivalent

I have already illustrated simple, non-phylogenetic models for which parsimony and likelihood are equivalent. The first attempts to establish this equivalence for phylogeny reconstruction were those of Farris (1973) and Felsenstein (1973). These models were different in that Farris's was basically Bayesian, with equal (flat) prior probabilities on all trees, whereas Felsenstein's was based on likelihood. Farris solved for the tree topology and character-state assignments at all points along branches, while including no assumption about rates of character-state change. Felsenstein's model summed over all possible character-state assignments, but required low rates of character-state change. According to Sober (Chapter 3), Farris's solution for topology *plus* character-state assignments (additional parameters) renders it inequivalent to likelihood, but that Felsenstein's parsimony model does achieve a likelihood equivalence. These considerations depend of course on the type of likelihood under consideration, for which there are several variants (Steel and Penny 2000; Goloboff 2003). To echo the point made by Steel and Penny (2000), both Farris's and Felsenstein's models are likelihood equivalents, just not for the same kind of likelihood.

Goldman's (1990) parsimony-likelihood formulation permits all branches to have the same length—a very simple model. However, Goldman, and indeed Sober (Chapter 3), assert its inequivalence with likelihood for basically the same reasons as for Farris's (1973) model: inference of the topology *plus something else*, in this case, ancestral character states. However, it is worth pointing out other views in the literature. According to Farris (1986) and Goloboff (2003a), ancestral states are not to be viewed as parameters:

Goldman (1990) decided that, even if the ancestral reconstructions are not parameters, they "could be treated as if they were." But they could also be treated (much more properly) as if they were not a parameter. The ancestral states are more like a kind of inferred observation (Farris, 1986). Parameters are instead those variables of the process that determine the conditions of the problem—the variables that determine the outcome of evolution, that is. Even if not observed, the ancestral states are (just like observed states) part of that outcome. [Goloboff 2003a, p. 100]

Goldman's formulation of parsimony assumes that each character type occurs with a probability equal to the pathway with highest probability, among all the pathways that lead to that character type. If the probability of change in each branch is low, this estimation produces probabilities that are roughly proportional to the actual probabilities (i.e., the ones obtained by summing); that is, all the resulting character types are ranked in the same order of increasing probability by both criteria. This, however, does not convert the calculations under Goldman's model into estimations of a parameter; if a reconstruction was indeed a parameter, there would be one of them which would confer to the corresponding character type its true probability of occurrence under the model, and there is none. Only the sum of all reconstructions provides the true value for a given type. [Goloboff 2003a, p. 100]

Thus, Goloboff argues, using *the most likely reconstruction* (instead of the sum of likelihoods for all reconstructions) produces a good approximation of the actual likelihood, which is not exact, but then again some likelihood methods are not exact either. Goloboff gives the example that the assumption of nucleotide state frequencies remaining constant over time also implies that likelihood calculations are approximate instead of exact, since reconstructions are then not truly independent (they must sum to the assumed frequencies) (Goloboff 2003a, p. 101).

Without debate as an equivalence between parsimony and likelihood (Sober, Chapter 3; Goloboff 2003a) is the formulation of Tuffley and Steel (1997). They provided a proof that parsimony was a maximum likelihood estimator under the assumption of no common mechanism (potentially unequal change probabilities) for each character with r states and a symmetric change assumption. With this formulation, the different rates can either be very, very small or very, very large: in fact, only 0 or infinity, and nothing in between. The Tuffley and Steel result has been considered by some a *complex* parsimony equivalent because of its numerous adjustable parameters (the lengths of each branch for each character, as estimated from a single datum). On the other hand, Goldman's formulation is an extremely

simple model: the fit of a tree to data is solely based on its topology and on state change/stasis probabilities. *As such, parsimony inference can receive a likelihood equivalence at both ends of the complexity spectrum, which has been interpreted to speak toward its generality as an inferential method* (see Goloboff 2003a). Of course, equivalence between parsimony and likelihood between a few models does not mean that equivalence extends to *all* models, or that it has to do so in order to justify use of parsimony methods.

1.2.4 A non-likelihood justification for parsimony

Not all users of parsimony analysis care about equivalencies between parsimony and likelihood under certain process models. Farris himself, who produced a series of statistical interpretations of parsimony (1973, 1977, 1978), later downplayed these in one of the most important philosophical papers on parsimony analysis (Farris 1983). He famously stated that:

A number of authors, myself among them (Farris, 1973, 1977, 1978), have used statistical arguments to defend parsimony, using, of course, different models from Felsenstein's [1973]....my own models, if perhaps not quite so fantastic as Felsenstein's, are nonetheless like the latter in comprising uncorroborated (and no doubt false) claims on evolution. If reasoning from unsubstantiated suppositions cannot legitimately question parsimony, then neither can it properly bolster that criterion. The statistical approach to phylogenetic inference was wrong from the start, for it rests on the idea that to study phylogeny at all, one must first know in great detail how evolution has proceeded. That cannot very well be the way in which scientific knowledge is obtained. [Farris 1983, p. 17]

Farris argued in favor of parsimony as a method that maximizes explanatory power among observations that could be expected to reflect genealogical relationships, i.e. potential homologies. He characterized most-parsimonious trees as the least falsified hierarchical hypotheses in the context of the philosopher Karl Popper's ideas on the treatment of observations (see Kluge, Chapter 2). However, Farris carried his argument into more general terms: *trees with minimal homoplasy (i.e. with minimal parallelism or reversal) must be preferred over trees that have more, because the latter require more observations to be dismissed for the sole purpose of protecting conclusions from 'offending' evidence.* This is a fundamentally different use of the parsimony criterion, *that which the data requires* (P_{DR} parsimony), and indeed, *this is a criterion that can be interpreted quantitatively.* Operationally, for a given set of independent pairwise similarity statements,

$$\min \sum_k H_k \iff \max \sum_k J_k$$

where H and J represent independent statements of pairwise homoplasy and homology, respectively, across k characters (see De Laet, Chapter 6). Minimization of H, with the addition of a tree-independent constant, is equivalent to minimizing total steps (character-state changes) as calculated by standard parsimony software.

With reference to the minimization of H, Farris also refuted the commonly held belief that parsimony assumes rarity of homoplasy by use of an analogy to linear regression analysis; although residual variation in a least squares fit is certainly minimized, there is no requirement that this variation be small. Likewise, minimization of H occurs in the context of all characters, and this involves no requirement that estimated homoplasy be rare. For further discussion of Farris's arguments, see De Laet (Chapter 6; including an interesting elaboration) and Kluge (Chapter 2).

1.2.5 Parsimony and statistical consistency

Although consistency enters into further discussion below, I will only briefly deal with its formalities. As Felsenstein (2004; p. 107) explains:

An estimator is *consistent* if, as the amount of data gets larger and larger (approaching infinity), the estimator converges to the true value of the parameter with probability 1. If it converges to something else, we must suspect the method of trying to push us toward some untrue conclusion. In 1978 I presented . . . an argument that parsimony is, under some circumstances, an **inconsistent** estimator of the tree topology. [italics in the original; bold emphasis is mine]

Farris (1983) rejected Felsenstein's 1978 model by arguing against its applicability to real data. He didn't object to the general idea of seeking a consistent estimator; he just felt that one was not available in practice. A decade ago, colleagues and I modeled consistency for sequence evolution and concluded that the 'Felsenstein zone' of inconsistency (under Felsenstein's own conditions) was small enough to be insignificant for real data (Albert *et al.* 1992, 1993). Felsenstein showed similar results himself (2004; see also Steel and Penny 2000), which echo our findings that as r states increase, the zone of inconsistency decreases. This will be seen to have bearing when gene-order data are discussed below.

1.2.6 Other practical considerations.

Parsimony analysis yields hierarchic results that are both fully diagnosable and interconvertible with the original data (Fig. 1.2). This is a very positive feature in terms of tree interpretation and for information storage and retrieval (Farris 1979). As stated above, most-parsimonious trees have branch lengths in terms of changes among the states of informative characters. This format is more intuitive than branches in terms of state-change probabilities, distances (via some metric), or those with no dimensions whatsoever. Indeed, many investigators have exploited parsimony's branch-length properties to optimize their original data on to trees derived from other methods; however, such comparisons are not interconvertible with the data matrix, rendering interpretations potentially *tree*-biased as opposed to *data*-biased (remember that the *data* that go into most-parsimonious trees provide the branch lengths that come back out; see Mishler, Chapter 4).

1.3 Genomic-scale data and parsimony

Current whole-genome sequences and projects underway represent the tip of the iceberg. Now that we have complete genome sequences for many prokaryotes, several eukaryotes, and numerous organellar DNAs, bioinformaticians

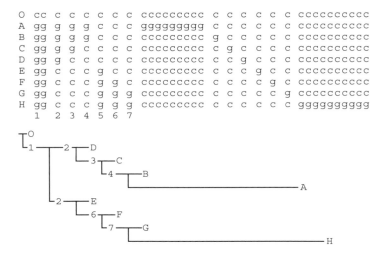

Figure 1.2 Phylogenetic trees based on parsimony are fully diagnosable. Similarities that had the capacity to bear hierarchical information, here characters 2–7, were those used to build the most parsimonious tree, and upon inspection it can be seen that inferred character-state changes can be readily optimized on to internal branches. All other characters shown in this example only show difference, as opposed to evidence for hierarchy. This fact can be readily appreciated by examination of the lengths of branches—two differences separate O (outgroup) from the other species or genes A–H; A and H have whole blocks of singular differences, B–G each have singular differences assigned, and all of this is reflected in the tree as differential branch lengths, not as arguments for a different overall hierarchy. Note also that the tree and its branch lengths are fully interconvertible with the original data matrix.

must find ways to reduce this complexity to provide meaningful fodder for biological hypotheses. One urgent need is for fast and predictive phylogenetic estimation of species and gene relationships. Parsimony is a method that has the logical and practical attributes discussed above, as well as, recently, the speed necessary to carry out massive topological calculations.

1.3.1 Sequence data and tree size

Parsimony analyses of sequence data for large numbers of species have been possible for a number of years (see Davis *et al.*, Chapter 7), but only recently have these become fast enough to be considered of use at the genomic scale. For genome comparisons, it will be important to use parsimony calculations to determine gene relationships within gene families or superfamilies. The issue of orthology vs. paralogy is important in the context of molecular evolutionary hypothesis testing (see Liberles, Chapter 10, and Rogozin *et al.*, Chapter 11). One current application, TNT (Goloboff *et al.* 2004), has the ability

to solve reliably for large most-parsimonious trees that were once thought to be intractable problems. For example, the 500-sequence *rbcL* data set for seed plants (see Davis *et al.*, Chapter 7) can now be solved for most-parsimonious trees in seconds (<42 s on my 1.6 GHz Pentium M laptop, in fact). This application has also been parallelized (Goloboff *et al.* 2003b), so for individual sequence alignments, the practical limits for genomic-scale sequence data will be the strength of such alignments.

In the case of *rbcL*, a highly conserved protein-coding gene with no introns, sequence start-stop and internal base alignment is unambiguous. However, this is certainly not a generalization that can be made for genes in general, not to mention (e.g.) non-coding regions in between genes. This begs an issue that I have avoided until now—according to Wheeler (Chapter 5) and De Laet (Chapter 6), the logic of *a priori* multiple alignment is erroneous and the results are incomplete at best. These authors argue persuasively for *optimization of sequences as whole, complex characters* under Sankoff parsimony (Sankoff 1975). However, practical limits rise greatly for such algorithms, which create compound NP-complete problems that must, by current necessity, be solved heuristically.

Likelihood- or distance-based phylogenetics will reach different sorts of complexity blockades in dealing with relationships among large gene families. For example, the R2R3-*MYB* gene family in *Arabidopsis* is composed of ca. 100 members, as opposed to only three found in human (Martin and Paz-Ares 1997; Romero *et al.* 1998). Imagine ca. 100 R2R3-*MYB*s across 500 plant genomes—perhaps 50 000 genes, which given advances in sequencing technology, isn't a far-fetched possibility in the not-too-distant future. If full diagnosability (tree/matrix interconvertability) and speed of execution together form the important criterion, distance matrix and likelihood methods will prove inferior. Distance matrix methods are fast, but they decompose information into pairwise estimates of path lengths between n_i and n_j and then, at best, try to reassemble them into a tree that somehow optimizes these lengths. Such an operation is fraught with error since pairwise path lengths may show no relationship to those on reconstructed trees. Moreover, the inherent information loss, especially as n gets large, is unacceptable (see above). Likelihood methods do not provide character-state diagnoses either, and are understandably slower than parsimony given that calculations are CPU-intensive, especially as n increases (even using the pruning algorithm, computational effort is proportional to $k(n-1)r^2$; Felsenstein 2004). Supercomputers and CPU clusters speed up likelihood calculations, but eventually a tradeoff will be reached. Besides, to quote Felsenstein (2004, p. 122):

If it escapes the clutches of long branch attraction [inconsistency], parsimony is a fairly well-behaved method. It is close to being a likelihood method, but is simpler and faster. It is robust against violations of the assumption that rates of change at different sites are equal. (It shares this with its likelihood doppelganger.)

Given the equatability of parsimony with likelihood under several models that range from simple to complex (see above), parsimony should be the method of choice as applied to genomic-scale questions that include enormous numbers of species or genes. It will no doubt arise in some readers' minds, however, that the issue of tree size, e.g. for sequences of 50 000 genes, could impact statistical consistency. Indeed, some have cautioned that inconsistency can occur more often as trees become larger and larger (Kim 1996). I will present a more positive outlook on parsimony and large trees below.

1.3.2 A conjecture on parsimony and large phylogenetic trees

Background

With reference to the largest phylogenetic analysis yet attempted (2 538 *rbcL* sequences for photosynthetic organisms), colleagues and I (Källersjö *et al.* 1999) observed that relatively rapidly evolving nucleotide sites, such as those in third positions of codons, provide the majority of tree structure despite initial estimates of saturation and high levels of homoplasy on most-parsimonious trees. We pointed out in reference to this analysis that, analyzed by themselves, third positions resolve 1 327 supported groups with an average parsimony jackknife frequency of 85%, whereas the first two positions together resolve only 431 groups, with an average frequency of only 75%. The groups recovered by third positions are also well supported by the full data and are spread over the tree, including both older and younger lineages. In contrast, the first two positions fail, for example, to recognize either land plants or flowering plants as monophyletic groups.

We also generated random subsets (10 for each) of $n = 100$ species, n, $2n \ldots 10n$, from the 2538-species matrix, and calculated the average retention index, position-wise within codons, for each subset. The retention index for individual characters—$(g - s)/(g - m)$, where g is the maximum number of steps, s is the most-parsimonious number, and m is the minimum number—measures the amount of initial similarity retained as homology on most parsimonious trees (Farris 1989a). As matrix size rose from 100 to 1 000, the retention index rose for third positions as matrix size increased. In contrast, the retention indices for first and second positions—those sometimes favored for molecular phylogenetics because they evolve more slowly—decreased. Our simple interpretation, also following the group support data reported above, was that third positions were performing better than first or second positions. Moreover, with respect to total homoplasy, the consistency index (m/s; Kluge and Farris 1969), which is inversely proportional to the total number of substitutions, gave the converse view: first, second, and third positions averaged 0.155, 0.178, and 0.046, respectively on the tree calculated from all positions—that is, *greatest homoplasy was discovered for the more rapidly evolving third positions, despite their better performance*. Given these results, our conclusion was that homoplasy can *increase* phylogenetic structure.

Conjecture

Homoplasy on trees can have a direct relationship with rates of change. It certainly does on large trees, such as those discussed above ($n = 100$–$2\,538$), for which enough branches exist to observe the products of high evolutionary rates. The branches of small trees (e.g. $n = 4$–6), such as those often used for simulation studies, would not be expected to reveal underlying rate differences as accurately for vast divergences, and such differences are precisely those that might lead to inconsistency.

In Chapter 9 of this book, Steel and Penny prove a common-mechanism equivalence between parsimony and likelihood when the number of character states, r, is large enough (we will get back to this issue later regarding gene-order data). They also show that under such conditions, few characters, k, are required to arrive at a most-parsimonious solution.

My conjecture is that a parsimony-likelihood equivalence can hold when r is much smaller than required by Steel and Penny's Theorem 9.6.2, e.g. in the r = 4 case as for nucleotide data, if n is large enough.

Erdős et al. (1999) proved that a compatibility tree, from which homoplasy is prohibited, requires at least $(n-3) * \log(n-3) - (n-3)$ informative characters (i.e., with grouping potential) to reconstruct a tree of n species or genes with at least 50/50 probability. For the $2\,538$ species case illustrated

above, this quantity is at least $17\,300$ informative sites. But, in reality, k was only $1\,428$—the number of bases in the *rbcL* gene. Erdős et al. (1999) also established that at least some phylogenetic methods should require, for a given constant c, only $c \log(n)$ or at worst a power of $c \log(n)$ characters. Steel and Penny's Conjecture 9.4.3 and Proposition 9.4.4 (Chapter 9) similarly suggest that k follows logarithmic growth on n, *and as n grows, so too would homoplasy for some characters, especially given any limitation put on r. This requirement for homoplasy echoes the empirical findings discussed above.* With Mike Steel's help, I provide below a mathematical formalization of my conjecture in light of these findings.

Conjecture 1.3.1. Consider the r-state symmetric Poisson model. For any $\epsilon > 0$ and constant $B \geq 1$ there exist constants h and c that depend only on r, ϵ, and B for which the following holds.

Suppose k characters are generated independently for this model on any fully resolved phylogenetic tree T with n species or genes for which all the branch lengths of T are at most h and the ratio of any two branch lengths of T is at most B. Then provided $k \geq c \log(n)/f^2$, where f is the smallest branch length in T, maximum parsimony will correctly recover T with probability at least $1 - \epsilon$.

Here, ϵ is any real number, and $B = 1$ is the case where all branch lengths are equal. As f, which is a function of n, gets smaller and smaller, the sequence length needed to 'detect' that branch has to grow (indeed quadratically with $1/f$). The role of B is to avoid inconsistency (as described by Felsenstein 1978a). However, note that the conjecture just says 'for any B', so one could take $B = 1\,000$ a priori, and thereby allow one branch to be $1\,000$ times as long as another. This will impact the values of c and h, but these are constants so far as n is concerned. Presumably also as n increases, the value of B may come down closer to 1, provided that adding to n does not create a branch that is too short.

The conditions stated mean that most of the characters will have a fair degree of homoplasy—indeed, the expected number of steps will go to infinity with n, since each branch length is bounded below by h, which is a positive number.

1.3.3 Strings, and more on *r*-state characters

Around the same time, Steve Farris and I developed methods that were intended to lower worries about parsimony and inconsistency. I will begin with my procedure (Albert *et al*. 1994), which was to accept strings of nucleotides (randomly selected) as unit characters instead of individual bases; these strings were recoded as presence vs. absence for data analysis. The intended effect was to reduce the probability of homoplasy given, e.g. that a six-base pair string is much easier to lose once it exists than it is to regain once lost.

My argument was based on investigations of Dollo parsimony (Farris 1977) and its use with DNA restriction-site data (Albert *et al*. 1992). In this context, Dollo *never* permits parallel gains of a restriction site (often a six-base recognition string), only multiple losses. In this earlier work, we concluded that the Dollo model was too severe and that despite the asymmetry in probabilities just discussed, parsimony with equal character-state weights (Kluge and Farris 1969; Farris 1970) was more appropriate. However, my work did not consider increasing string length. Felsenstein (2004, p. 236) has examined the parallel gain case more thoroughly, solving for the probability (under the Jukes–Cantor model) that two species or genes and their common ancestor each have/had (+) a particular nucleotide string of *k* sites after *t* units of branch length:

$$\Pr(+++) = \left(\frac{1}{4}\right)^{k} \left(\frac{1+3e^{-\frac{4t}{3}}}{4}\right)^{2k}$$

This equation clearly demonstrates that as strings grow longer and longer, and as time since common ancestry increases, the probability of parallel gain becomes very small.

This is precisely why I developed the string character method; if one were to code *only* those completely matching strings beginning at certain nucleotide positions, especially larger and larger ones, then these should be rather conservative characters for deep branchings within phylogenetic problems. As such, the string character concept need not be restricted to nucleotide data; amino acid data, exons/introns, or even genomic regions could be so coded (see below).

Now on to *r*-state characters. Farris and Källersjö presented a related method, supersites, at the 1999 meetings of the Willi Hennig Society in Göttingen, Germany. With supersites, strings of nucleotides are recognized beginning at nucleotide *W* and then parsed downwards through the matrix, recognizing as many character states as necessary to account for differences within the strings. Supersites can therefore generate considerable character-state space among fewer informative characters. However, Steel and Penny (2000) suggest that such procedures may not avoid inconsistency because probabilities of change along branches increase → 1 as a function of *k*, where $r = d^{k}$ and *d* is the number of possible character states.

1.3.4 Gene content

Genomic-scale phylogenetic studies based on gene content are reviewed by Rogozin *et al*. (Chapter 11). Two approaches have been used: (1) estimate species trees from orthologous gene presence vs. absence among whole genomes, or (2) optimize these data on to a predetermined species tree. In my string character method, above, I permitted string presence-absence to have equal weight, whereas the probabilities modeled above imply asymmetric weights favoring losses. So which character-state weights to use? Rogozin *et al*. (Chapter 11) discuss Dollo analyses based on whole genes, which are of course nucleotide strings themselves (see above). The Dollo assumption is the asymptotic case, and with reference to the equation above this should be entirely appropriate as string size increases, say, to 1 428 bases. Use of Dollo optimization onto species trees incorporates the same state-change asymmetry. Moreover, Huson and Steel (2004) have shown that Dollo parsimony compares very favorably with a genesis-loss likelihood model they constructed to analyze gene-content data.

1.3.5 Gene order

The growing rate of whole-genome sequencing, particularly for the relatively small and circular genomes (prokaryote, chloroplast, and most mitochondrial), has been accompanied by heightened interest in determining phylogenetic relationships based on gene order (synteny). The problem is not a simple one, at least in terms of encoding the data. For one, Steel and Penny point out in Chapter 9 that the order of G genes in a signed (oriented) circular genome can display any of $2^G(G-1)!$ combinations. Nonetheless, their proof of equivalence between likelihood and parsimony for large character-state space states bodes well for use of computationally simpler parsimony calculations in this genomic arena.

Parsimony analyses of gene order are related to analyses of string data (as discussed above), but differ in their attempt to account for *adjacency vs. non-adjacency* of strings. Coding methods for use with parsimony analysis have already been under investigation, e.g. Maximum Parsimony on Multi-state Encodings (MPME; Wang *et al.* 2002, as suggested by Bryant 2000) methods. This method produces signed, multistate circular permutations of gene adjacency on circular genomes (see discussion in Steel and Penny, Chapter 9). In one simulation study, MPME has been shown to have greater accuracy in comparison with a method incorporating neighbor joining (Wang *et al.* 2002), a distance matrix method that inherently incorporates less information from the data (see above). Still other coding methods exist or are under development, including a technique that utilizes Dollo parsimony on tightly linked gene pairs that are then binary-recoded (Wolf *et al.* 2001; see Rogozin *et al.* Chapter 11). This method is directly related to the string recoding method of Albert *et al.* (1994).

A limitation encountered by Wang *et al.* (2002) and mentioned by Steel and Penny (Chapter 9) is the relatively small number of character states permitted by the most rigorous parsimony software (e.g. TNT) and required by the multistate coding methods. In other words, getting anywhere with the gene-order issue will require algorithmic advances regarding state space.

1.3.6 Microarray data

Hierarchic analysis of microarray expression data has become routine. However, almost all methods used are those of phenetic clustering (Eisen *et al.* 1998), which only supplies dimensionless levels of difference based on a distance matrix of log-transformed fluorescence intensities. Clusters of genes or 'treatments' (including tissue types) are formed based on shared patterns of up- vs. down-regulation of gene expression. Such analyses have proven of some use in (e.g.) tumor classification by gene-expression profiles, as well as, by inverting the matrix, identification of genes active in different tumor types. The analyses do not intend to be phylogenetic. However, phylogenetic methods, such as parsimony analysis, can be brought to bear on microarray data, at least when these data could be expected *a priori* to show evidence for hierarchy (e.g. through hereditary relationship). A parsimony approach to microarray analysis has been developed (Planet *et al.* 2001; Sarkar *et al.* 2002) and applied to the tumor classification problem. However, classification of different tumor types may not fit a phylogenetic model; gene regulation can be hierarchical and is certainly heritable, but it may also be networked, and tumor types do not share clear evolutionary relationships. A less stringent view on fit to model might be worth adopting for exploratory studies, since Sarkar *et al.* did identify gene-expression events that had also been identified by phenetic clustering.

There is one study of which I am aware that explicitly used parsimony analysis to reconstruct heritable relationships (they cited Planet *et al.* 2001). Uddin *et al.* (2004) used genome-wide expression profiles from primate brains to perform a parsimony analysis of organismic relationships; echoing other substantial evidence (e.g. Salem *et al.* 2003), the chimpanzee was identified as *Homo sapiens'* closest relative. Another classic study of heritable gene expression relationships within and between species was that of Oleksiak *et al.* (2002) on *Fundulus* fish populations. These authors used the phenetic clustering methods of Eisen *et al.* (1998) to group populations by genes, as well as genes by populations. Although Oleksiak *et al.* were studying population differentiation and not phylogeny *per se*, it would have been possible to use parsimony methods that incorporate population-level information. Sarkar *et al.*'s Characteristic Attribute Organization System (CAOS) is closely related to Population Aggregation Analysis (PAA; Davis and Nixon 1992), which identifies patterns of discrete features that unambiguously mark groups (i.e. that have gone to fixation). The CAOS approach further identifies characteristic expression patterns found in some members of a group and never outside that group, those found in all members of a group but never found together outside that group, and those found in some samples within a group but never outside that group.

CAOS illustrates the major advantage of using parsimony analysis for microarray data: diagnosability (see above). State changes that identify groups (of genes or treatments) and changes among members of groups have far greater predictive value than dimensionless clustering.

1.4 The future: some predictions

It is difficult to predict future modes and rates of genomic-scale data acquisition, but with Moore's law, computer capacity should open up previously inaccessible data-analysis possibilities in less than a decade. Parsimony will remain an indispensable part of the phylogenetics and genomics tool kit, particularly to estimate enormously large trees with full diagnosability, and also for data with large character-state space (e.g. geneorder information). Perhaps a proof for Conjecture 1.3.1 can be given, establishing that particular amounts of homoplasy on large-enough trees can render another parsimony-likelihood equivalence. A massive increase in whole-genome sequencing will no doubt permit refinements to estimations of ancestral gene content. To mention an area barely discussed in this chapter, optimization of whole sequences as complex characters will also become a practical and everyday tool with large numbers of species or genes. The gene-order issue, which will no doubt develop further with different encoding methods, will also include similar approaches to optimization of whole genomes as complex characters (see De Laet, Chapter 6). Finally, parsimony analyses of microarray data should become commonplace for gene-expression data with underlying hereditary relationships, such as for phylogenetic and population genomics.

1.5 Acknowledgments

I thank Pablo Goloboff for his extremely helpful comments on the manuscript, Mike Steel for his interest in helping me formalize Conjecture 1.3.1, and Steve Farris for the example in Fig. 1.2. Again, the work of Elliott Sober is acknowledged for useful examples. I also thank the authors of the various chapters of this book; their insights and topical reviews helped make this introductory chapter possible.

I

Philosophical aspects of parsimony analysis, including comparison with model-based approaches

What is the rationale for 'Ockham's razor' (a.k.a. parsimony) in phylogenetic inference?

Arnold G. Kluge

Anyone suggesting a justification for a method of inference—be it parsimony or anything else—should be careful to distinguish sufficiency from necessity. It is one thing to show that a set of assumptions suffice to justify a method; it is much more difficult to show that those sufficient conditions are also necessary. I have suggested a [likelihood] framework, which, if true, suffices to justify parsimony. But what assurance can there be that this framework is necessary for the method to be justified? The possibility always remains that different or more meager assumptions will suffice to legitimize the method.

Although this possibility cannot be ruled out in principle, there is an indirect test that provides some indication of whether the suggested justifying principles are fundamental. If those principles provide a general framework that allows one to characterize and investigate other aspects of phylogenetic inference, this is some indication that the framework proposed is not only sufficient, but fundamental.

Elliott Sober (1986, p. 41)

2.1 Introduction

Philosophers continue to debate the meaning and rationale of 'Ockham's razor.' For instance, Sober (1994) concluded that parsimony is not a global principle of theory evaluation because it has no subject-matter-invariant applicability. He also maintained (p. 77) "parsimony, in and of itself, cannot make one hypothesis more plausible than another," a position that obtains under the anti-quantity principle (AQP). On the other hand, according to Barnes (2000, p. 370), interpreting parsimony as an anti-superfluity principle (ASP), parsimony in and of itself does make a theory more plausible because it "releases a theory from its commitment to components unsupported by the relevant data." Moreover, Barnes went on to conclude that ASP is a global principle of theory

evaluation because it does not depend on any subject-matter-specific assumption. Adding yet another dimension to this recent debate, Baker (2003; see also Nolan 1997) justified ASP parsimony hypotheses on strictly quantitative grounds. Thus, parsimony has been rationalized in terms of AQP and ASP, and for which there are different quantitative rationales.

Most empirical scientists have shown little interest in these debates and the newer meanings and justifications for Ockham's razor, not that empiricists have ever exhibited much concern for the philosophical. Indeed, it is common to find no argument whatsoever for parsimony being cognitively virtuous in the evaluation of a set of competing theories or hypotheses, and if an argument is provided at all it is usually made in operational terms. But, in the absence of a concept, most kinds of justification are difficult, if not impossible, to make (Grant 2002). One of the underlying themes of this chapter is the importance of making the distinction between concept and operation, and where the former precedes the latter (see also Farris 1967, p. 44). Another important theme is identifying incoherence among nested conceptualizations.

Some unconcerned empiricists, when pressed, will fall back on syntactic simplicity, such as the conventionalist argument that stresses aesthetic value (e.g. W. C. Wheeler, personal communication, XVI Annual Meeting of the Willi Hennig Society). Such appeals to simplicity cannot, however, save conventionalism from arbitrariness, since choice of a convention is arbitrary (Popper 1959). Others have laid claim to the pragmatic, where for example the most parsimonious hypothesis is judged a "necessary property of methods of analysis" (Patterson 1988, p. 79). Still others have embraced simplicity/parsimony for its descriptive efficiency (Farris 1979; e.g. see Brower 2000; Frost 2000), but for which there is no epistemology. A few empiricists have spoken of a statistical, goodness-of-fit, kind of justification when applying a parsimony method, where for example the most probable or plausible hypothesis is claimed to be most predictive. A best explanation kind of justification has also been mentioned by a few scientists, but usually without stating the relevant cause and effect. To merely assume 'some kind of explanation' is to treat explanation as a primitive term.

The ontological status of that to which a parsimony criterion is applied is not always considered, but without which the application cannot be judged. For example, the goodness-of-fit kind of justification can, and often does, involve an explicit connection between parsimony and probability, and that in turn supposes there are multiple instances of a kind with which to statistically estimate or model fit. However, not all sciences are concerned with class concepts and lawful regularities, i.e. not all are nomothetic sciences (Wrinch and Jeffreys 1921; Grant 2002), and the justification for parsimony has to be sought elsewhere (see below). For example, it has been argued that there

can be no probabilification when the entities of interest—objects and events—are necessarily unique, as they are in a historical science like phylogenetics (Grant 2002; Kluge 2002).

Also relevant to parsimony is the question of the meagerness of the conditions legitimizing a choice among hypotheses. What are the minimally *sufficient* assumptions required to make an inference? And, of those justifications for parsimony that are ontologically sound and minimally sufficient, are there any that can be judged *necessary* (see epigraph)? I begin my evaluation of these questions in the inference of phylogeny with a brief discussion of the newer classifications of parsimony, and their most general justifications. To set the stage for my evaluation of the kinds of parsimony that have been, or are likely to be, employed in phylogenetic inference, I briefly explicate phylogenetic inference, as I see it, focusing on the ontological status of what is being inferred, as well as the scientific approach that is consistent with achieving that kind of knowledge. I conclude with some examples of how the fundamental nature of parsimony in phylogenetics, as well as theory unification, might be judged. I will only discuss uses of parsimony applied in the inference of phylogeny *per se*, i.e. relative recency of common ancestry. Parsimony as it pertains to networks (undirected graphs), methods of algorithmic efficiency, data exploration (Grant and Kluge 2003), optimization, and indexes of support will not be considered. Mention of parsimony in relation to outgroups and additivity, the study of adaptation, coevolution, and biogeography will only be made in passing. Each of these topics deserves a separate forum. While it is fair to say that I seek a necessary and sufficient justification for parsimony in phylogenetic inference, I will be satisfied if I am only able to convince a few more empirical scientists to become involved in debating the meaning and rationale of Ockham's razor (a.k.a. parsimony) in phylogenetic inference.

2.2 Parsimony: classifications and justifications

Syntactic simplicity has long been distinguished from ontological simplicity. Some philosophers refer to the former as elegance (Walsh 1979), while the term parsimony is applied to the latter.

According to Baker (2003), syntactic simplicity involves the number and complexity of hypotheses, where justification for that kind of minimization is sought in descriptive efficiency, subjective epistemology, such as aesthetic value, or instrumentalism. In addition to these, syntactically simple hypotheses can have practical consequences, since the simpler theory can be more clearly expressed and thereby more easily understood (McAllister 1996). "Such a structure, simply as a structure, is intrinsically perspicuous" (Walsh 1979 p. 243). This subjective, 'easily understood,' argument can also be interpreted objectively in terms of testability—a simpler theory being more logically improbable than a less parsimonious hypothesis (Popper 1959, p. 119; see below). Obviously, with this kind of understanding, we have passed from the syntactic to the ontological.

Some continue to restrict the domain of parsimony to the theoretical, usually the theoretical hypotheses that posit the fewest entities, objects, events, or processes, or ascribe certain properties to objects (Barnes 2000, p. 354). Appealing to a most-parsimonious hypothesis of species relationships in any of these senses certainly gives the appearance of assuming evolution is parsimonious. However, parsimony can be seen only as a rule of inference, not an empirical assumption of reality. It need not be an ontological claim that evolution, wherever it applies in the universe, is really parsimonious. Enjoining parsimony is then consistent with the background knowledge assumption of 'descent, with modification' (see below). Having said that, however, any phylogenetic method that denies the independent evolution of similarities would seem to be making a claim on the minimization of the evolutionary process.

Justification for parsimony is usually sought in explanatory power, realism, objective epistemologies, or objective cognitive (epistemic) values. Those who use explanatory power to justify parsimony are considered realists, not subjectivists, when they argue from cognitive (epistemic) criteria in favor of empirical assumptions. Moreover, in the sense of realism, there are good reasons to treat the epistemologically preferred hypothesis as tentatively true, i.e. as an objectively optimal knowledge claim.

As mentioned in the introduction, Barnes (2000) distinguished two kinds of parsimony justification, AQP and ASP. The former principle recommends positing as few theoretical components as possible, whereas the latter recommends against positing the superfluous. As Barnes (2000) exemplified (p. 354):

The two principles are clearly not equivalent: consider two competing theories, A and B, which both fit the relevant data equally well. Theory A contains more components than B, and is thus less parsimonious than B by the lights of the AQP. But while A contains no components that are not required (within A) to explain the data, theory B posits one or more superfluous components—i.e. one or more components which could be deleted from B without impairing B's ability to explain the data. Thus B is less parsimonious than A by the ASP.

The fact that AQP entails ASP, but not vice versa, leads to the conclusion that whatever justifies B may not justify A. Barnes (2000) thought ASP to be at the heart of what is generally known as inference to the best explanation.

As this discussion suggests, there are many ways to define Ockham's razor, as well as to classify its various kinds. My only reason for choosing Barnes' (2000) general background knowledge, pragmatic, unification, anti-free parameters and local background knowledge as the kinds of parsimony categories that I recognize in relation to phylogenetic inference is that they expose justifications that have been previously lumped together in this field. I have added testability and ASP categories to complete that exposure. As will become apparent, the justifications that I exploit in these discussions are the usual philosophical contrasts, between theory-based and narrative explanation, descriptive efficiency and explanatory power, subjective and objective epistemologies, instrumentalism and realism, and the subjective values of aesthetics and the like and the objective cognitive (epistemic).

2.2.1 General background knowledge

A general background knowledge kind of parsimony supposes the universe is naturally parsimonious in some way. Certainly one of the most general arguments for its justification is due to Sir Isaac Newton (Thayer 1953, p. 3), who claimed, "nature is pleased with simplicity and affects not the pomp of superfluous causes." This kind of justification is more than a curiosity, as will be illustrated below with examples from phylogenetic inference.

2.2.2 Pragmatic

Truth is tested by its practical consequences with this kind of parsimony, where the more parsimonious of two hypotheses stands a better chance of confirmation. According to Quine (1963, p. 105), for example, the more parsimonious hypothesis, "the one with fewer parameters, is initially the more probable because a wider range of possible subsequent findings is classified as favorable to it." This is an instrumentalist justification, and for which there can be no straightforward report as to what is explained (Walsh 1979, p. 244).

2.2.3 Unification

Friedman's (1983; see also Greene 2004) lengthy discussion of parsimony in relativistic physics underscored the importance of unified theories. As Barnes (2000, p. 356) put the issue inductively, "unified theories are multiply confirmed by the various empirical phenomena they explain, while a competing theory with less unifying power can only be confirmed by its smaller class of explanada—thus unified theories tend to be better confirmed than their disunified competitors." Eliminating those entities that are indeterminate or have no unifying power in the particular context is where parsimony is said to come into play in unification. Isomorphism and reduction are the traditional techniques used in the analysis of theory unification. For example, two or more theories are said to be isomorphic if there is a one-to-one mapping of their structures; e.g. if the properties attributed to them by their respective theories are the same. Reduction of theories focuses on the microstructure of phenomena, rather than on their physicalities, such as properties.

"According to scientific realists, the unification of theories reveals common causes or mechanisms underlying unconnected phenomena" (McAllister 2000, p. 538). Realists also point out that the relative ease with which theories of different domains

are unified would have to be considered fortuitous according to instrumentalism, which is currently popular in phylogenetics. The unification of evolutionary and genetic theories in the neo-Darwinian synthesis is an example that should be familiar to most contemporary biologists.

2.2.4 Anti-free parameters

This kind of parsimony concerns a preference for hypotheses with few adjustable parameters. For many years the focus was on minimizing parameters, and now, with so much emphasis placed on models in inference (Burnham and Anderson 1998), Akaike's (1973) approach to evaluating models is receiving most of the attention. Akaike argued that the predictive accuracy of models provides a reason for choosing among models, and he proved how the predictive accuracy of a model is estimated (Forster and Sober 1994; Sober 1996). Simply, old data are used to estimate the maximum likelihood of the parameters of the model, and this fitted model is in turn used to predict future data. Parsimony is involved when a penalty is given to the number of adjustable parameters, which is subtracted from the log-likelihood estimate of the best-fitting model. The proof of the theorem is not, however, without its difficulties (Forster and Sober 1994). For example, it is assumed that: (1) nature is uniform, i.e. the old and new data sets involved in the definition of predictive accuracy come from the same underlying distribution (Forster 2000), (2) the likelihood function is asymptotically normal, where the likelihood is a function of the parameter values, and (3) the sample size (amount of data) is large enough to ensure the likelihood function approximates asymptotic normality.

2.2.5 Local background knowledge

This anti-quality kind of parsimony is firmly grounded in local background knowledge. Several different rationales have been advanced that depend on subject-matter-specific assumptions. This is even true in a narrowly defined field like phylogenetic inference, as will be illustrated below.

2.2.6 Testability

Testability places a premium on the improbability of the hypothesis, not its probability (Popper 1959, p. 119, 1983, pp. 283–240; Kluge 2001b; contra de Queiroz and Poe 2001, 2003). Testability is defined objectively, as the power of an hypothesis to explain the evidence, in light of the background knowledge, where the data consists of reports of the outcome of sincere attempts to refute the hypothesis, not of attempts to confirm it (Popper 1959, p. 414; see also Salmon 1966, p. 46; Kluge 2001b; contra de Queiroz and Poe 2001, 2003). In maximizing severity of test, explanatory power and degree of corroboration are maximized. In Popper's logic of scientific discovery (Popper 1959, p. 145; see also below), enjoining parsimony protects the falsifiability of the system from going to zero.

2.2.7 Anti-superfluity

Being opposed to the superfluous has been justified in various ways. For example, Nolan (1997, p. 339) justified it in terms of plausibility; however, I believe his argument, that it is "Better, rather, to have quantitative parsimony expressed as a different principle to the independently plausible principle about explanatory parsimony, rather than tying them together in this way," unnecessarily emphasizes plausibility in determining extravagance (see also Baker 2003, p. 248). Barnes (2000) appealed to the subjectivism of Bayesian inference as his justification, while Baker (2003) made it clear that he interpreted ASP as pertaining to the class of cases that are demonstrably additive, i.e. that involve the postulation of a collection of qualitatively equivalent individual entities in the relevant respects, be they objects or events (see below). In this, Baker presumed an analysis of equivalent singular causal statements of one kind, of the form '*e* causes *e'*.' This is not a holist justification for parsimony in the sense that the strengths and weaknesses of one kind of science carry over to another kind. Its virtue is that collectively such equivalent entities can explain some particular phenomenon. "The explanation is 'additive' in the sense that the overall phenomenon is

explained by totaling the individual positive contributions of each object" where "quantitative parsimony tends to increase the explanatory power of hypotheses compared to their less quantitatively parsimonious rivals" (Baker 2003, p. 248). A less quantitatively parsimonious hypothesis can only match the most quantitatively parsimonious proposition in explanatory power by adding auxiliary claims. "Thus the preference for quantitatively parsimonious hypotheses emerges as one facet of a more general preference for hypotheses with greater explanatory power" (p. 258). However, it is also clear from Baker's lengthy discussion that he defined explanatory power as a primitive term. For example (p. 258), he simply concluded, "quantitatively parsimonious hypotheses allow the explanation of more things." In other words, Baker did not define quantitative parsimony specifically in terms of explanation in relation to evidence. The ASP from which this justification obtains is not strictly equivalent to Sober's (1981, p. 145) more general principle that entities should not be postulated beyond those that have explanatory power (see also Farris 1983).

2.3 The ontological status of phylogeny: what is ideographic science is not nomothetic science

The historical science of phylogenetic inference is ideographic (Grant 2002). The word ideographic, in this context, springs from the idea that relative recency of common ancestry can be represented directly as a concrete, spatio-temporally restricted, explainable thing, the phylogenetic hypothesis, cladogram, or tree, as can the accompanying transformation of an inherited trait or homologue. For all such things there is orderliness to their unfolding, a transformation series, and for more complex (highly integrated) traits there is usually an assumable direction to the sequence of change, based on such things as ontogeny. At the 'quantum level' of systematics, where physico-chemically identical nucleotide states (A, G, C, T) substitute for one another indeterministically, direction of transformation may not be assumed. In either case, there is an asymmetry between the past and the future. As an historical science, phylogenetics is then retrodictive, but not pre-dictive; it does not predict speciation events or evolutionary changes that have yet to happen. While the laws of relativity may govern time itself, the direction or arrow of time does not appear to be governed by relativity. It seems that time's arrow was merely conditioned at the birth of the universe, at a starting point of low entropy (Greene 2004, Fig. 6.3).

Phylogeny defined in terms of Darwin's (1859, p. 420) principles of "descent, with modification"—relative recency of species common ancestry (monophyletic entities or clades) and the trans-formation of the phenotype/genotype (homo-logous characters or character states)—is an evolutionary concept, where only heritable things can evolve. The event of transformation is the same factor that gives absolutely the same result in all places and at all times, and it is on this basis that the law of inheritance is argued.

Phylogeny is judged to be a lineage system, one consisting of "bundles" of characters transforming through time as part of the evolution of species (Hennig, 1966). Within a species lineage is located the lineage histories of the organisms, tokogeneti-cally related in the case of biparentals. Within each of those parts are located the ontogenetic histories of the more complex phenotypes. At the least inclusive level of spatio-temporal restrictedness there is the transformation series—what the evo-lutionary systematist claims as evidence (Grant and Kluge, 2004). This part/whole system is spatio-temporally restricted at each level. Phylogenetic inference is then devoted to the deduction of those sister lineages and the explanation of the singular, non-recurrent, heritable events in each bundle that mark such points of phylogenesis. Species diver-sification is the result of historical contingency and all of the propensities acting at the time of divergence (Kluge, 2002).

Inference in such an ideographic system is lim-ited to what can be observed of organisms, observations that are then used to retrodict some objectively defined part of the system's history to which they belong. That inference explains the heritable variation observed among lineages by identifying the series of necessarily unique trans-formation events that occurred in the lineage immediately subtending a lineage splitting event. Each stage of occurrence in a hypothesized trans-formation series is tested simultaneously against all other such evidence, thereby maximizing severity of test (Kluge 2003a). Phylogenetic hypotheses are chosen for their explanatory power, on the basis of the number of transforma-tion events they explain. The only things that are explanatorily relevant in this system are the cla-distic and patristic (Farris 1967; not the patristic of Sokal and Camin 1965; see below), which is to say that only the inherited patristic things can provide a critical test of competing cladistic hypotheses. The spatio-temporally unrestricted can be rejected, because it has no meaning in the part/whole sys-tem of organism history. The a priori testable and a posteriori reciprocally illuminated hypotheses are the statements of relative recency of species com-mon ancestry and homology, respectively (Kluge 2003a). As Grant and Kluge (2004) pointed out, however, the number of possible hypotheses of homology is defined a priori by pure logic, as a function of the number of inherited parts identified for each terminal taxon, just as all possible hypotheses of phylogeny are predefined as a function of the number of those terminals (Siddall and Kluge 1997). As such, no special procedure is required to generate hypotheses of homology, nor hypotheses of relationships, since they already exist.

Popper (1957, pp. 105–122, 143–147; see also Scriven 1959; Goudge 1961, p. 63; Hull 1974, 1982; Sober 1993, pp. 14–18) discussed the dis-tinction between historical things and lawful generalizations—the ideographic and the nomo-thetic, respectively. For the most part, the ideographic and the nomothetic are readily dis-tinguished in terms of being concerned with the spatio-temporally restricted and unrestricted, res-pectively. Nomological necessity pertains to rela-tions that are repeatable in an indefinitely recurrent way, or to sequences of variable phe-nomena, which are invariable under the same conditions. The historical entity—object or event—is firmly grounded in objective reality, whereas laws explain what is inherent in the abstract—classes (kinds) or sets of particulars (Grant 2002). No frequency-based probability exists for the necessarily unique. There is no basis for modeling the probability of error in the ideographic science of phylogenetic inference (Kluge 2002). As will be argued below, that homoplasy is a universal concept, the complement, 'not-a,' of homology means that it can play no role in the conceptualization of phylogeny or in its inference.

Hull (1977, 1989) and Sober (1980) forcefully argued that evolutionary theory *precludes* the conception of taxa, including species, as classes or sets. I agree, and would add that the end game or goal of identifying species and the monophyletic taxa of which they are a part involves historical questions, and which cannot be answered with nomothetic means (Kluge 2002; see however Rieppel 2004). Why is it then that some phylogeneticists use the lawful 'if/then' means of modeling historical relationships (e.g. Felsenstein 2004)? I believe the answer lies in "distinguishing *means* from *ends*," and from separating the nomothetic and ideographic sciences according to that distinction (Sober 1993, p. 14). As Sober (1993, p. 15) illustrated:

The astronomer's problem is a historical one because the goal is to infer the properties of a particular object; the astronomer uses laws only as a means. Particle physics, on the other hand, is a nomothetic discipline because the goal is to infer general laws; descriptions of particular objects are relevant only as a means.

The problem to ponder, as I see it, is how the means (say, of modeling) can justify the end (of identifying species and their relationships). By focusing first on the kind of question involved, as dictated by the ontological status of what is being inferred, as I argued above, you are effectively saying the opposite—that it is the end that justifies the means.

That there really is a general truth in this aphorism, consider what is often presented as one of the greatest challenges to Darwin's theory of natural selection—the origin of sex, where as a consequence the individual female wastes half her energy producing males. The significance of the problem is usually stated as a function of the number of biparental relative to uniparental species, n_b and n_u, respectively, and the solution to the problem is sought with lawful, if/then, means. These are the standards of neo-Darwinian science, such as modeling and frequentist statistical inference applied within and among populations, where

modeled accuracy and prediction are the goals. Alternatively, however, one can identify the minimum number of independent character-state transformations of sex, inferred from the most parsimonious hypothesis of species relationships, assuming for example only $n_b \leftrightarrow n_u$, and proceed to explain each of those relatively few unique past instances of evolution on a case-by-case basis. To apply the standard nomothetic means of analysis at the population level, as if the problem is necessarily one of lawful optimality, 'due to natural selection,' may be arguable, but to treat the n_b and n_u observations as if they are all independent in those analyses does not give consideration to the historical nature of the problem, including the uniqueness of each instance of transformation, and Darwin's other major principles of evolution, "descent, with modification." To think of Darwin's contribution only in terms of the lawful regularities of natural selection and adaptation is to miss the significance of his theory of propinquity of descent.

Popper (1957) also used the distinction between lawfulness and explanatory retrodiction, the nomothetic and the ideographic, as part of his argument that laws cannot predict history, nor explain trends in history, because you cannot use what is spatio-temporally unrestricted to retrodict what is spatio-temporally restricted. Even familiar law-like evolutionary statements, such as "all swans are white," can be accorded a singular, spatio-temporally restricted, object explanation, because "all white swans" can be hypothesized to be parts of an historical individual or monophyletic taxon (Kluge 1999). In addition, there is the argument that "evolutionary theory contains no reference to particular taxa, just what one would expect if taxa are actually individuals and not classes. According to this view, 'All swans are white' could not count as a scientific law even if it were true" (Hull 1977, p. 83). As Simpson (1964, p. 128) succinctly concluded, "The search for historical laws is . . . mistaken in principle."

Nomothetic science is not the domain of phylogenetics, not only because each instance of common ancestry is a spatio-temporally restricted unique part of history, but because each species is part of a replicator system (Lidén 1990) that renders it "necessarily unique," which is uniqueness in the strictest sense (Goudge 1961; Simpson 1964, p. 186; Kluge 2002; see however Hull 1974, p. 47, 97–98). The phylogeneticist cannot meaningfully practice estimation because there is no set of instances with which to assess a frequentist probability statement of species relationships. Not only is each event of common ancestry necessarily unique, so is each transformation event that is used as evidence (a proposition of homology) in the inference of phylogenetic relatedness and the explanation of observed biological diversity. Whether inherited variation is identified with DNA substitutions or the modification of a complex (highly integrated) phenotypic character, each hypothesized transformation or proposition of homology involves a necessarily unique event—an historical individual—just as are species and monophyletic groups of species (Grant and Kluge 2004).

The rationale one uses in defense of parsimonious inferences in phylogenetics must not rely on assumptions that violate the logic imposed by the ontological status of history. But what is assumed is often disputed. For example, there is the popular claim that uses of unweighted parsimony in phylogenetic inference are dependent, at least implicitly, on a model that assumes a constant rate of evolution in which all character transformations are equally likely to occur, and therefore parsimony is unable to identify any patterns of relationships other than those kinds. The error in this argument is obvious upon inspection of empirical results—parsimony methods often identify heterogeneous rates of character evolution in the unweighted most-parsimonious phylogenetic hypothesis. In other words, there is no basis for modeling data, either in light of, or independent of, the hypothesis.

Even this brief discussion indicates why those who apply parsimony must be careful to evaluate the ontological status of what is being inferred, as well as the nature of the evidence used in those inferences—which are the normative aspects of parsimony. No scientist should feel safe in appealing to parsimony without first assessing the ontological status of what he/she is making an inference. Even the instrumentalist's heuristic use of probability or likelihood as the basis for historical prediction, and not explanation, cannot be founded on an illogical thesis (Ariew 1998).

2.4 Causality and scientific practice in phylogenetic inference

Given the ideographic generalities discussed in the previous section, I can now explicate the specifics of the causality involved and the kind of scientific method that can be practiced in the inference of the necessarily unique parts of history. To begin with, the causal event of heritability is the sufficient condition for claiming a homology or historical identity (**H**), where that relation specifies a part of phylogeny (**P**), that which is ostensively defined in terms of common ancestry. The point is that an event of heritability is precisely located in relation to the object or character state that is inherited (Hennig, 1966, fig. 21). Although the part/whole relations of ontogenesis/tokogenesis/phylogenesis may be necessary to the conceptualization of causality in phylogenetic inference, it is the event of heritability or transformation that fixes the nature of the causality involved.

This logic does not tell us what things exist; it only suggests how to determine what things a theory claims to exist. Only **H** is relevant in the study of **P**, which in the simplest case of three terminal taxa, A, B and C, can be represented as (A,B)C, (A,C)B and (B,C)A. Since neither **P** or **H** are observable, *typically* a hypothesized shared derived state, a synapomorphy (**S**), is used as the unit of *empirical* evidence to test among the logically possible **P**, a choice of which in turn hypothesizes at least some **S** being explained by **H**. Regarding linguistic conventions (Kluge, 2003a, p 236), we can say that $P_{(A,B)C}$ causally explains $H_{(A,B)C}$, as inferred from $S_{(A,B)C}$, but not $H_{(A,C)B}$ or $H_{(B,C)A}$, as inferred from $S_{(A,C)B}$ or $S_{(B,C)A}$, respectively. Likewise, $P_{(A,C)B}$ causally explains $H_{(A,C)B}$, as inferred from $S_{(A,C)B}$, but not $H_{(A,B)C}$ or $H_{(B,C)A}$, as inferred from $S_{(A,B)C}$ or $S_{(B,C)A}$; or $P_{(B,C)A}$ causally explains $H_{(B,C)A}$, as inferred from $S_{(B,C)A}$, but not $H_{(A,B)C}$ or $H_{(A,C)B}$, as inferred from $S_{(A,B)C}$ or $S_{(A,C)B}$.

Two attendant considerations turn this ideographic kind of causal explanation into a *historical* kind of scientific operation (following the general outline provided by Goudge 1961). For the practice of phylogenetic inference to be scientific (1) the evolutionary principles of "descent, with modification," must contain concepts that do not necessarily correlate with what is hypothesized, the **P** and the **H**. These concepts are called theoretical constructs, which is what heritability is in the present case. These constructs may not be directly observable, but nonetheless play an important role in the framework of the theory. Their scientific admissibility depends on the fact that they occur in statements that have a deductive connection with statements that refer directly to the inherited object, i.e. the empirical data. It is because of this connection that theoretical constructs have scientific meaning conferred on them. In the practice of phylogenetic inference (Kluge 2003a), so-called observation statements, **S**, stand as a hypothesis of **H**, if at best a weak one (see below), which in turn provide the means whereby the theory, **P**, is tested to ascertain its falsity. (2) In addition, there are reasons for holding that a theory, **P**, is properly called *scientific* provided it entails observation statements, which are capable of being refuted by any empirical data (Kluge 1999). This guarantees that the theory, **P**, is potentially falsifiable. Character congruence, and the reciprocally illuminating process of character re-analysis, provides the test of such statements in phylogenetic inference (Farris *et al*. 1970, pp. 177–178; Kluge 1997b). There is no vicious circularity in this scheme of causality and testing (for further discussion see Hull 1967; Kluge 2003b, p. 365), because the observation statements, such as **S**, are not perfectly correlated with **H** (Farris *et al*. 1970). The logical proof of this obtains from the familiar argument that while all **H** are **S**, not all **S** are **H** (Farris *et al*. 1970, p. 187). Thus, the scientific quality of **P**, as inferred from **H**, is maintained.

Goudge (1961) also formulated two kinds of historical explanation, integrative and narrative. While neither supposed genuine scientific laws, they are not to be confused with the deductive historical explanation described immediately above, which I will consider further in my explication of phylogenetic inference. In the case of Goudge's integrative explanation, similarity relations and spatial patterns observed among organisms are explained in light of a phylogenetic hypothesis, showing the relations and patterns to be the outcome of, or partly dependent on, past *sequences* of historical phenomena, which have continuity and direction. Although Goudge argued that integrative explanations can be

causally and critically evaluated, most such analyses are fraught with issues of non-independence and enumerative confirmation, which I believe cannot be part of a scientific philosophy.

According to Goudge (1961), why particular historical events have occurred requires a narrative explanation. As he summarized (p. 77):

What we seek to formulate is a temporal sequence of conditions which, taken as a whole, constitutes a unique sufficient condition of *that* event. This sequence will likewise never recur, though various elements of it may. When, therefore, we affirm 'E because s', under the above circumstances, we are not committed to the empirical generalization (or law) 'Whenever s, then E'. What we are committed to, of course, is the *logical principle* 'If s, then E', for its acceptance is required in order to argue 'E because s'. But the logical principle does not function as a premise in an argument; the affirmation, 'E because s', is not deducible form it . . . Both s and E are concrete, individual phenomena between which an individual relation holds.

Critical to Goudge's thinking on integrative and narrative explanations (p. 174), and the part with which I do not take exception, is the idea that if we envisage a transformation series "as a unique sequence of historical events, extending from the past into the present, then it is irreversible in the sense of being *irrevocable*. What has happened cannot be altered, and *a fortiori* cannot be reversed."

The particulars involved in a historical narrative explanation are akin to the central subjects of literary narratives, and they take their place in the chronicle as a consequence of interpretative or explanatory writing. Under this interpretation, explanation is achieved through closure, that which contributes to the cohesiveness and conclusiveness of the chronicle. The analogue to closure in phylogenetics is Hull's (1975, 1981) notion of integration, where explanation is achieved through integrating ontological individuals, i.e. by making them wholes.

Narrative explanation is, however, not without significant problems. As O'Hara (1988) pointed out, not all clades exhibit closure; the more inclusive ones remain open, to the extent that any included lineage is extant. Also, O'Hara argued that the literary interpretation tends to emphasize linearity, which in reading the phylogenetic hypothesis promotes unnatural, paraphyletic, groups. O'Hara (1988, p. 153) concluded that these "false concepts arise out of our expectation that the central subject of an evolutionary history is a linear individual, instead of a branched tree." Then there is the disanalogy between the divergent nature of phylogenetic hypotheses and the reticulate nature of at least some aspects of cultural evolution, such as the "tree of knowledge." In addition, it remains unclear how historical integration (*sensu* Hull 1981; O'Hara 1988) would involve those properties exhibited by the central subjects, as well as those processes in which central subjects participate. These are the cause–effect relations (the sufficient conditions) that connect the central subjects, such as evolving species. Indeed, the historical narrative does not have the form of theory-based explanation, as in science, where an hypothesis is sought that has the power to explain the evidence (Popper 1957, 1962a, b; see however, Ruse 1971). Additional criticisms of Goudge's narrative explanation can be found in Ruse (1971; see also R. Laudan 1990).

2.5 Parsimony: justifications in phylogenetic inference

With the ontological status of phylogeny, the nature of historical causality, and the scientific practice of phylogenetic inference having been explicated, we are now in a position to more critically evaluate the justifications for parsimony in phylogenetics in light of these details.

2.5.1 General background knowledge

The Camin—Sokal parsimony method, popular for only a brief time in phylogenetics, was argued in terms of the parsimonious nature of the evolutionary process, such as the probability of character state change being rare. As Camin and Sokal (1965, pp. 311–312) stated:

Comparison by Camin of these various schemes with the "truth" led him to the observation that those trees which most closely resembled the true cladistics invariably required for their construction the least number of postulated evolutionary steps for the characters studied.

However, this justification for parsimony falls short of general background knowledge, because it

has been falsified empirically. Any unweighted most-parsimonious hypothesis of species relationships on which character states cannot be optimized as unique and unreversed disconfirms this justification, assuming the absence of systematic error. The best falsifiers in this regard are physicochemically *identical* nucleotide states. Moreover, this kind of falsification has long been considered commonplace (Felsenstein 1979, p. 60). At best, the probability of character-state change being rare is an auxiliary conditional of the kind one expects to find in a model. In fact, Felsenstein (1979) took Camin and Sokal's statement to be the model for a likelihood argument—the hypothesis of maximum likelihood being the rooted branching pattern that requires the fewest character-state changes to explain the observed data, *assuming* the absence of systematic error and processes that lead to reversals of character evolution, $1 \rightarrow 0$.

A similar auxiliary conditional, again not of the quality of general background knowledge, forms the basis for Farris' (1977a, b) Dollo method. Here, the opposite of the Camin and Sokal model is assumed. (1) Forward changes $(0 \rightarrow 1)$ are allowed, but are considered very rare. (2) As many reversions $(1 \rightarrow 0)$ are permitted to occur as are necessary to explain the data. As Farris discussed (1977a, p. 86): "A useful way of assessing the significance of a probability ratio between two trees is to compare it to the likelihood ratio between null and alternative hypotheses attained when the null hypothesis can be rejected in favor of the alternative at exactly error-rate a in large-sample normal statistics." Basically, the phylogenetic hypothesis with fewest reversions is preferred under that model of evolution (see review by Blackburn 1984).

Earlier references to a general background knowledge kind of minimum evolution assumption in phylogenetic inference can be found in Edwards and Cavalli-Sforza (1963, 1964), and that assumption continues to be explicitly assumed or implied (e.g. Dayhoff and Eck 1968; Dayhoff and Park 1969, p. 7–16; Crisci and Stuessy 1980; Cartmill 1981; Kumar *et al.* 1993; Pritchard 1994; de Queiroz 1996; Larson and Losos 1996; Gee 2000, p. 6–7). This class of justifications for discrete character-state change, as well as those for minimum distances (e.g. see Kidd and Sgaramella-Zonta 1971; Farris 1972; Rzhetsky and Nei 1992; Swofford *et al.* 1996, p. 451), not only generally fail in their presumption that evolutionary change is rare, but they exceed the sufficient general background knowledge premise of "descent, with modification" (see below).

2.5.2 Pragmatic

Many pattern cladists appeal to a kind of pragmatic justification for their use of parsimony. Their usual interpretation of pragmatic is like Friedman's (1983, p. 269), where the descriptively most efficient hypothesis is sought because it is argued to be the most predictive (Farris 1979). The kind of prediction most commonly mentioned is that of other characters, even in the situation where an evolutionary explanation of homology is explicitly denied (Brower 2000)! A frequently repeated justification for why pattern cladists exorcise evolutionary assumptions is that the most-parsimonious pattern of relationships, once determined, can then, and only then, serve as evidence of the basic principles of evolution. Brady (1985) went even further, arguing that the pre-Darwinian standing patterns of natural history—common plan, homology, ontogenetic parallelism, and the hierarchy of groups—are recoverable with *the* parsimony method, without presuming "descent, with modification." In other words, it is parsimony's inductive confirmation of patterns that is critical, not its evolutionary explanation. And, like fitting the simplest curve to a set of points, pattern cladists argue that an hypothesis will be produced in the long run that is predictively efficient and *arbitrarily* close to the truth. Of course, Brady's argument runs afoul of reification, i.e. it is illogical to interpret an abstract pattern as a historical thing (Kluge 2003b; see however Rieppel and Kearney 2002).

There are a number of additional points that can be made regarding the pattern cladists' pragmatic, theory-free, justification for parsimony when using it to predict phylogeny (see review by Kluge 2001a). (1) To begin with, the evidential basis for grouping species (taxa) is overall similarity, which is notoriously deficient when it comes to

identifying the relative recency of common ancestry required of all approaches to phylogenetic inference. As will be further discussed below, it is also imperfect when using shared-derived similar states (Farris *et al.* 1970, p. 187; Kluge and Farris 1999). Even physico-chemically identical states of nucleotide characters are well known to be an imperfect index to common ancestry. (2) Without a minimal *a priori* assumption of 'descent' there is no reason to presuppose a nested, hierarchical, pattern of relationships; it might just as well be a circular array, a reticulate pattern, or a periodic order. (3) There is also no reason to assume that the pattern of relationships is *necessarily* dichotomous, since the difference between the reticulate pattern of tokogeny and the increasingly divergent pattern of species relationships is theory-dependent. (4) There is no reason to exclude compatibility/clique analysis as a kind of parsimony method because the largest clique consists of a completely congruent set of characters, one in which all the evidence is unique and unreversed, without exception. (5) Without an assumption, such as "descent, with modification," there is no justification for optimizing character states at the internodes of a pattern of relationships. The fact that Wagner and Prim networks provide most-parsimonious hypotheses among the terminals without entailing internodes should make them the methods of choice for the pattern cladist. (6) Without a limiting assumption, such as 'descent,' i.e. one history, there is no reason to seek *one* most-parsimonious hypothesis of relationships. In other words, pattern cladists have no basis for using the phylogeneticists' optimality criterion of *the* most parsimonious tree hypothesis.

2.5.3 Unification

Some might consider systematics, including most of its sub-disciplines, to be the "poster-child" for disunification, not unification. For example, there is no consensus as to phylogenetic method and relevant evidence, let alone a theory of inference. Consider, phenetics was born out of a concern for similarity relations, shared states, not shared steps (Farris *et al.* 1970, pp. 178, 187), and pattern cladism continues to be an argument for phenetics

(Kluge and Farris 1999). Further, Bayesian and likelihood inference of species history involve excess assumptions and subjectivism, where there is little to connect what is observed to what is analyzed and explained as character data. Moreover, there is as yet little published that convincingly indicates theory unification within any of the disciplines that relate to phylogenetic inference, such as the study of adaptation, vicariance biogeography, coevolution, and taxonomy (see below).

However, I believe there is one overarching consideration that holds promise for the unification of all the historical sciences—which is to eliminate or reformulate the theories (and methods) of those disciplines and sub-disciplines that are inconsistent with the ontological status of what is being inferred. The history of species relations being what it is clearly defines the criteria for systematics more generally and phylogenetic inference in particular—an ordered set of historically contingent and necessarily unique events of common ancestry. As will be further discussed below, such a unification of this kind would eliminate those theories (and methods) that are indeterminate or inconsistent and cannot therefore have any unifying power in this kind of historical science. Such a unified theory in its ultimate form could then be legitimately judged an ideographic science, as distinct from the nomothetic. That is to say, historicism would finally be removed from phylogenetics (Popper 1957). This ideographic science would not only be recognized for its power to unify, but to simplify and explain the particulars of species diversity. A step in the direction of this kind of unification has already been taken in questioning the similarity-based theory of character and replacing it with an evolutionary concept of "transformation series" or "stages of expression" (Hennig 1966, p. 91; Kluge 2003b; Grant and Kluge 2004; see however Rieppel and Kearney 2002). This is considered significant because all of the important entities in phylogenetics, the species relations, as well as the statements of homology, conceptualized as spatio-temporally restricted objects or events, are interpretable in terms of evolutionary theory (Hull 1977, 1989; Sober 1980). This is not to say that another form of theory unification cannot be achieved according to other, or alternative, principles. If competing forms of phylogenetic unification should actually be formulated then they can be evaluated in terms of that which has the greater generality or scope, the more general and broader in scope being more vulnerable to refutation (see epigraph).

2.5.4 Anti-free parameters

Numerous authors have examined the premium placed on free parameters in terms of the models assumed by parsimony and likelihood methods of inference (e.g. see Farris 1973b; Felsenstein 1973, 1979, 1981a,b, 1982, 1983, 1988; Sober 1985, 1988a; Felsenstein and Sober 1987; Goldman 1990; Steel et al. 1993, 1994; Yang et al. 1995a; Yang 1996; Lewis 2001; Steel and Penny 2000; Steel 2002). According to Tuffley and Steel's (1997: 599, italics in the original; Steel and Penny 2000, Theorem 2) Theorem 5, "*Maximum parsimony and maximum likelihood with no common mechanism are equivalent in the sense that both choose the same tree or trees.*" No common mechanism in this theorem refers to the absence of constraints on edge parameters from site to site. This theorem does require, however, the simplest type of substitution model at a particular nucleotide position, a Poisson model, where each of the possible substitutions occurs with equal probability. Given just a tree and a single character (and no information as to edge lengths), the maximum likelihood estimate of the state at any internal node is precisely the maximum parsimony state.

More recently, Goloboff (2003) provided an example of anti-free parameter justification for parsimony in phylogenetic inference, where he proved that parsimony assumes fewer model parameters than does likelihood. Thus, that unweighted most-parsimonious hypothesis of species relationships must necessarily be included in any likelihood ratio test to decide whether the simpler model should be rejected.

However, I believe the fact that Tuffley and Steel assumed more model parameters than did Goloboff illustrates the futility of attempting to understand a method based on the model it supposedly implies. If exactly the same hypothesis can be understood as being derived from a maximally complex model (in terms of free parameters) or a maximally simple model (in terms of free parameters), or just assuming "descent, with modification," as background knowledge, then this demonstrates that focusing on free parameters is a meaningless exercise.

Another anti-free parameter justification for parsimony in phylogenetic inference is Akaike's framework (e.g. see Posada and Crandall 1998, 2001a, b; Sober and Steel 2002), where the goal of model selection is predictive accuracy, and parsimony is employed in hypothesis evaluation. However, the 'uniformity of nature' assumption disqualifies that framework (see above) when it comes to phylogenetics. To assume that the old and new data sets evolved according to the same underlying distribution is a counter-factual conditional, and one that is recognized among phylogeneticists as being generally false. More importantly, while this framework may provide instrumentalism with a kind of justification, its appeal to frequentism is denied in the study of phylogeny, because the events of interest and the relevant evidence are necessarily unique (Kluge 2002; Grant and Kluge 2004).

Minimally, any phylogenetic method, parsimony or likelihood, that assumes a model, can be criticized. First, models assume counter-factual conditionals. There is also the issue that models are usually statistical, and to relate them to the necessarily unique hypotheses of phylogeny is illogical. Further, to employ a model is to assume more than background knowledge, that which is minimally sufficient to provide a causal explanation of historical individuality (Kluge 2002; Grant and Kluge 2004; see also below).

2.5.5 Local background knowledge

This justification for parsimony has taken a variety of forms in phylogenetic inference, including weighting (Wheeler 1986; Goloboff 1993b, p. 83). The several kinds of weighted parsimony analysis—*a priori*, successive (iterative, *a posteriori*), implied (heaviest tree), support, and strongest evidence—attempt to correct for instances of homoplasious similarity, which is assumed to lead to a better-supported, more reliable, hypothesis

(e.g. Farris, 1966, 1969, 1979, 2001; Le Quesne 1969; Goloboff 1993b; Mindell and Thacker 1996; Penny *et al.* 1996; Salisbury 1999). All of the more explicitly stated arguments for differential character weighting assume some concept of conservatism/constancy (uniformatarianism; e.g. Goloboff 1993b; see Kluge 1997b).

More specific criticisms of weighting that must also be rebutted before proceeding with any such kind of practice include the following: (1) Weighting leads to suboptimal, less-parsimonious, not more-parsimonious, phylogenetic hypotheses when it comes to the data of observation (Kluge 1997b; see however Farris 1983). Thus, to weight is to be logically inconsistent with parsimony's goal of maximizing explanatory power and finding the best-supported hypothesis given the evidence (*sensu* Grant and Kluge 2003). As an aside, by the same argument, weighting does not maximize descriptive efficiency (Farris 1979; Kluge 1997a). (2) Assuming a conservative/constancy model of evolution also diminishes severity of test (Kluge 1997a). (3) Weighting contributes to a loss of character independence; there is a loss of independence by virtue of the fact that the members of any weighted-class of characters (the more or less conservative classes) are weighted the same (Kluge 1997b). (4) There is also a potential loss of information because an incongruent character state can in fact increase phylogenetic structure; e.g. a reversed state can be diagnostic of a monophyletic group (Källersjö *et al.* 1999).

Sober (1988a, 1994, p. 85; see also Felsenstein 1973) provided quite a different kind local background knowledge justification for parsimony— where parsimony impacts on likelihoods in terms of a common causal explanation. Many phylogeneticists, including some who call themselves cladists (e.g. Nelson and Platnick 1981), have implied that their inference extends from such an explanation. As developed by Sober (1988a), Bayes' theorem summarizes the plausibility of the common causal explanation of homology (**cc**) in relation to the separate 'explanation' of homoplasy (**sc**) in light of shared-derived character-state similarities or synapomorphies (**e**), $p(\mathbf{cc}, \mathbf{e}) = p(\mathbf{e}, \mathbf{cc})$ $p(\mathbf{cc})/p(\mathbf{e})$, and $p(\mathbf{sc}, \mathbf{e}) = p(\mathbf{e}, \mathbf{sc})$ $p(\mathbf{sc})/p(\mathbf{e})$. Sober (1988a, p. 79) traced the concept of common causal explanation to the idea of improbable coincidences developed by Russell (1948) and Reichenbach (1956): "If two events are similar in ways that would be immensely improbable if they had separate causes, we may reasonably hypothesize that they trace back to a common cause." Sober's 'Smith/Quackdoodle theorem' formalized a common causal explanation for three taxa (Sober 1988a, p. 239). In the present case, a hypothesis of homology is considered more plausible than a separate cause of homoplasy when the independent origin of similar shared-derived (synapomorphic) character states is relatively unlikely. Sober (1988a) argued, by analogy, that the genealogical relatedness of two people listed in a phone book as Smith is not as plausible as two people named Quackdoodle.

Given the same denominators in $p(\mathbf{cc}, \mathbf{e})$ and $p(\mathbf{sc}, \mathbf{e})$, $p(\mathbf{e}, \mathbf{cc})$ $p(\mathbf{cc}) > p(\mathbf{e}, \mathbf{sc})$ $p(\mathbf{sc})$, or $p[(A_1, B_1), \mathbf{cc}] > p[(A_1, B_1), \mathbf{sc}]$, where the hypothesis of homology is considered more plausible than the separate cause of homoplasious similarity observed in two species, A_1, B_1. For three taxa, the likelihood terms, $p(\mathbf{e}, \mathbf{h})$, are $p[1\ 1\ 0, (A,B)C] > p[1\ 1\ 0, A(B,C)]$, or $p[1\ 1\ 0, B(A,C)]$. It is in this context that set of terms was considered "a *prima facie* plausible inference principle" (Sober 1993, p. 174). While there appears to be explanatory power in Sober's Bayesian approach to phylogenetic inference, where shared-derived similarities are explained as homologues, his justification depends on a frequentist assumption of character-state occurrence (the frequency of Smiths and Quackdoodles in a phone book), as well as a causal explanation of homoplasy (see above). Those who assert that parsimony is a kind of likelihood have some form of plausibility parsimony in mind (e.g. Swofford *et al.*, 1996; de Queiroz and Poe 2001, 2003), not an unweighted parsimony analysis (see also Kluge 1997b, 2001b).

A phylogenetic hypothesis *cannot* tell you whether a given character relation is expected or not. Each transformation is necessarily unique, and any hypothesis of such change can only be true ($p = 1$) or false ($p = 0$); some frequentist probability value in between true and false has no meaning when applied to the concept of historical individuality of transformation (Kluge 2002; Grant and Kluge 2004;

see however Sober 1988a; Felsenstein 2004). To be sure, phylogeneticists can be wrong in their choice of data used to test some part of species history, but that concerns the uncertainty of the operational issues required to identify transformation series. There is no logic that says phylogenetic hypotheses are able to say how probable the observations are as evidence of common ancestry "if we append *further* assumptions about character evolution" (contra Sober 1994, p. 88; my italics). Yes, there is uncertainty in the observations systematists employ as evidence of phylogenetic relationships, a normal part of operationalism; however, as we will see below, there is no uncertainty in the relationship between the ideographic character concept of transformation series and the nested hierarchy concept of species relationships—the two concepts are perfectly coincident (e.g. see Hennig 1966, Fig. 21).

Further, the concept of support is important in the inference of species relationships (Grant and Kluge 2003), as measured by the relative degree of corroboration of the competing hypotheses, not their probability/plausibility. Assessing truth is subjective, founded on probabilities, statistics, or likelihood, where what is being inferred, and the evidence used to infer it, are misapplied class concepts. As argued elsewhere (e.g. Kluge 2002), it is illogical to treat historical individuals as class concepts, and to do so leads unnecessarily to over-reductionism (Frost and Kluge 1994).

2.5.6 Testability

Phylogenetic inference has long been cast in Popperian terms (Wiley 1975; see also Bock 1973), where testability is a function of the improbability of a hypothesis of relative recency of common ancestry, not its frequentist probability. In the case of phylogenetic inference (Kluge 2003a), assuming only "descent, with modification," as background knowledge, the evidence for competing hypotheses of sister-group relationships should be equally likely. Thus, in the simplest case of three terminal taxa, the possible hypotheses are $\mathbf{P}_{(A,B)}$, $\mathbf{P}_{(A,C)}$, and $\mathbf{P}_{(B,C)}$, and the expected data of observation equally likely, $\mathbf{S}_{(A,B)} = \mathbf{S}_{(A,C)} = \mathbf{S}_{(B,C)}$. However, if a large majority of one of those possible kinds of data were to be observed in an unbiased sample, say $S_{(A,B)}$, which counts against $P_{(A,C)}$ and $P_{(B,C)}$, but counts for $P_{(A,B)}$, then this is improbable given the background knowledge alone, but not under that background knowledge *plus* the postulated rooted cladogram $P_{(A,B)C}$ (Kluge 1997a). It follows from this improbability argument, where only incongruent data count as a falsifier, that severity of test increases with the number of those tests that have been carried out, and the more severe the test, as supporting evidence of the corroborated hypothesis, the greater power the hypothesis has to causally explain the data. The corroboration of the hypothesis by the evidence is simply the measure of the degree of support given by the evidence to the hypothesis, and explanatory power and degree of corroboration are maximized by minimizing the data on the hypothesis, where *"enjoining parsimony protects the falsifiability of the phylogenetic system from going to zero"* (Kluge 2003a, p. 237).

In this system, the unweighted most-parsimonious phylogenetic hypothesis requires the fewest character-state changes or steps. The application of this kind of parsimony in a total evidence analysis of equally weighted data minimizes the total number of hypotheses of character transformation required to explain the heritable variation observed among species and, as such, the unweighted most-parsimonious cladogram represents the objectively optimal phylogenetic theory (Grant and Kluge 2004; Kluge 2004). Moreover, support can be defined objectively in this system (see also above), as the "degree to which critical evidence refutes competing hypotheses. A hypothesis is unsupported if it is either (1) decisively refuted by the critical evidence or (2) contradicted by other, equally optimal hypotheses (i.e. the evidence is ambiguous), otherwise it is supported. That is, rational hypothesis preference is based on the *relative* degree of corroboration of competing hypotheses, where the hypothesis that is the least refuted by critical evidence is preferred" (Grant and Kluge 2003, p. 383). While some have argued that Popper (1959) interpreted testability only in nomothetic terms, there still appears to be no reason why it, and his epistemological principle of explanatory power, do not apply to the ideographic science of phylogenetic inference (Goudge 1961; Popper 1980; Kluge 2003a, p. 238).

2.5.7 Anti-superfluity

Farris' (1983) minimization of *ad hoc* hypotheses is a well-known ASP justification for most parsimonious phylogenetic hypotheses. The value of minimizing *ad hoc* hypotheses is unassailable, as philosophers and scientists alike acknowledge (e.g. Popper 1962b, p. 288; Farris 1983, p. 18), because such hypotheses are adopted only for the purpose of saving a theory from difficulty or refutation, in the absence of any independent rationale. Without such minimization there would be no way to distinguish personal belief from evidence in choosing among competing theories. Moreover, ad hocisms can be explanatorily empty. As Popper (1957, p. 103) pointed out, "the ad hoc hypothesis that the laws have changed would 'explain' everything," but in doing so would explain nothing at all.

Farris (1983) was clear that it was *ad hoc* hypotheses of a particular kind that are explanatorily superfluous. As he stated (p. 18; my italics), "the explanatory power of genealogy is...measured by the degree to which it can avoid postulating *homoplasies*," where (Farris 1989b, p. 107) "A postulate of homology explains similarities among taxa as inheritance, while one of homoplasy requires that similarities be dismissed as coincidental, so that most parsimonious arrangements have greatest explanatory power." Contrary to homologues, homoplasious similarities are then minimized in phylogenetic inference, according to Farris, because they do not constitute propositions of similarity that identify monophyletic groups.

There are, however, significant problems with the different ways homoplasy has been explicated. First, there is the issue of interpreting homoplasy as independently evolved instances of a similar kind (for a review see Kluge 2003b). Suffice it to say, similarity in this context is being treated as a class concept, one tied to lawfulness or natural necessity, where one or more immutable properties constitute the basis for intensionally defining any particular class or kind. As such, similarity is an abstraction, and so too is any group of organisms defined in terms of having properties of that kind (see however Sober 1988b). Aside from the arguments that have been lodged against using similarity in the inference of phylogeny (Hennig 1966; Farris *et al.* 1970; Kluge 2003b; see also below), homoplasy cannot be explained in terms of evolution, when homoplasy is intensionally defined as an immutable set of similarity relations. Only ostensively defined, spatio-temporally restricted, things have the potential to evolve according to Darwin's principles of "descent, with modification" (contra Sanderson and Hufford, 1996).

When propositions of homology are tested with character congruence, and from which homoplasy is deduced, homology and homoplasy become a complementary relation, a, and not-a, respectively. As the not-a relation, homoplasy is nominal (everything that a is not) and as such it cannot be causally explained. Of course, any one of the *independently* evolved instances of homoplasy might be explained in its own right as homology (Kluge 1999). None of this argument denies the lawfulness of natural selection, only as it applies to a set of independently evolved similar things (Kluge 2003b). Homoplasy *per se* can have no common causal historical explanation because the independently evolved instances of similarity are spatio-temporally unrestricted. If nature has taught us anything, it would be that living things respond to the same selective pressures in any number of ways, a lesson that is an anathema to inductive reasoning in comparative biology (T. Grant, personal communication).

Lastly, it was Farris' (1983) position that homoplasy is merely investigator "error" in the inference of homology. However, there is no *natural* causal explanation for such error. Although homoplasy as systematic error may be defined intensionally as a class concept, it cannot be modeled as if it were a historical law. Of course, increasing precision by minimizing error is a worthwhile endeavor in all sciences, but it has no epistemological standing itself. As Popper (1979, p. 356–357) recognized, a "precise statement can be more easily refuted than a vague one, and can therefore be better tested." However, as he went on to note, the theoretical or the explanatory has

logical priority over the practical or "instrumental" tasks of science, such as precision.

Phylogeneticists are concerned with the ideographic, patterns of inherited things that can be deduced from a common ancestral state. That being the case, I assert that homoplasy can be nothing more than a *description* of inferred transformation events; effectively, it is a description of explanations. Just as referring to something as similar is acausally descriptive, referring to something as homoplasious is acausally descriptive. Explanation of the observed, independently evolved heritable variation is achieved through the inference of transformation events, and nothing explanatory is added by referring to them as homoplasies.

Arguably, homoplasy is an example of Aristotle's "fallacy of accident," where distinct differences between the essential and the accidental are assumed, i.e. that independent transformations result in a set of "similar," causally accidental, things. As Ghiselin (1966, p. 148) argued, we may want to compare similar things, but it is an error to subsume one relation within the other, because homology involves some kind of similarity between organisms (Farris *et al.* 1970).

It is true that incongruent transformations can be made useful, both in the sense of Hennig's (1966) reciprocal illumination and also as a heuristic in developing and testing adaptive/selectionist explanations of particular transformations, but even in these cases there is nothing at all explanatory in the term homoplasy. Although further explanation may in principle be achieved for each of those transformation events, e.g. by establishing the selective basis for their origin and retention, those conditions may be determined to be causally the same (cf. homoplasy) or causally different, and in neither case does this impinge on phylogenetic explanation which is concerned with the spatio-temporally restricted, i.e. historical individuals.

I conjecture that it has been the inductionists' preoccupation with homoplasy (e.g. see Sober 1988, p. 32), with the possibility of interpreting *similar* character states as repetitions of a kind, that has given license to the myriad of methods concerned with which hypothesis is most likely to be true (Bayesianism), which hypothesis is statistically the most probable (frequentism), or which hypothesis confers the highest likelihood on the data (likelihoodism), where counter-factual auxiliary assumptions are entailed in an attempt to model the course of such a history of independent evolution. Repetitions, like repeated trials, may instantiate a class concept of some kind of similarity. That set of instances may even be used to generate a frequency profile that is interpreted as approximating a probability distribution relevant to some method. That concept, distribution and method may even be thought of as governed by a universal law or propensity. While this may be nomothetic science at its best, it bears no relationship to the practice of ideographic science.

This is the same argument that denies the use of weighting against instances of homoplasy, from *a priori*, successive (iterative, *a posteriori*), implied (heaviest tree), support, to strongest evidence (see above). It is true that the multiple hypotheses required to explain the variation that we describe as similar can lead to the reciprocally clarifying elimination of operational error (Kluge 1999). That, however, does not offer an *historical* epistemological argument for minimizing instances of independent evolution. The study of homoplasy may be of interest to students of function, but that research has no special meaning for historical biologists, independent of the separately evolved states being homologous. To be so concerned, the phylogeneticist is engaged in fallacious reasoning.

A quantitative kind of ASP has yet to be articulated as a justification for parsimony in the inference of phylogenetic relationships. Wheeler's (1996) direct optimization approach to gene-sequence alignment may be an analog of ASP, but an epistemological justification has yet to be provided for it (see below). To be sure, few phylogeneticists have shown much interest in the philosophical, and Baker's (2003) discussion of the quantitative justification was published only recently (however, see preliminaries by Nolan 1997). Also important is the additivity requirement—a collection of qualitatively equivalent individual objects or events in the relevant

respects—which would appear to make this justification an unlikely candidate for phylogenetic inference. Indeed, none of the character concepts usually referred to in phylogenetics suggest that kind of individuality, where character states can be interpreted as additive instances of one kind. Recently, however, Grant and Kluge (2004) made that connection with their definition of an ideographic character. As they stated (p. 29; my italics),

the application of phylogenetic parsimony in a total evidence analysis of equally weighted evidence minimizes the total number of hypotheses of transformation required to explain the *heritable* variation observed among species and, as such, the most parsimonious cladogram represents the objectively optimal phylogenetic theory.

In fact, it was this treatment of Hennig's (1966) transformation series character concept that I consider fundamental to my quantitative parsimony rationale—where the conceptualization of history determines the operational and methodological means used in its inference—and which in turn is critical to my attempt to focus phylogenetic inference only on the ideographic.

As Grant and Kluge (2004) pointed out, most of the character concepts currently in use emphasize kinds and degrees of similarity among terminal taxa as evidence of their relationships (for details see Kluge 2003b; see also Rieppel and Kearney 2002). For example, it is usually stated that "Derived similarity *is* evidence of propinquity of descent" and "Ancestral similarity is *not* evidence of propinquity of descent" (Sober 1994, p. 87). Aside from the problem of not being able to provide an evolutionary epistemology for "similarity"—because the properties of organisms to which similarities refer are spatio-temporally unrestricted, abstract, and immutable (see above)—there is no basis on which to claim an additive accounting. To begin with, the plesiomorphic and the apomorphic states of a single heritable transformation may entail any number of properties (Hennig 1966, pp. 92–93), with the total number of properties being infinite. Moreover, logic dictates that a similarity relation according to one kind of property cannot be equivalent to one based on another kind, because each kind has its own intensionally defined necessary and sufficient set of conditions. Thus, similarity is without a common currency, because it is one of degree as well as kind.

Grant and Kluge's (2004) ideographic definition of character is an event concept, events being things that happen, such as phylogenesis and transformation. That definition is not a material object concept, those being things to which physical features are attributed, like volume, mass, and being containable and storable, even though the object is the thing that systematists claim to observe when operationalizing the concept character, and it is the thing geneticists currently use to measure heritability, i.e. the proportion of the variance in a trait among individuals that is attributable to differences in genotype. What is it then that allowed Grant and Kluge to argue that their transformation series character concept is concerned with heritability when the ontological distinctions between event and object imply their incommensurability? To begin with, the problem is simplified by virtue of the fact that the transformation event(s) and the transformed object(s) form a spatio-temporally restricted, historically contingent, transformation series (Hennig, 1966, fig. 21). That is, the locatability and mobility of the event is not a problem with reference to the object, they are causally related, and consequently, paraphrasing Woodger (1929, pp. 301–302), it can be stated that the perceptual object we also call the character state is expressive of certain of the knowable characteristics of the event that can be exemplified in sense-experience. That is, the character state is the event and the event is the character state, or, in a word, the event and the object are *coextensive*. Thus, the ontological distinctness of mutation (event) and mutant (object) concepts does not deny their causal continuity and their comparability in such terms as heritability. Moreover, it is because the transformation events occupy the same place in the causal sequence, i.e. have the same causes and the same effects, and that they are identical with events described in the causal law of inheritance, that they can be considered identical and additive (Davidson, 1991; Baker, 2003).

With regard to the systematists' and geneticists' character operationalisms, it is important to recognize that the phenotypic states that are attributed to an organism are, at best, only proxies for the actual "stages of expression" in the transformation series (Hennig, 1966, p. 91). For example, no one should be fooled into thinking that the states of eye color and handedness in humans are things that literally pass from parent to offspring. An investigator would do well to sample nucleotides if precision in heritability is of particular concern.

As already mentioned above, Hennig's (1966, Fig. 21) transformation series character concept, assuming just "descent, with modification," as background knowledge, does not entail the contradictions and inconsistencies with respect to evolutionary theory as do similarity-based definitions. More importantly in my formulation of a quantitative parsimony rationale, adopting Grant and Kluge's (2004; see also Kluge 2003b) ideographic explication of Hennig's concept, insofar as it is relevant to phylogenetic explanation, each inferred transformation is metaphysically the same kind of process, with each such event counting equally as heritable evidence in the analysis of singular causal statements of that kind (Bach 1981; Davidson, 1991).

How this quantitative ASP parsimony rationale is connected to Grant and Kluge's (2004) ideographic character concept is clarified by Farris' (1967) definition of evolutionary relationship. In that seminal, but largely overlooked, paper, Farris distinguished phenetic and evolutionary or phylogenetic systems, and in doing so he made distinctions and identified other relevant parameters sufficiently rich in ideas to restrict the permissible meanings of relationship. As he pointed out, in distinguishing phenetic relations (pp. 45–47; my italics), "The best one can do is to study the form of the measure of overall phenetic *similarity* . . . until the meaning of overall similarity is standardized." Whereas, with four axioms, he precisely defined *a priori* the evolutionary form of the measure of phylogenetic relationship.

Axiom 1: The objective of the [phylogenetic] system is to place [species] in such a way as to describe their patristic and cladistic relationships as completely as possible.... Axiom 2: The patristic difference between [species] is a function of the displacement in all the unit characters of the [species] along the phyletic line connecting the [species].... Axiom 3: The phylogenetic relationship between two given [species] is a fixed value.... Axiom 4: The measures of patristic and cladistic difference are non-negative real numbers.

And from these axioms, Farris characterized two phylogenetic relationship functions, cladistic and patristic (not the patristic of Sokal and Camin 1965), with phylogenetic relationship being the negative of the corresponding cladistic or patristic differences. For example, as he stated:

The overall patristic difference is the sum of the patristic unit character differences. Each patristic unit character difference is the summation of the changes of that character from point to point over the phyletic line between the [taxa] compared.

Likewise, Farris defined cladistic difference as the sum of the number of lineage divergences between any two taxa and their most recent common ancestor, which means that both phylogenetic relationship functions have the same properties of historical individuality, where each divergence and transformation is spatio-temporally restricted and necessarily unique. As Farris (1967, p. 47) concluded, "The two components of evolutionary difference thus have similar properties, and this fact lends a certain unity to the concept of [phylogenetic] relationship."

Let there be no mistake, Farris' definition of phylogenetic relationship and relationship functions are genealogical and not phenetic. As Darwin (1859, p. 420) stated:

All the foregoing rules and aids and difficulties in classification are explained, if I do not greatly deceive myself, on the view that the natural system is founded on descent with modification; that the characters which naturalists consider as showing true affinity between any two or more species, are those which have been inherited from a common parent, and, in so far, all true classification is genealogical; that community of descent is the hidden bond which naturalists have been unconsciously seeking, and not some unknown plan of creation, or the enunciation of general propositions, and the mere putting together and separating of objects more or less alike.

While Kluge and Farris (1969; see also Farris 1970) provided an heuristically efficient, if not an effective, algorithm in their Wagner method for choosing a best fitting hypothesis, it was Farris *et al.* (1970) who further clarified and extended the meaning of cladistics and patristics (Farris 1967). They abstracted four premises from Hennig's (1966) "Phylogenetic Systematics," and from these they derived three theorems, plus corollaries, which they used to explicate evolutionary tree hypotheses in accordance with Hennigian phylogenetic principles. Their specific points that constitute the basis for my ideographic interpretation of phylogenetic systematics are as follows. (1) Their Axiom I described Hennig's transformation series concept of character (see also Grant and Kluge 2004), which defined the evolutionary ordering of character-states as plesiomorphous and apomorphous in the simple case or as a character-state tree when the transformation series consists of more than two stages of expression, and they allowed reversals and any state to be potentially permissible as the most ancestral state for some restricted part of the tree (their Axiom I'). (2) According to their Axiom II (Farris *et al.* 1970, p. 173), all monophyletic groups are distinguished by *sharing* one or more apomorphous "stages of expression," whether the group has an apomorphic state x or a state apomorphous relative to state x (as determined by the predefined character-state tree). (3) Transformation series or stages of expression were characterized in terms of "steps" or "derived steps," where emphasis was put on sharing "stages of expression." As Farris *et al.* (1970, p. 174; italics in the original) summarized, "two [taxa] with states y and z *share a step*, x, if and only if y [is derived from] x and z [is derived from] x. (4) In their Axiom III (Hennig's auxiliary principle), in the absence of evidence to the contrary, any state corresponding to a *step shared* by a group of taxa is assumed to be unique and unreversed (at least locally). (5) Their Axiom IV measured the strength of the evidence for a monophyletic group—the more characters certainly interpretable as apomorphous the better founded is the assumption the group is monophyletic. (6) Their Theorem I provided the basis for describing the common ancestral state of a monophyletic group, i.e. the most derived state from which the sister lineages are derived. In this theorem they identified homologous states, at least in the unambiguously optimized case. (7) As a corollary, their Theorem II described the same relation for taxa, i.e. the common ancestor for a monophyletic group is the most derived hypothetical taxon from which the sister lineages are derived. In these two theorems, they had characterized both monophyletic taxa and transformation series as spatiotemporally restricted. (8) Their Theorem III stated, in terms of derived steps, the evidential basis for a taxon being excluded from a monophyletic group. Summarizing, Farris *et al.* identified a close connection between Hennigian phylogenetic systematics and unweighted most parsimonious hypotheses of species relationships, and they found the Wagner method for inferring those hypotheses (Kluge and Farris 1969; Farris 1970) to be consistent with their generalization of the Hennigian axioms. Effectively, they made the connection between amount of evidence (Axiom IV) and the minimum number of steps required of the unweighted most-parsimonious hypothesis of species relationships, which is measured by the additive requirement of quantitative parsimony.

The virtue of minimizing the quantitatively superfluous—the patristic difference—is that collectively the hypothesized heritable stages of transformation (**T**, not syanpomorphy or **S**) can explain the particular phenomenon of homology (**H**) as a nested series of such statements. The explanation is 'additive' in the sense that the overall phenomenon of phylogeny (**P**) is explained by totaling the individual positive contributions of each transformation, where quantitative parsimony tends to increase the explanatory power of phylogenetic hypotheses compared to their less quantitatively parsimonious rivals. Less quantitatively parsimonious hypotheses can only match the more quantitatively parsimonious propositions in explanatory power by adding auxiliary claims of one sort or another.

The following injunction summarizes these details—*choose the hypothesis of cladistic relationships that minimizes the overall patristic difference, because that hypothesis has the greatest power to explain the independently heritable transformation*

events as propositions of homology. In keeping with Farris *et al.*'s (1970, p. 172) reference to a "quantitative analog of phylogenetic systematics," I believe it is fitting to designate this ideographic kind of phylogenetic inference *quantitative phylogenetic systematics* (QPS), and I name its quantitative parsimony rationale *Farris parsimony* (FP), in recognition of James S. Farris' many significant contributions to the theory of phylogenetics. Wagner parsimony remains an efficient, if not an effective, method for operationalizing FP (Kluge and Farris 1969; Farris 1970).

I underscore the fact that not one of the parameters of QPS (*sensu* Farris 1967; Kluge and Farris 1969; Farris *et al.* 1970) is conceptualized in terms of similarity, nor is FP's rationale identified with the minimization of *ad hoc* hypotheses of homoplasy. Setting aside similarity in these conceptualizations means more than discounting overall similarity, which subsumes symplesiomorphy, synapomorphy, and independent evolution (Hennig 1966; Kluge 2003b). It also means setting aside synapomorphic similarity, s(A,B), i.e. "shared-derived character states," when it comes to the conceptual. At the very most, conceptually speaking, one might say that QPS is left with a kind of similarity "owing to ancestral states," as per Farris *et al.*'s (1970, p. 187) formal distinction between s(A,B) and s_E(A,B). And, as they pointed out:

The actual choice of a phyletic tree is left to an algorithm that effectively constructs the evolutionary hypothesis most in accord with available data. Thus only a weak connection between s or s_E and relationship is assumed.

Unfortunately, their distinction between the conceptual and the operational when it comes to similarity in phylogenetic inference (see also Kluge 2003b; Grant and Kluge 2004) has been largely overlooked, even in the most recent literature (e.g. see Mayr and Bock 2002; Ghiselin 2004; Padian 2004).

Although there may not be an absolute criterion for knowing the truth, as I stated above, specifying the conditions for truth is no more burdensome in QPS than it is in other approaches to phylogenetic inference. For example, given three terminal taxa, the statement "(A,B)C is true" *if and only if* A and B share a more recent common ancestor than either does with C; "(A,C)B is true" *if and only if* A and C share a more recent common ancestor than either does with B; "(B,C)A is true" *if and only if* B and C share a more recent common ancestor than either does with A. In turn, a transformation series characteristic of A and B is presumed to be homologous *if and only if* the stage of expression observed in A and the stage of expression observed in B can be derived eventually from the ancestral state of the group (A,B), not including C; a transformation series characteristic of B and C is presumed to be homologous *if and only if* the stage of expression observed in B and the stage of expression observed in C can be derived eventually from the ancestral state of the group (B,C), not including A; a transformation series characteristic of A and C is presumed to be homologous *if and only if* the stage of expression observed in A and the stage of expression observed in C can be derived eventually from the ancestral state of the group (A,C), not including B (Theorem I, Farris *et al*. 1970, p. 175).

As suggested above, the virtue of ASP is that collectively qualitatively equivalent things can explain some particular phenomenon, like transformation series, whose stages of expression constitute the basis for inferring the parts of species history. It is the application of FP—choosing the unweighted most parsimonious hypothesis of species relationships—that maximizes explanatory power, i.e. the stages of expression form a nested series of homology statements. I maintain that the difference is conceptually significant between minimizing steps, $s_E(A,B)$, in order to maximize explanatory power, and minimizing *ad hoc* hypotheses of homoplasy in order to explain shared-derived character state similarities, $s(A,B)$, as homologues. Transformation has a common causal explanation in this historical science—heritable change—whereas similarity and homoplasy do not.

2.6 Judging the ontological consistency and sufficiency of parsimony justifications

Each parsimony justification discussed in the previous section was examined for its ontological and epistemological consistency in the inference of phylogeny—inconsistency being any apparent negation or contradiction among the concepts and operations that are claimed to lead to advances in objective knowledge (Grant 2002). In these evaluations I was primarily concerned with the normative aspect of parsimony, what justifies the minimization, and not with the descriptive or how the related optimality criteria are scored (Sober 1983). The limited evaluations undertaken are an example of normative naturalism, where rules or criteria are identified for picking the theories or concepts according to the aims of the discipline at hand (L. Laudan 1990).

Leaving the theory unification justification for parsimony until the final section, it is in this sense of evaluation that all but the testability and quantitative justifications were judged to be inconsistent, having failed ontologically and/or epistemologically. For example, the failures of the general background-knowledge justifications were simply a function of the falsity of the assumptions they make. And, claiming an advance in knowledge is not possible according to the pragmatic justification because there is no way to evaluate either 'best fit' or explanation. The principal reason for the failures of the other justifications was not distinguishing between the ideographic and nomothetic—not taking account of the ontological status of what is being inferred—each instance of common ancestry being necessarily unique and not a class or set of things.

What constitutes a minimally sufficient justification for parsimony is another important kind of evaluation. In this I am guided by the 'principle of less is more,' a conditional of the usual form, if *p* then *q*, if the less of something then the more of something else. For example, according to Sober (1988a, p. 11; my italics), the "less we need to know about the evolutionary process to make an inference about pattern, the more *confidence* we can have in our conclusions. From the point of view of an evolutionary theory that is used to uncover phylogenetic relationships, the best outcome would be that minimal process assumptions suffice to identify that pattern. If on the other hand a detailed understanding were required of why evolution proceeded in the way it did, then an inference about pattern would have to await a detailed understanding of process," or be unattainable if the understanding of process is entirely dependent on knowledge of the historical pattern.

How to exploit the 'less is more' conditional is suggested by Sober's reference to "confidence", i.e. there being a basis in the logical condition of being necessary/sufficient: if p is a *necessary* condition of q, then q cannot be true unless p is true. If p is a sufficient condition of q, then given that p is true, q is also true. However, as Sober recorded, there are some potential difficulties with this logic when applied to current topic. (1) What is the minimum set of assumptions required to make an inference of phylogeny? (2) At some point, pattern will become irretrievable, because process assumptions will have become too meager, and how is that point to be recognized (Sober 1988a)?

In phylogenetic inference, the principles of "descent, with modification," are widely considered the minimal set of assumptions. To assume "descent" alone is too meager, since it does nothing more than provide an assumption of common ancestry (contra de Queiroz 1992, p. 305), albeit an important assumption in the argument against a creationist interpretation of pattern. To assume absolutely nothing about evolution, as pattern cladists claim, is obviously vacuous because it provides no basis whatsoever for an empirical evaluation of competing hypotheses of relative recency of common ancestry. The simplicity justification that Goloboff (2003) attributed to parsimony pertains to model parameters, i.e. auxiliary assumptions, which are in addition to "descent, with modification."

Only premises that are not known to be false can serve as background knowledge, and it is in this sense that assuming just "descent, with modification," is not considered problematic (Siddall and Kluge 1997, p. 320). Further, the hierarchy of relative recency of common ancestry, given this particular example of background knowledge, cannot be judged an *a priori* truth (*pace* Brady 1994, p. 22). On the other hand, models are problematic, because they are counterfactual conditionals, of the general form "if p were to have happened q would have happened, where the supposition of p is contrary to the known fact not-p." It is for this

reason that assumptions used in the inference of phylogeny, other than "descent, with modification," should be looked upon with skepticism, if not outright rejected.

The 'less is more' principle can also be interpreted as saying "the less one assumes the more one can test, and thereby explain." For example, in assuming a model of homogeneity of rate of evolutionary change in phylogenetic inference, one cannot in turn use the phylogenetic hypothesis to measure that rate or its homogeneity. Likewise, if the temporal order of fossils is used to polarize a character then that temporal record cannot serve as the basis for testing competing hypotheses of species relationships (Donoghue *et al*. 1989). Consider further, to *a priori* down-weight transversions relative to transitions leaves no opportunity to judge that inequality most critically, i.e. historically. In Popperian terms, the simplest (unweighted most-parsimonious) set of assumptions maximizes severity of test, and in turn explanatory power and degree of corroboration (Kluge 2003a). Historical scientists cannot afford to lose these opportunities to critically evaluate hypotheses of relative recency of common ancestry, because that is the basis for providing objective knowledge.

I conclude that the testability and quantitative justifications for parsimony are judged ontologically consistent, as well as minimally sufficient, in that they rely solely on the non-problematic background knowledge of "decent, with modification." Not only are model assumptions of some of the other justifications ontologically inconsistent, they violate the principle of less is more. The question remains, however, whether the testability and quantitative justifications are rival, complementary, or coextensive in the inference of phylogeny. The basic issue is whether or not QPS is inclusive of testability.

Having a skeptical research ethic, as formally provided by testability, is certainly to be commended in the historical sciences, because nothing inferred can be proven to be true, or even probably true (Kluge 2002). As already discussed above, testability places a premium on the improbability of hypotheses, not on their probability, where evidence consists of reports of the outcome of sincere attempts to refute a hypothesis, not of attempts to verify it. Thus, falsificationism is distinguished from verificationism, deduction from induction (Kluge 1997a). In maximizing severity of test, given the total relevant available evidence, explanatory power, and degree of corroboration are maximized, and objective knowledge gains can be claimed, as can an objective measure of support (Grant and Kluge 2003; Kluge 2004). All that is required of testability is that the investigator exhibit no *a priori* bias towards one or more of the competing phylogenetic hypotheses, as in the simplest case of $P_{(A,B)}$, $P_{(A,C)}$, or $P_{(B,C)}$. Thus, when a large majority of one of the kinds of data, $T_{(A,B)}$, $T_{(A,C)}$, or $T_{(B,C)}$ in QPS, is observed in the unbiased sample, say that which counts for $P_{(A,B)}$, then the phylogeneticist can argue that this is improbable given only "descent, with modification," but not given that background knowledge plus the rooted cladogram $P_{(A,B)C}$. That a parsimony algorithm maximizes severity of test is what is important, not that its justification is sought in a particular argument, such as testability or the quantitative interpretation of ASP.

I believe testability benefits QPS in another way, besides providing for severity of test. As Grant and Kluge (2003) pointed out, testability is accompanied by an objective concept of support, which QPS is not provided with in the quantitative justification for parsimony. While it can always be argued that such a concept will eventually be defined for that justification, I fail to see that there is any room for it. By this I mean FP is limited to maximizing explanatory power, without the accompanying severity of test that underlies the concept of support in testability. A concept of support might be formulated for QPS but I don't see how it could be anything but "explanatory power = support." Having defined QPS as including testability, this problem may be considered mute. While adopting the philosophy of testability in the practice of QPS has great merit, it is FP that provides a sufficient justification for advancing our knowledge of species relationships (Kluge 2003a, p. 237).

2.7 The fundamental nature of Farris parsimony in phylogenetic inference

While FP certainly qualifies as sufficient in the inference of phylogeny, and I believe minimally so

as well, some may doubt those conditions are also necessary. Taking my cue from Sober (1986, p. 41; see epigraph), I will now judge how necessary FP is by examining some of its sufficient conditions for fundamental contributions—those justifying principles that provide a general framework for characterizing and investigating the empirical aspects of phylogenetic inference and related fields of inquiry.

In this regard, I believe the empirical nature of phylogenetic inference benefits significantly from parsimony being defined in terms of transformation series. For example, according to that definition, FP then provides an evolutionary epistemology. That argument, either the empirical or the epistemological, cannot be made when the concept of character is defined in terms of similarity, s(A,B) (Hennig 1966; Farris *et al.* 1970, p. 187). Moreover, similarity is neither predictive nor projectible (Kluge 2003b).

Further, it is equally important to re-emphasize the fact that FP departs significantly from similarity valued comparisons, where the phylogeneticist is faced with having to argue relations in terms of natural kinds and properties, concepts that are contradictory, if not antagonistic, to evolutionary theory (Frost and Kluge 1994; Kluge 2003b; Grant and Kluge 2004). The ideographic character concept defined by Grant and Kluge (2004), with its unambiguous reference to the concept of transformation series (Hennig 1966), and in turn heritability is not only evolutionary, it focuses directly on what it is the phylogeneticist is concerned with—parts of species history and homology—and more precisely on the congruence of two kinds of historical things, monophyletic taxa and transformation or heritable change (Farris *et al.* 1970). That FP provides a criterion for choosing among competing hypotheses of these parameters may be interpreted as a fundamental conceptual advantage over all those methods that rely on counting instances of similarity of a kind, including those similarity-based uses of parsimony, s(A,B).

Character coding is an area where similarity has been mistakenly involved in a variety of ways. For example, Lipscomb (1992) advocated that *all* multistate characters be treated non-additively, even though that kind of coding discards both the form and direction of the character-state tree, preserving only the identity of the character states. Whereas, additive coding preserves the form and the direction of evolutionary ordering, and as Farris *et al.* (1970, p. 181) concluded, only "Additive coding corresponds directly to the operations employed in the phylogenetic system," even though an analysis of non-additive characters results in a hypothesis of relationships equal to or of fewer steps than an additive analysis (Grant and Kluge 2003). Curiously, Lipscomb (1992, p. 51; italics in the original; my bold) asked the correct question, *"If parsimony is minimizing some assumptions and this is to be used to derive hypotheses of multistate character* **transformation***, we must decide what types of assumptions are most important to minimize,"* but then proceeded to operationalize it incorrectly. Incorrectly I say (p. 52), because the "order of the states is postulated so that states that are most similar are adjacent to each other" was the first step in her "transformation series" method. To assume the non-additivity of complex morphological characters is in principle to embrace a pattern cladistic kind of phenetics, i.e. to be content with similarity relations or *shared states*, s(A,B) (Kluge and Farris 1999). Moreover, Lipscomb's (p. 54) elaborate method for distinguishing "the incongruence in the character that is due to the character state order from that caused by non-homology of states" becomes a non-issue. As Farris *et al.* (1970, p. 181) clearly had in mind, the *shared steps* ordering relation of taxa and character states is integral to the phylogenetic system, of which FP is a part, whereas "in phenetic practice this need not be."

An unweighted most-parsimonious phylogenetic hypothesis is chosen because it counts more transformation events as homologues than does any competing hypothesis, and when the concept character is defined ideographically, as an historical individual, all such events are necessarily independent. "The dependency between such transformation series is non-problematic, because it merely reflects the transformation event(s) they share, i.e., the shared portion of their history" (Grant and Kluge 2004, p. 26). The functional and developmental dependence that occurs at the level of the organism is not problematic either, because it merely reflects the integrated nature of the organism, as a whole. However, a category error is committed when these two kinds of the non-independence are conflated (e.g. see Naylor and Adams 2001, 2003), and the problem is unavoidable when the concept character is defined in purely operational terms, as it was in Rieppel and Kearney's (2002) similarity definition of character. Dependence in QPS is not a problem, providing the ideographic concept of character is primary, and from which the operationalisms of character analysis follow. To observe functional and developmental dependence may effect how the systematist *a priori* defines the unit character, but the possibility of such dependence does not invalidate choosing the unweighted most-parsimonious hypothesis.

The importance of the distinction between $s_E(A,B)$ and $s(A,B)$, a patristic difference and synapomorphic similarity difference, is no more evident than in the analysis of homologous nucleotide sequences that differ in length. Typically, in multiple sequence alignment, gaps are introduced so that base correspondences can be interpreted as shared similarities. Alternatively, there is Wheeler's (1996) direct optimization approach, which was founded on the idea that the number of DNA sequence events is provided directly by phylogeny, with character-optimization procedures finding the minimum number of those events on the competing tree hypotheses. Ignoring the fact that Wheeler's actual method used weighted (cost) functions (relying ultimately on frequentist probability arguments), the relevant conceptual feature of direct optimization is that it analyzes *all* events as transformations, $s_E(A,B)$, insertion and deletion (indels), as well as substitutions, rather than the implied similarity relations that obtain from multiple sequence-alignment methods. However, Wheeler's (p. 1) justification for that minimization was descriptive efficiency, direct optimization providing "more efficient (simpler) explanations of sequence variation than does multiple alignments." But as Frost *et al.* (2001, p. 354) pointed out, even that interpretation is only consistent when setting all substitution costs and the unit gap cost equal. Still, the bottom line remains that unweighted direct optimization can find a pattern of character state change that is more parsimonious than those based on maximizing pair-wise statements of alignment similarities, s(A,B) (e.g., compare figs. 1B and D to 1C and E in Simmons, 2004). New methods that use s(A,B) and claim to avoid the problem of suboptimal hypotheses have yet to be justified epistemologically (e.g., see DeLaet, 1997; DeLaet and Smets, 1998).

Lastly, on the subject of similarity, QPS does not provide a basis for distinguishing 'good' from 'bad' data, unique and unreversed from homoplasy, where the number of instances of independent evolution supposedly marks the relatively weakness of the evidence. Indeed, FP calls into question the whole issue of weighting (Kluge 1997b). Instead of weighting, Hennig (1966, p. 148; Hull 1967, p. 186; Farris *et al*. 1970; Kluge 1997b) emphasized the importance of research cycles in his empirical concept of reciprocal clarification (reciprocal illumination). Basically, the incongruence of different kinds of observations, in light of the unweighted most-parsimonious phylogenetic hypothesis, suggests the need for further study, *a posteriori*, and further testing of the incongruences may lead to a reinterpretation of the data, such as a redefinition of the characters and character states, and ultimately to a more severely tested and better-supported hypothesis. There is no vicious circularity of reasoning at work in QPS under these conditions; however, one must be careful to maintain testability at all levels of analysis and reanalysis, because it is easy for reciprocal clarification to be born out of utilitarianism. Obviously, this argument for research cycles favors including testability in QPS, because ASP does not advocate any particular scientific scheme of inference, either deductive or inductive. Being able to claim the potential for research cycles in testability also draws attention to how little potential of that objective kind there is in the Bayesian and likelihood kinds inference, where models and claiming to know the truth are critical (subjectively speaking, of course).

I believe QPS is also relevant to the issue of character-state polarity (rooting). In theory at least,

QPS argues against Nixon and Carpenter's (1993, p. 413) "unconstrained, simultaneous analysis of all terminals," which those authors judged to be sufficient with respect to the discovery of monophyletic groups according to the parsimony criterion. The problem with their conclusion is that they asserted an operational imperative, global parsimony, without regard for the epistemological argument that justifies the concept of parsimony in phylogenetic inference. A convincing justification is required of global parsimony, just as it is of FP. To appeal to descriptive efficiency in these circumstances won't do, because it is without an epistemological foundation of its own.

The importance of FP maximizing explanatory power goes beyond the philosophical and theoretical, with the practical and the heuristic being covered as well. For example, the fact that the unweighted most-parsimonious hypothesis of species relationships maximally explains the relevant available evidence in terms of discrete and incontrovertible homology statements means the evidence can then be used to diagnose monophyletic groups. Such information is practically important to the community of scholars responsible for identifying and systematizing museum/herbarium collections. It is also possible to use the history of each character, described in light of that optimal hypothesis of species relationships, as the basis for formulating testable population-level hypotheses, such as in the study of adaptation. That not all methods of phylogenetic inference are explanatory, and do not therefore have these extra research benefits, is clear. For example, the histories of individual characters in a maximum likelihood analysis can only be estimated subjectively, as probabilities. Effectively, there is no distribution of real valued (other than abstract) character states on a maximum likelihood tree, a distribution that is often obtained from an *a posteriori* most-parsimonious mapping of discrete character states on the tree of greatest likelihood (e.g. Smith *et al*. 2004, Fig. 9)!

2.8 Ideographic theory unification

While FP may be both necessary and sufficient in the inference of phylogeny (see previous two sections), the question remains whether QPS addresses more than the empirical in the evaluation of scientific hypotheses. Can that ideographic theory make significant contributions to the philosophical—to metaphysical system building?

In addressing this question from the point of view of theory unification (Friedman 1983; McAllister 2000), I briefly survey a small sample of relevant areas of comparative biology to determine to what extent they multiply disconfirm ideographic theory. I will also consider using 'Ockham's razor' to eliminate other approaches to phylogenetic inference, other than those based on unweighted evidence, on the grounds that their contributions are indeterminate, or have less unifying power.

The latter exercise is certainly not the first attempt to unify phylogenetic theory. Initially, there were the debates that resulted in the elimination of phenetics and the evolutionary systematics of the neo-Darwinian synthesis. Recently, the indeterminate nature of phenetics was identified in pattern cladistics, which I believe has marginalized, if not eliminated, the influence of that approach. Even more recently, the excess of assumptions and the subjectivism of evolutionary systematics have been identified in the currently popular Bayesian and likelihood approaches to phylogenetic inference. Thus, the ideographic continues to be tested, as it should be if unification is to be a scientific exercise.

Fink (1982) was the first to explicate development in the context of a cladogram. He argued that to interpret ontogenetic processes requires such a hypothesis. For example, a paedomorphic condition can resemble the plesiomorphic state, and the only way to distinguish the two is in light of a phylogenetic hypothesis, the former often being described as an "evolutionary reversal." This is a good example of the well-known principle that to explain any pattern of interspecific variation in terms of "descent, with modification," that pattern must be in accord with ideographic theory. Certainly, the best-known example of this principle is the explanation of homologues. Not all theories of historical inference, however, such as represented by Bayesian and maximum likelihood methods,

provide these kinds of causal explanation on their own (e.g. Smith *et al.* 2004, Fig. 9). Many systematists are especially interested in discrete evolutionary changes in the phenotype and genotype, and not being able to deduce that kind of history, in light of the most probable or likely phylogenetic hypothesis, must be considered a significant shortcoming of the methods employed. I eliminate Bayesian and likelihood approaches from ideographic theory because they are indeterminate when it comes to the causality and explanation of the evidence employed. Phylogenetic inference is about more than a probable or likely classification, assuming those subjective conditions can be determined.

Also relevant to ideographic theory unification is reliance on a character concept of transformation series, assuming "descent, with modification," and not on one of similarity. Not only does QPS avoid being logically inconsistent with evolutionary theory in this conceptualization, it is directly relevant to all fields of comparative biology that assume heritable objects/events. This consideration is relevant even to areas of historical research outside biology. For example, the study of illuminated manuscripts (Platnick and Cameron 1977; Cameron 1987) assumes a concept of a transformation series that involves a kind of "heritability," a change in what is copied and passed on to subsequent illuminators, and for which there is the objective of maximizing the explanatory power of the event—explaining mistakes in copying in terms of the history of manuscript. The same cannot be said however for the studies of languages that depend on class concepts of historical evidence (Rexová *et al.* 2003), and here there is an explicit basis for their elimination from ideographic theory.

Taxonomy has always been an important area of systematics, and one where phylogenetics is now being promoted with increased vigor. Not only has there been an emphasis on the individuality of taxa in at least some of these efforts (e.g. Kluge 2005), giving license to the distinction between the classification of classes and the systematization of historical things, but the use of testability has been advocated as the basis for ruling on the content (wholeness) of taxa and their names, as opposed to handing over such important responsibilities to international commissions of nomenclature which reach decisions according to legalistic, non-scientific, conventions (ICZN 1999; Grueter *et al.* 2000). I interpret this example to be an issue of metaphysical system building, reaching out into areas that heretofore had not been considered scientific. Perhaps only a small point, but a point nonetheless, taxonomic names at all levels of taxonomy are proper names, and therefore the things to which they apply are required to be spatio-temporally restricted, as they are in QPS. Most other approaches to phylogenetic inference treat most or all of the parts of species history as class concepts, as is evident in names being preceded by the article "the," as in *the Homo sapiens*. As noted earlier, evolutionary theory precludes the conception of taxa, including species, as classes or sets.

Most of the current interspecific studies of adaptation, coevolution, and vicariance biogeography entail a phylogenetic hypothesis. Moreover, most of the analyses in these areas use parsimony as an optimality criterion. Still, I doubt they multiply confirm ideographic theory. For example, Lauder *et al.*'s (1993) parsimony-based lineage (homology) method for inferring adaptation was tested for its ability to deliver increased knowledge, but sufficient failures were found in its theory, especially in its presuppositions, to call into question its credibility in the inference of adaptation. Harvey and Pagel's (1991) comparative (convergence) method, although statistical, fails for the same kinds of reasons.

The historical treatments of coevolution and biogeography fare no better. As Sober (1988b) put it, there is a disanalogy between phylogenetic inference and those sciences accompanied by "dispersal" theory, such as coevolution and vicariance biogeography. In phylogenetic inference, only ancestor-descendant (vertical) relations are *assumed* and counted as heritable events in choosing between competing hypotheses. In coevolution and vicariance biogeography, deciding among competing hypotheses is a complex function of the kinds of vertical *and* horizontal (dispersal) events the researcher is prepared to accept. I believe assuming a shared history (Brooks 1988)

is ontologically indecisive in making that decision, and increasing information density (Brooks 1981), which is just another phrase for increasing descriptive efficiency, provides no epistemological basis for choosing. However, as Sober (p. 252; see also Kluge 1988, p. 316) contemplated, "If horizontal transmission and vertical transmission are unequally probable or make different predictions, that may provide a reason for preferring some ...hypotheses to others." But, here we are being asked to use a nomothetic means to address an ideographic end, which is the instrumentalist scheme that I rejected earlier in this paper. Thus, given these issues, and aside from the possibility of their heuristic potential, I doubt the historical theories of adaptation, coevolution, and vicariance biogeography can confirm ideographic theory unification.

Like the evolutionary systematics of the neo-Darwinian synthesis, the nomothetic sciences practiced in the inference of phylogeny, such as Bayesian inference and likelihood, make excessive assumptions, and thereby have relatively little unifying power. As already discussed, these model assumptions are counterfactual conditionals. Nor can those assumptions be tested because testing requires a phylogenetic hypothesis in the first place. In addition, nomothetic inference relies on a frequentist interpretation of history. However, as Kluge (2002) argued, the adequacy of a probabilistic interpretation must be judged according to the nature of the event, or object, being inferred, and he found a conditional (frequency) interpretation of probability fails all the tests of adequacy—admissibility, ascertainability, and applicability. Dilemmas of this sort are becoming especially evident in emerging fields of comparative biology, like genomics, where the impressive technicalities of nomothetic science seem to count more than an unweighted parsimony analysis that can objectively identify hypotheses of species relationships and character history, and do so with philosophical, theoretical, and methodological consistency.

It is significant that QPS assumes only background knowledge, "descent, with modification," whereas Bayesian and likelihood methods of inference depend on choosing from among an infinite number of *additional* auxiliary or model assumptions. Not only is the assumption of "descent, with modification," not known to be false, unlike model assumptions, it is uniformly required of all the comparative biological sciences, including the nomothetic. Surely, choosing from among an infinite variety of models contributes relatively little to any kind of theory unification.

Obviously, the details involved in this most recent attempt at unification of the historical sciences have only just begun to be exposed. At least some of the important bases for distinguishing the ideographic from the nomothetic have been identified, and all that remains is to continue to critically evaluate those research programs that are comparative and claim to assume historical change. As for future studies in this area, I see no reason to *necessarily* exclude any kind of research, including those at the populational level, such as illustrated by the nomothetic sciences of phylogeography (Avise 2000) and conservation genetics (Moritz 2002). While I am not optimistic that confirmation can be obtained from any theory that does not focus on historical individuality, the unification of the phylogenetic and the tokogenetic may yet be possible by reformulating the latter fields in ideographic terms, by divesting them of all their references to class concepts and kinds. If successful, the result would at least be equivalent in scope, although not in kind, to the neo-Darwinian synthesis, i.e. the unification of evolutionary and genetic theories (McAllister 2000).

2.9 Acknowledgments

I dedicate this paper to James S. Farris. His justification of parsimony—that the minimization of *ad hoc* hypotheses of homoplasy maximizes explanatory power—is certainly among the best known of his many theoretical contributions to phylogenetic inference. This mantra was argued early in the phylogenetics revolution (Farris 1983) and it has guided more than one generation of systematists in their preference for most-parsimonious hypotheses, including myself. I especially want to thank Taran Grant for a most critical review of the diverse ideas expressed in this paper. I also thank Taran for bringing Nolan's and Baker's ASP justifications of parsimony to my attention. Prior to knowing about their works I appealed either to Farris' mantra or, more recently (Kluge 2003b), to Popper's (1959) falsificationist justification for parsimony. Elliott Sober also read an early draft of this paper and he provided several trenchant comments. It was Ward C. Wheeler's candor in acknowledging aesthetic value that caused me to search more widely for a sufficient justification for parsimony in theory evaluation. I take full responsibility for the final positions taken in this paper. This manuscript was written and revised at the Cladistics Institute, Ann Arbor, MI, USA.

CHAPTER 3

Parsimony and its presuppositions

Elliott Sober

3.1 Introduction

The use of a principle of parsimony in phylogenetic inference is both widespread and controversial. It is controversial because biologists who view phylogenetic inference as first and foremost a statistical problem have pressed the question of what one must assume about the evolutionary process if one is entitled to use parsimony in this way. They suspect not just that parsimony makes assumptions about the evolutionary process but that it makes highly specific assumptions that are often *implausible*. That it must make *some* assumptions seems clear to them because they are confident that the method of maximum parsimony must resemble the main statistical procedure that is used to make phylogenetic inferences, the method of maximum likelihood.[1] Maximum likelihood requires the *explicit* statement of a probabilistic model of the evolutionary process. Parsimony does not; you can calculate how parsimonious different tree topologies are for a given data set without stating a process model. Likelihoodists suspect that parsimony nonetheless involves an *implicit* model. The question, for them, is to discover what that model is.

Cladists who defend the criterion of maximum parsimony often reply that parsimony *does* make assumptions about evolution, but that those assumptions are modest and unproblematic. For example, cladists sometimes claim that parsimony assumes just that descent with modification has occurred. This suggests that the disagreement between critics and defenders of parsimony is

not about whether parsimony makes assumptions about the evolutionary process, but concerns what those assumptions are and whether they are troublesome. But perhaps more important is the fact that critics and defenders also disagree about how those assumptions should be unearthed and evaluated. Defenders of maximum likelihood approach this problem by embedding the principle of parsimony in a statistical framework; they evaluate parsimony by examining it through the lens of probability. Defenders of parsimony often reject the use of statistics and probability as a criterion for evaluating parsimony; as Farris (1983/1994, p. 342) says, "the modeling approach was wrong from the start." His preferred alternative is to evaluate (and justify) parsimony in terms of what he takes to be the more basic idea that the best phylogenetic hypothesis is the one that has the most explanatory power.

There are many dimensions to this dispute—too many to discuss in the brief compass of the present chapter. What I wish to concentrate on here is the relationship that exists between maximum likelihood and maximum parsimony. Felsenstein (1973, 1979) and Tuffley and Steel (1997) have each identified models of the evolutionary process that suffice to insure that the two methods always agree on which hypothesis is best supported by a given data set. I will argue that these results provide only negative guidance concerning what parsimony presupposes. They allow one to establish that this or that proposition is *not* assumed by parsimony, but do not allow one to conclude that any proposition *is* an assumption that parsimony makes. To discover what parsimony presupposes, another strategy is needed. I suggest that parsimony's presuppositions can be found by examining simple

---

[1] For the purposes of this chapter, I will treat 'the maximum likelihood approach' as an umbrella term that covers both frequentist and Bayesian implementations. The difference between them is discussed later.

examples in which parsimony and likelihood disagree. My arguments will assume a broadly likelihoodist point of view, but will not require the assumption that any evolutionary model is correct.

3.2 Preliminaries

The principle of parsimony does not provide a *rule of acceptance*; rather, it provides a *rule of evaluation*. That is, the principle does not tell you to *believe* the phylogenetic hypothesis that requires the fewest changes in character state to explain the data at hand. After all, if the most-parsimonious tree requires that there be at least 25 changes, and the second and third most-parsimonious trees require that there be 26 and 27 respectively, the most you should conclude is that the most-parsimonious tree is *better supported* than the others; you are not obliged to conclude that the most parsimonious tree is *true*. In other words, parsimony would be a sound principle if the parsimony ordering of phylogenetic hypotheses and the support ordering of those hypotheses came to the same thing:

(1)   For any data set *D*, and any phylogenetic hypotheses H_1 and H_2, *D* supports H_1 more than *D* supports H_2 if and only if H_1 is a more parsimonious explanation of *D* than H_2 is.

If (1) is always true, I will say that parsimony is 'correct' in what it says. And if (1) is true in some restricted domain, I will say that parsimony is correct in what it says about that domain.

Likelihood likewise seeks to provide a rule of evaluation, not a rule of acceptance. If one hypothesis confers a higher probability on the data than another hypothesis does, it does not follow that the first hypothesis is true; in fact, it doesn't even follow that the first has the higher probability of being true. The fact that $\Pr(\text{Data} \mid H_1) > \Pr(\text{Data} \mid H_2)$ does not entail that $\Pr(H_1 \mid \text{Data}) > \Pr(H_2 \mid \text{Data})$. Rather, the virtue that has been claimed for the likelihood concept is that it provides an indication of which hypotheses are better supported by the data. The following principle has come to be called the *Law of Likelihood* (Hacking 1965, Edwards 1972, Royall 1997):

(2)   For any data set, and any phylogenetic hypotheses H_1 and H_2, the Data support H_1 more than they support H_2 if and only if $\Pr(\text{Data} \mid H_1) > \Pr(\text{Data} \mid H_2)$.

Proposition (2) restricts likelihood to hypotheses that describe phylogenetic relationships. I state the Law of Likelihood in this way to preserve its symmetry with (1), even though likelihood is supposed to be a perfectly general criterion for evaluating the direction in which the evidence points. Proposition (2) is not a consequence of the axioms of probability; it is not a mathematical truth, but rather is a philosophical thesis—that the epistemological concept of support is adequately represented by the mathematical concept of likelihood. Just as one can ask whether, or in what circumstances, (1) is true, the same questions can be posed about (2). If (2) is always true, then I will say that likelihood is 'correct' in what it says. And if (2) is true in some restricted domain, I will say that likelihood is correct in what it says about that restricted domain.

How are (1) and (2) related? If there is a data set and a pair of hypotheses H_1 and H_2 such that the parsimony ordering and the likelihood ordering do not agree (e.g. where H_1 is more parsimonious than H_2, but $\Pr(\text{Data} \mid H_1) < \Pr(\text{Data} \mid H_2)$), then (1) and (2) cannot both be true. On the other hand, if parsimony and likelihood agreed about the relative support of any two hypotheses for any data set you please, then (1) and (2) would be perfectly compatible. The fact that one is stated in terms of likelihood and the other in terms of parsimony would be no more significant than the difference between measuring distance in meters and measuring it in feet.

Proposition (2) gives a somewhat misleading picture of what it means to apply the Law of Likelihood to phylogenetic hypotheses. The problem is that phylogenetic hypotheses that describe the topology of a tree—not the times of branching events, or the amount of change that has taken place on branches, or the character states of interior nodes—do not, all by themselves, confer probabilities on the data. In the language of statistics, phylogenetic hypotheses are *composite*, not *simple* There are two possible solutions to this problem. The first is Bayesian; one represents the likelihood of a phylogenetic hypothesis *H* as a *weighted average*. *H* will vary in its likelihood, depending on which process model is considered, and depending also on what the values are for the parameters that occur in a process model. The 'full' Bayesian approach is to take all these possibilities into account, weighting each by its probability, conditional on H:

$\Pr(\text{Data} \mid H) = \sum_i \sum_j \Pr(\text{Data} \mid H \ \& \text{ process model } i \ \&$ values j for the parameters in model $i) \times \Pr(\text{process model } i \ \& \text{ values } j \text{ for the parameters in model } i / H)^2$

Although biologists are starting to explore Bayesian methods in phylogenetic inference (see e.g. Huelsenbeck *et al.* 2001), no one has proposed to represent a hypothesis' likelihood by averaging over all possible process models;[3] rather, Bayesians have tended to adopt a single process model M and to average over the different values that the parameters in M might take. According to this 'attenuated' Bayesian approach, the likelihood of H should be written as $\Pr_M(\text{Data} \mid H)$, not as $\Pr(\text{Data} \mid H)$, where

(B) $\Pr_M(\text{Data} \mid H) = \sum_j \Pr(\text{Data} \mid H \ \& \text{ model } M \ \& \text{ values } j$ for the parameters in $M) \times \Pr(\text{values } j \text{ for the parameters in model } M \mid H\&M)$

Whereas Bayesians treat the likelihood of H (once a model has been adopted) as a weighted average over the likelihood that H would have under the different possible settings of the model's parameters, frequentists treat the likelihood of H (given an assumed model) by finding $L(H\&M)$, where $L(H\&M)$ is the likeliest special case of the conjunction $(H\&M)$; it is found by setting the adjustable parameters in M to values that maximize the likelihood of $(H\&M)$.[4] For them, the appropriate quantity is

(F) $\Pr_M(\text{Data} \mid H) = \Pr(\text{Data} \mid L(H\&M))$.

Whereas likelihood means *average likelihood* for a Bayesian, likelihood means *best-case likelihood* for a frequentist. We will return to this difference between Bayesian and frequentist treatments of

likelihood in a moment. For now, let's focus on what they have in common—both evaluate the likelihood of H by assuming a process model M.[5]

How should this recognition of the model-relativity of likelihoods be incorporated in (2)? If different models are correct for different data sets and different taxa, there won't be a single 'master model' that should be used to evaluate the support of all phylogenetic hypotheses. Rather, what we need is the following:

(2*)   If M is the correct model for how the characters described in a data set evolved in the taxa described in phylogenetic hypotheses H_1 and H_2, then the Data support H_1 more than they support H_2 if and only if $\Pr_M(D \mid H_1) > \Pr_M(D \mid H_2)$.

Just as Proposition (2) is a philosophical thesis, not a mathematical truth, the same point holds for (2*). If (2*) is true, then I'll say that likelihood$_M$ is correct for the taxa and data set in question.

I earlier described how (1) and (2) can come into conflict. What would it take for (1) and (2*) to conflict? You need the same ingredients as before, plus a model M that is correct for the taxa and characters involved. That is, consider a pair of phylogenetic hypotheses, a data set, and a model M, where the parsimony ordering of the hypotheses differs from their likelihood$_M$ ordering. *If you accept (2*) and also think that model* M *is correct, then you are obliged to accept the judgment of likelihood$_M$ and reject the judgment of parsimony concerning which hypothesis is better supported by the data.* Notice that there are two if's in this italicized statement. This means that if you do *not* reject what parsimony says about the hypotheses, there are *two* options available, not just one. You can reject model M *or* you can reject (2*). That is, cladists are not obliged to reject model M; they also have the option of rejecting the Law of Likelihood as it is embodied in (2*).

I so far have described how a model can lead to a conflict between (1) and (2*). However, it is equally true that there are models of the evolutionary process that lead to a perfect harmony

---

[2] As an expository convenience, I represent H's average likelihood as a discrete summation, rather than as a continuous integration.

[3] Hulesenbeck *et al.* (2004) average over all of the many different time-reversible models by assigning them equal prior probabilities. Since some of these models are nested inside others, this prior distribution is questionable.

[4] Note that $L(H\&M)$ is a proposition, not a number between 0 and 1.

[5] In Sober (2004a) I argue that model selection criteria such as Akaike information criteria (AIC) permit phylogenetic inference to proceed by considering any number of process models without one's having to commit to any of them.

between (1) and (2*). Such models lead parsimony and likelihood$_M$ to be *ordinally equivalent*:

(OE)   Parsimony and likelihood$_M$ are ordinally equivalent if and only if, for any data set D, and any pair of phylogenetic hypotheses, the parsimony ordering of that pair is the same as the likelihood$_M$ ordering.

If a model M induces ordinal equivalence, what does that establish about the legitimacy of parsimony and likelihood$_M$? If you accept the model and regard one method as legitimate, then you should regard the other method as legitimate as well. In this circumstance, likelihoodists will say that M provides a likelihood justification of parsimony, whereas friends of parsimony will say that M provides a parsimony justification of likelihood$_M$. On the other hand, if you do not accept the model that induces ordinal equivalence, the status of the two methods is left open; for example, both could turn out to be unsatisfactory methods for evaluating the support of phylogenetic hypotheses. The point to notice here is that (OE) says nothing about whether parsimony is correct; it merely says what it means for parsimony and likelihood$_M$ to be in the same boat; if parsimony and likelihood$_M$ are ordinally equivalent, then both are correct or neither is. Two broken thermometers can be ordinally equivalent in what they say about the temperatures of different objects.

In summary, the model-relativity of likelihood entails that we are asking the wrong question when we ask "what is the relationship between likelihood and parsimony?" The word 'the' is where the trouble lies; there are many likelihood concepts (one for each possible model of the evolutionary process) and so there are many relationships between the different likelihood concepts and parsimony. More specifically, if we adopt (2*), the following two lines of reasoning are valid.

• If model M is correct, then likelihood$_M$ correctly evaluates support. If likelihood$_M$ has this property, and moreover is ordinally equivalent with parsimony, then parsimony also correctly evaluates support.
• If parsimony and likelihood$_M$ are not ordinally equivalent and parsimony correctly evaluates

support, then likelihood$_M$ does not. If likelihood$_M$ does not correctly evaluate support, then M cannot be correct.

The first line of reasoning describes what would be true if likelihood$_M$ and parsimony were not just in perfect agreement, but additionally had the property of correctly evaluating support. The second describes how a failure of ordinal equivalence can help uncover a presupposition of parsimony—if parsimony correctly evaluates support, then process model M must be false. Notice that both lines of reasoning require (2*). A more thorough investigation would address the question of why one should accept this formulation of the Law of Likelihood. This is a topic I will not take up here; I'll assume (2*) without trying to justify it.

3.3 How to determine what parsimony does *not* presuppose

A number of writers have attempted to find models that induce ordinal equivalence. Three have succeeded, Felsenstein (1973, 1979) and Tuffley and Steel (1997). In Felsenstein's model, characters are constrained to have very low probabilities of changing state, but there is no requirement that the probability of a character's changing from state i to state j on a branch is the same as its probability of changing from state j to state i. In Tuffley and Steel's, characters can have high probabilities of changing state (though they need not), but the probabilities of change must be symmetrical.[6] The models are very different, but each entails ordinal equivalence.

Both Felsenstein, and Tuffley and Steel, evaluate the likelihoods of phylogenetic hypotheses by using the frequentist approach (F) for assigning values to the parameters in the models they discuss. For example, consider a single site in the aligned sequences that characterize four species W, X, Y, and Z. Suppose that W and X are in state G and that Y and Z are in state A. The most parsimonious unrooted tree is (WX)(YZ). Under the

---

[6] The two models agree that different traits on the same branch can have different probabilities of changing; this also applies to the same trait on different branches.

symmetrical model that Tuffley and Steel assume, the highest likelihood this tree can have, relative to this character, is $(\frac{1}{4})(\frac{1}{4})(1)(1)(1)(1) = 1/16$, and this is the likelihood that Tuffley and Steel take the unrooted tree to have.[7] A Bayesian would want to consider the *average* likelihood of (WX)(YZ), not the *maximum*.

There are other attempts in the literature to establish ordinal equivalence. Farris (1973) tried to prove this result by using a model that makes very weak assumptions about the evolutionary process. Goldman (1990) sought to do the same thing by using a model in which the probability that a character will change state on a branch is independent of the branch's duration. Both these efforts fail to establish ordinal equivalence because both interpret parsimony as inferring not just the topology of a tree but something more inclusive. Goldman viewed parsimony as a procedure for inferring the topology plus an assignment of character states to interior nodes; Farris took parsimony to output the topology plus an assignment of character states to all points along the branches. Why does this vitiate the arguments that Farris and Goldman present? The reason is that even if $H_1 \& X_1$ is more likely and more parsimonious than $H_2 \& X_2$, and H_1 is more parsimonious than H_2, it doesn't follow that H_1 is more likely than H_2 (Felsenstein 1973; Sober 1988; Steel and Penny 2000). The likelihood of a tree must sum over all possible assignments of character states to points in the tree's interior.

When a model induces ordinal equivalence, what does this reveal concerning parsimony's presuppositions about the evolutionary process? It most certainly does not show that parsimony assumes that the model is true. The models of Felsenstein (1973, 1979) and of Tuffley and Steel (1997) are simply *sufficient* conditions for ordinal equivalence. No one has shown that either of these models is *necessary* for ordinal equivalence. And, obviously, neither of them is; if each of two models suffices, neither is necessary. We must be careful to

distinguish what a *modeler* assumes from what the model reveals concerning what *parsimony* assumes (Sober 1988, 2004a).

Still, if we accept the instance of the Law of Likelihood given by (2*), these results about ordinal equivalence provide a partial test for whether parsimony assumes this or that proposition about the evolutionary process (Sober 2002, 2004a). As noted earlier, parsimony and likelihood$_M$ can be ordinally equivalent even if both are wrong in what they say about support. However, if they are ordinally equivalent *and* model M is true, then both are correct, given (2*). Consider a model M that induces ordinal equivalence; M might be Felsenstein's model, or the one described by Tuffley and Steel, or some third model that no one has yet identified:

$$M \to \text{parsimony is correct } \to A$$

If model M is true (where M induces ordinal equivalence), then parsimony is correct in what it says about support (and so is likelihood$_M$, of course). What does parsimony assume? The assumptions (A) of parsimony are just those propositions that must be true, if parsimony is correct in what it says about support. Notice that any proposition that is entailed by the claim that parsimony is correct also must be entailed by model M. However, the converse is not true—if model M entails a proposition, that proposition may or may not be entailed by the thesis that parsimony is correct. This means that the results of Felsenstein (1973, 1979) and of Tuffley and Steel (1997) provide the following test concerning whether parsimony assumes that proposition X is true:

- If model M entails X, then X may or may not be an assumption of parsimony's.
- If model M does not entail X, then X is not an assumption of parsimony's.

Applying this partial test yields some surprising results. First, many biologists have suspected that parsimony assumes that changes in character state are very improbable and that homoplasies are rare; from a likelihood point of view, this suspicion is *provably* mistaken. The reason is that the Tuffley and Steel model does not entail that changes are

---

[7] Of the two occurrences of one-quarter in this expression, one of them is the prior probability of the root's being in a given state; the other is the probability of a change in state in the tree's interior.

improbable or that homoplasies are rare. Second, it follows that parsimony does not assume that change is symmetrical; the reason is that the Felsenstein model does not assume this. These results depend on using the Law of Likelihood (2*); but once that interpretative framework is adopted, these results are secure.

As illuminating as these results are, they still have the limitation of being purely negative. The partial test can show that this or that proposition is *not* an assumption that parsimony makes, but the test isn't able to demonstrate that a given proposition *is* assumed by parsimony. Results that demonstrate that a model induces ordinal equivalence have this inherent limitation. In order to obtain positive results concerning what parsimony assumes about the evolutionary process, a new strategy is needed.

3.4 How to determine what parsimony presupposes

Mathematical arguments for ordinal equivalence are necessarily general; they must show, for any data set and for any pair of phylogenetic hypotheses (which may describe an arbitrarily large number of taxa), that parsimony and likelihood$_M$ agree about the support ordering. In contrast, an argument that demonstrates a *failure* of ordinal equivalence need not be general; it can just take the form of a simple example. All that is needed is a model M, a single data set, and a pair of hypotheses such that the parsimony ordering differs from the likelihood$_M$ ordering. If parsimony is right in what it says, then likelihood$_M$ is wrong. And if likelihood$_M$ is wrong, so is the model M (assuming that 2* is true). Such cases therefore help reveal parsimony's presuppositions.

3.4.1 Example 1

Let's begin with a simple example in which the hypotheses being evaluated don't describe tree topologies, but rather assign character states to ancestors in trees that are taken as given. Imagine a bifurcating tree in which all the tip species are observed to have the same character state a. Parsimony asserts that the best-supported estimate of the character state of the most recent common ancestor

A of those tip species is that A was also in state a. Parsimony's solution to the problem would remain the same if we were talking about a star phylogeny. In fact, distilled to its simplest form, the problem and parsimony's solution to it can be formulated by considering a single lineage that ends with a descendant D that is in state a; the problem is to infer what the character state was of the ancestor A that existed at the start of the lineage. Parsimony says that the best estimate is that $A = a$.

What would a likelihood analysis of this problem look like? If the character in question is dichotomous (with character states 0,1), the standard approach from the theory of stochastic processes (Parzen 1962) is to divide the lineage into a large number of brief temporal intervals. In each, there is a probability u that the lineage will change from state 0 to state 1, and there is a (possibly different) probability v that the lineage will change from state 1 to state 0. Each of these instantaneous probabilities are assumed to be small (at least less than one-half). They allow us to describe the probability $\mathrm{Pr}_N(i \to j)$ that a lineage that is N units of time in duration will end in state j, given that it starts in state i. These *lineage transition probabilities* are as follows:

$$\mathrm{Pr}_N(0 \to 1) = u/(u+v) - [u/(u+v)](1-u-v)^N$$
$$\mathrm{Pr}_N(1 \to 1) = u/(u+v) + [v/(u+v)](1-u-v)^N$$
$$\mathrm{Pr}_N(1 \to 0) = v/(u+v) - [v/(u+v)](1-u-v)^N$$
$$\mathrm{Pr}_N(0 \to 0) = v/(u+v) + [u/(u+v)](1-u-v)^N$$

This is the two-state Markov process model. Each of these transition probabilities averages over all possible scenarios consistent with the specified initial and end states. For this reason, it would be misleading to say that the two transition probabilities of the form $\mathrm{Pr}_N(i \to i)$ describe the probability of *stasis*; $\mathrm{Pr}_N(i \to i)$ encompasses the possibility that there has been no change in the lineage but also the possibility that the lineage has flip-flopped an even number of times. There is no assumption in this model as to whether $u = v$. If $u = v$, the lineage undergoes an unbiased process of drift. If $u > v$, there is a directionality or bias in the evolutionary process, favoring state 1 over state 0. One possible source of this bias is natural selection; however, mutation and migration also can induce a bias in how the lineage tends to evolve.

When N is very small, the two probabilities of the form $\text{Pr}_N(i \rightarrow i)$ are close to unity and the two probabilities of the form $\text{Pr}_N(i \rightarrow j)$ (where $i \neq j$) are close to 0. When N is infinite $\text{Pr}_N(i \rightarrow j) = \text{Pr}_N(j \rightarrow j)$; the lineage has the same probability of ending in state j, regardless of what the state was in when the lineage began. Thus, when a lineage has a very short duration, its initial condition virtually determines its final state and the relationship of u and v doesn't matter; when a lineage is very old, it is the process that occurs during the lineage's duration (represented by u and v) that matters; the initial condition is forgotten.

It is a property of this model that a *backwards inequality* obtains: $\text{Pr}_N(j \rightarrow j) \geq \text{Pr}_N(i \rightarrow j)$, with strict inequality when N is finite (Sober 1988). Don't confuse the backwards inequality with the *forwards inequality* $\text{Pr}_N(j \rightarrow j) > \text{Pr}_N(j \rightarrow i)$. An instance of this forwards inequality (e.g. that $\text{Pr}_N(1 \rightarrow 1) > \text{Pr}_N(1 \rightarrow 0)$) will be true for some values of u, v, and N, but not for others. The backwards inequality says that if a descendant is in state j, that outcome is made more probable by the hypothesis that its ancestor was in state j than by the hypothesis that the ancestor was in state i. This provides a likelihood solution to our problem: *if the descendant is in state* a *of a dichotomous character, the hypothesis of maximum likelihood about the state of the ancestor is that the ancestor was in state* a *as well*. This result holds regardless of what the values of u, v, and N are; even if these values entail that the expected number of changes in the lineage is large, the most-parsimonious assignment of character state to the ancestor is *still* the assignment of maximum likelihood. Parsimony and likelihood$_M$ therefore agree when M is the two-state Markov process model and the problem is to infer an ancestor's character state from the character state of a descendant.[8]

In analyzing this simple problem, I used the Bayesian method (B), not the frequentist procedure (F), for taking account of the fact that the two-state Markov model can have different values assigned to its parameters u, v, and N. I didn't focus exclusively on the values for these parameters that would maximize the likelihood of each hypothesis about the state of ancestor A; that would have led to the conclusion that the two likelihoods are as close together as you please, since

$$\text{Pr}(D=1 \mid L(A=1 \,\&\, u=a_1 \,\&\, v=a_2 \,\&\, N=a_3)) = 1$$

and

$$\text{Pr}(D=1 \mid L(A=0 \,\&\, u=a_1 \,\&\, v=a_2 \,\&\, N=a_3)) \rightarrow 1 \text{ as } a_3 \rightarrow \infty$$

Rather, my argument is that for each value of u, each value of v (each less than 0.5), and for each finite value of N,

$$\text{Pr}(D=1 \mid A=1 \,\&\, u=a_1 \,\&\, v=a_2 \,\&\, N=a_3)$$
$$> \text{Pr}(D=1 \mid A=0 \,\&\, u=a_1 \,\&\, v=a_2 \,\&\, N=a_3)$$

From this it follows that

$$\sum_{i,j,k} \text{Pr}(D=1 \mid A=1 \,\&\, u=i \,\&\, v=j \,\&\, N=k)$$
$$\times \text{Pr}(u=i \,\&\, v=j \,\&\, N=k \mid A=1)$$
$$> \sum_{i,j,k} \text{Pr}(D=1 \mid A=0 \,\&\, u=i \,\&\, v=j \,\&\, N=k)$$
$$\times \text{Pr}(u=i \,\&\, v=j \,\&\, N=k \mid A=0)$$

if the settings of u, v, and N are independent of the character state of the ancestor A. In this instance, the Bayesian weighting terms $\text{Pr}(u=i \,\&\, v=j \,\&\, N=k \mid A=1)$ and $\text{Pr}(u=i \,\&\, v=j \,\&\, N=k \mid A=0)$ are *innocuous*; the last stated inequality holds, no matter what their values are.

I now want to consider the same problem—that of inferring the character state of the ancestor A from the observed character state of the descendant D—when the character in question is a quantitative phenotype (e.g. the average length in the species of a particular bone), not dichotomous. It remains true, of course, that if the descendant is in state a, then the most-parsimonious hypothesis about the state of the ancestor is that it was in state a as well. To see what likelihood says about this problem, we need to construct a probabilistic model of the

---

[8] The same result holds when we pose this question about a star phylogeny or a bifurcating tree. If branches are conditionally independent of each other, the support for $A = a$ (as measured by the likelihood ratio) is greater when there are several descendants than when there is just one.

evolution of the quantitative character. Let's begin by setting limits on the values of the character in question; suppose it can't go below zero or above 100. We can think of u as the probability of the lineage's increasing its character state by a very small amount during a brief interval of time, and v as the probability of the lineage's reducing its value during that instant. Since there are upper and lower bounds on the character state, u and v cannot remain constant over the full range of the lineage's possible states; for example, u must have a value of zero when the lineage is in state 100, though of course it can have a nonzero value when the lineage has a value less than 100.[9] In addition, we want to allow for the possibility that the lineage is evolving towards a stable equilibrium; for example, perhaps a trait value of 75 is optimal, and selection is pushing the lineage towards that value. This means that $u > v$ when the lineage's trait value is less than 75, but that $u < v$ when the population has a value greater than 75. In addition, the degree to which $u > v$ must decline as the population approaches 75 from below.

When a biased process (such as natural selection) is pushing a lineage towards a single attractor state, the lineage's probability of reaching that equilibrium is greater, the closer its initial state is to that attractor. Similarly, the equilibrium value has a higher probability of being attained, the more time there is in the lineage. When the lineage has a very short duration, stasis is almost certain; as the lineage is given a longer duration, the biased process takes over and the initial condition recedes in its impact on the lineage's final state. In the limit of infinite time, the initial condition is entirely forgotten and the lineage's probability of attaining a given end state is the same, regardless of what the state was in which the lineage began.

How should we conceptualize a pure drift process for continuous phenotypic characters? In this case, $u = v$, except when the lineage is at the limit values of 0 and 100. If the ancestor has a given trait value, that trait value is the expected value of its descendant. With very little time, the expected value of the descendant is tightly peaked around the lineage's initial state. As time goes on, this low variance bell curve flattens and spreads out. With infinite time, there is a flat distribution—each character state has the same probability. Whereas selection in a finite population involves both the shifting and the squashing of a distribution, the process of pure drift involves only squashing.[10]

Now let's return to the inference problem. If the descendant D is in state a of a quantitative phenotypic trait, which assignment of character state to the ancestor A has the highest likelihood? Since the backwards inequality holds for dichotomous characters, one might expect the model for continuous phenotypic characters to have the same unconditional consequence: that A $= a$ has maximum likelihood. This is not always correct (Sober 2002). If D $= a$, then A $= a$ is the maximum likelihood assignment when the process is one of pure drift (W. Maddison 1991). And if D $= a$ and a is the optimal character state towards which natural selection is pushing, then A $= a$ is again the maximum-likelihood assignment. However, if the descendant has a character state of, say, 40 and selection is pushing the lineage towards a value of 50, then the maximum likelihood assignment to the ancestor A will be *less than 40*; how much less than 40 the maximum likelihood value is depends on how long the lineage has been evolving, on how strong the directional force is, and on the character's heritability (Sober in press). This means that parsimony and likelihood$_M$ conflict when the model says that there is a directional process whose attractor is some state different from a, the observed character state of the descendant. Thus, to defend the parsimonious assignment of A $= a$ without rejecting the Law of Likelihood, one must reject this model. Parsimony assumes that the trait either evolves by pure drift or by a selection process in which the descendant's character state is optimal.

---

[9] I do not conceptualize its maximum and minimum values (0 and 100) conceptualized as *absorbing states*. The same will be true in the drift model to be discussed shortly.

[10] See Lande (1976) and Sober (2005) for further details concerning these phenotypic models for selection and drift. Let me emphasize that my discussion of 'drift' in this paper is not about random *genetic* drift, but concerns change in the population average of a quantitative phenotype.

3.4.2 Example 2

Consider two extant species A and B and their most recent common ancestor C. Suppose that A = 1 and B = 0; parsimony says that C = 1 and C = 0 are equally parsimonious. In what circumstances do these two assignments of character state to the ancestor have the same likelihood? That is, when will it be true that $Pr_A(0 \rightarrow 1)Pr_B(0 \rightarrow 0) = Pr_A(1 \rightarrow 1)Pr_B(1 \rightarrow 0)$? Here the subscripts A and B represent which of the two lineages the transition probabilities describe. It is helpful to rewrite this equality as

$$\frac{Pr_A(0 \rightarrow 1)}{Pr_A(1 \rightarrow 1)} = \frac{Pr_B(1 \rightarrow 0)}{Pr_B(0 \rightarrow 0)}$$

If the two lineages experience the same evolutionary processes (i.e. are characterized by the same pair of u, v values), then this equality holds if and only if the duration N of the lineages is 0, or infinity, or $u = v$. That is, parsimony assumes that if the two lineages are of finite duration and have experienced the same evolutionary process, then that process is pure drift.

3.4.3 Example 3

The next problem is just like the previous one, except that the two lineages have unequal temporal durations. When A = 1 and B = 0, parsimony says that C = 1 and C = 0 are equally well-supported estimates of the ancestral character state even when A is a present-day species and B is a fossil. This temporal difference between A and B means that the lineage leading from C to A lasted longer than the lineage leading from C to B. The two-state Markov model views this difference as evidentially significant; recall that N, the duration of a lineage, figures in the expressions for the lineage transition probabilities. If the processes in the two lineages are characterized by the same values of u and v, then B provides stronger evidence (in the sense of a larger likelihood ratio) about the state of C than A does; likelihood will then favor C = 0 over C = 1. Parsimony denies this. Parsimony therefore assumes that the u and v values that characterize one lineage must differ

from the u and v values that characterize the other. But not just any difference between the two pairs of values will suffice for C = 0 and C = 1 to have the same likelihood. The lineage with the longer duration (the one leading to A) must have a smaller value for $u + v$; in fact, the degree to which its value for $u + v$ must be smaller is determined by the two lineage durations. This has the embarrassing consequence that the lineage leading to A must change its values for u and v in a very precise way as it gets older. At one point the lineage leading to A and the lineage leading B were of equal duration. But then B went extinct while A continued to exist, so the lineage leading to A got longer while the one leading to B did not. According to parsimony, A's values for u and v must evolve in precise response to its duration and to the duration and u and v values that attach to the lineage leading to B.

3.4.4 Example 4

The next example in which a process model M induces a conflict between likelihood$_M$ and parsimony concerns a rooted tree in which the character state of the root is specified. It is a familiar property of parsimony in this context that shared derived characters are said to provide evidence of common ancestry, but that shared ancestral characters do not. If three tip species A, B, and C, are in states A = 1, B = 1, and C = 0, with 0 taken to represent the ancestral character state, then (AB)C will be more parsimonious than A(BC). However, if the polarity is reversed, with 1 now representing the ancestral condition, then (AB)C and A(BC) will be equally parsimonious.

Consider the two-state Markov model given before on which an additional constraint is imposed, namely that the probability of one branch's ending in state i if it begins in state j ($i,j = 0,1$) is the same as another branch's doing the same, if the two branches are simultaneous. This model has the consequence that (AB)C has a higher likelihood than A(BC), given the observation that A = 1, B = 1, C = 0, if all branches have finite duration, *regardless of what the polarity of the character is* (Sober 1988, pp. 206–212). This contradicts what parsimony asserts, when 1 is the ancestral state.

Parsimony's interpretation of the observations therefore requires a rejection of the process model just described.

3.4.5 Discussion of the examples

If the Law of Likelihood, as formulated in (2*), is correct, then parsimony assumes the falsehood of any model M for which likelihood$_M$ and parsimony are not ordinally equivalent. A summary of the models discussed in this chapter that parsimony assumes are false is provided in Table 3.1. These descriptions do not lay out the full details of the models that parsimony must reject. This is an important point, since these models are each conjunctions of several propositions. If parsimony assumes that model M is false, this means that at least one of the constitutive propositions that specifies the model must be false, not that all of them must be. So don't take the table's brief description of a model to mean that the detail described must be false. Also, I have described, for each inference problem, *a* model that parsimony must regard as false; don't assume that this is the *only* model that parsimony must reject when it addresses that inference problem.

Though each example requires that parsimony reject a process model, a model that parsimony needs to reject in one inference problem doesn't necessarily have to be rejected in another. In Example 1, parsimony requires a nontrivial assumption when the character is quantitative, but no such requirement is imposed when the trait is dichotomous. In Example 3, parsimony assumes that the two lineages experience different evolutionary processes. In Example 2, parsimony does not require this assumption; rather, it assumes that if the same process is at work in the two lineages, then that single process is drift. In Example 4, parsimony leaves open whether selection or drift is operating within a branch, but requires that different simultaneous branches be characterized by different pairs of values for u and v. These results suggest, but do not demonstrate, that parsimony may impose different assumptions about the evolutionary process when it addresses different inference problems.

Although I think these examples make clear at least some of the assumptions that parsimony makes about the evolutionary process, I have not commented on whether those assumptions are innocuous or implausible. I have emphasized that my arguments are predicated on the assumption that the Law of Likelihood, as formulated in (2*), is true. This is so general a proposition that it can hardly be said to be an assumption about *evolution*. Even so, if it is rejected, we are back to square one. If it is retained, the question becomes more specifically biological, but here again, there are choices to consider. For example, in problem 3, a likelihoodist may wish to maintain that the state of the fossil B provides more evidence about the state of the most recent common ancestor C than the extant organism A does. If so, parsimony's solution to this problem is mistaken. But it is open to the defender of parsimony to take the opposite

Table 3.1 Summary of the models discussed in this chapter that parsimony assumes to be false

	Inference problem	A model that parsimony assumes is false
1	For a quantitative character evolving in a lineage, infer the character state of the ancestor when the descendant has character state a.	Selection is pushing the lineage towards an optimum that differs from the character state of the descendant.
2	When two descendants alive now exhibit different states of a dichotomous character, infer the character state of their most recent common ancestor.	The two lineages are characterized by the same pair of values for u and v, where $u \neq v$.
3	When two descendants (one extant, the other a fossil) exhibit different states of a dichotomous character, infer the character state of their most recent common ancestor.	The two lineages are characterized by the same pair of values for u and v.
4	When two species share a symplesiomorphy not exhibited by a third, infer the rooted tree topology.	Simultaneous branches in the tree have the same pair of values for u and v.

position; however, I think it is not enough just to insist that parsimony is right in what it says about this example and to conclude from this that the model that leads likelihood to a contrary verdict must be mistaken. An additional argument is needed concerning why the Markov process model should not be taken seriously. This point generalizes to the other examples. All these examples can be handled in the same way by attacking the entire Markov process framework. I don't say that this framework is beyond criticism. Rather, I suggest that criticisms of this framework must be *biological* in their content. This is an important point: once the Law of Likelihood is accepted, both criticisms and defenses of parsimony must be based on biological, not purely methodological, considerations.

I have not discussed the issue of statistical consistency in this essay, but there is a part of the debate about that matter that bears on the present discussion. Felsenstein (1978) described a model of evolution and an assumed true phylogeny that together lead parsimony to converge on a false phylogeny as the data are made large without limit. Farris' (1983/1994) reaction was to reject Felsenstein's model as unrealistic; after all, Felsenstein's model says that all traits in a given branch have identical transition probabilities and that the probability of reversion from the derived to the ancestral state is always zero. Felsenstein said at the outset that the model he describes is unrealistic; Farris emphatically agreed, and took this point to cancel whatever criticism of parsimony the demonstration of statistical inconsistency might be thought to imply. Farris apparently was reasoning that the correctness of parsimony requires parsimony to be statistically consistent; thus, if model M entails that parsimony is *not* consistent, then the correctness of parsimony requires that model M be *false*.[11] I have reasoned

similarly about the examples in this essay, except that I have focused on ordinal equivalence with respect to finite data sets, not on statistical consistency, which describes what will happen if you have an infinite data set. This difference aside, I am hardly the first to suggest that parsimony's *failing* to have some property elucidates what its biological assumptions are.

3.5 Acknowledgments

I thank Joe Felsenstein and Michael Steel for useful comments on the present paper. I also want to acknowledge the considerable influence that Steve Farris has had on my thinking about the role of parsimony in phylogenetic inference, and in a wider scientific context. It is a pleasure to contribute this chapter to a volume honoring his work.

---

[11] In Sober (1988, pp. 166–171), I argue that a method's statistical consistency is not a necessary condition for one to be rational in using that method. As it happens, there are parameter settings of the Tuffley and Steel (1997) no common mechanism model (N) that have the consequence that parsimony and likelihood$_N$ both fail to be statistically consistent. However, I don't see why that forces one to decline to use N as one process model, possibly among several, in phylogenetic inference; for discussion of the use of multiple process models, see Sober (2004a). Furthermore, if using parsimony required the rejection of any model whose parameters can be assigned values that render parsimony statistically inconsistent, then the Tuffley–Steel model has both of the following properties: (1) it suffices for likelihood and parsimony to agree, and (2) its falsity is presupposed by parsimony. This illustrates how fundamentally different likelihood and statistical consistency are as tools for thinking about parsimony. Bayesians see the Law of Likelihood as fundamental; frequentists such as Felsenstein see consistency as the fundamental *desideratum*.

II

Parsimony, character analysis, and optimization of sequence characters

CHAPTER 4

The logic of the data matrix in phylogenetic analysis

Brent D. Mishler

4.1 Introduction

The process of phylogenetic analysis inherently consists of two phases. First a data matrix is assembled, then a phylogenetic tree is inferred from that matrix. There is obviously some feedback between these two phases, yet they remain logically distinct parts of the overall process. One could easily argue that the first phase of phylogenetic analysis is the most important; the tree is basically just a re-representation of the data matrix with no value added. This is especially true from a parsimony viewpoint, the point of which is to maintain an isomorphism between a data matrix and a cladogram. We should be very suspicious of any attempt to add something beyond the data in translating a matrix into a tree!

Paradoxically, despite the logical preeminence of data matrix construction in phylogenetic analysis, by far the greatest effort in phylogenetic theory has been directed at the second phase of analysis, the question of how to turn a data matrix into a tree. Extensive series of publications have been elaborated to attempt to justify such tree building approaches as neighbor-joining, maximum likelihood, and Bayesian inference, while ignoring entirely the nature of the data matrix that must underlie any analysis. The reasons for this asymmetry in research on phylogenetic theory are not entirely clear, but it probably has to do with the fact that the problem of tree building may appear simpler, more clear-cut. Perhaps it is just a matter of research fashions. For whatever reason, relatively little attention has been paid to the assembly of the data matrix, and it is high time to examine this all-important part of systematic research. At stake are each of the logical elements of the data matrix: the rows (what are the terminals?), the columns (what are the characters?), and the individual entries (what are the character states?).

The tree of life is inherently fractal-like in its complexity, which complicates the search for answers to these questions. Look closely at one *lineage* of a phylogeny (defined as a diachronic connection between an ancestor and a descendent) and it dissolves into many smaller lineages, and so on, down to a very fine scale. Thus the nature of both the *terminal units* (TUs; the twigs of the tree in any particular analysis) and the characters (hypotheses of homology, markers that serve as evidence for the past existence of a lineage) change as one goes up and down this 'fractal' scale. Furthermore, there is a tight interrelationship between TUs and character states, since they are reciprocally recognized during the character analysis process.

This chapter will deal with logical issues involving the elements of the data matrix in light of the nested and interrelated nature of TUs and characters. I will argue at the end that if care is taken to construct an appropriate data matrix to address a particular question of relationships at a given level, then simple parsimony analysis is all that is needed to transform the matrix into a tree. Debates over more-complicated models for tree building can then be seen for what they are: attempts to compensate for marginal data.

4.2 What exactly is a terminal branch on a tree (that is, a row in the data matrix)?

People who publish phylogenetic analyses are usually cavalier about what their terminal branches represent. One often sees species or other taxon names, or even geographic designations of populations, attached to terminal branches of published trees without explanation. Larger-scale units might indeed be a well-justified TU, but they need to be justified, not assumed *a priori*. Taxa or populations are never the fundamental things from which phylogenies are actually built. Not even individuals are the TUs (contra Vrana and Wheeler 1992). As was carefully elaborated by Hennig (1966), the fundamental terminal entity in phylogenetics is the *semaphoront*, an instantaneous time slice of an individual organism at some point in its ontogeny. A tube of extracted DNA and its associated museum voucher specimen—a semaphoront—should be considered the ultimate TU.

This realization helps conceptually, but doesn't solve all of the empirical problems that arise in assembling a matrix. In practice, TUs (i.e. rows in a data matrix) are usually not semaphoronts. Especially in larger-scale studies, TUs are usually a complicated assemblage of semaphoronts, and sometimes even include data removed from any connection with its original semaphoront. Many specimens often need to be examined for relevant character information (not all of which can be gathered from all semaphoronts because of their sex, life stage, or state of preservation). Information from the literature or a database such as GenBank is often included in the matrix, based on a taxon identification alone without reference to a voucher specimen. This process of assembly of such composite TUs needs careful examination.

Similar sorts of terminals have been called operational taxonomic units (OTUs) in the past, but I think a refined concept of TUs, as referred to above, is necessary, one designed specifically for phylogenetics. The original concept of OTU was defined by pheneticists as a minimal cluster in a Euclidian distance sense. Cladists need instead to refer to specific, potentially homologous and discrete-state characters in a Manhattan distance

sense. An additional flaw of the original concept of OTU is that, by using the word 'taxonomic,' it implies that one can do taxonomy before an analysis is completed. This view, by confounding the logical precedence of analysis before classification, has led to major mistakes in systematics research, both phenetic and cladistic, most acutely in the development of phylogenetic species concepts (see the debates framed in Wheeler and Meier 2000).

So how can we define a TU that is suitable for use in phylogenetics? Epistemologically speaking, *a TU is a set of semaphoronts that are homogeneous for the informative character states currently known* (as explained in detail below). A TU is essentially a pile of semaphoronts that cannot currently be subdivided by character data, and thus it is a pragmatic unit, always subject to change as knowledge of characters progresses. Ontologically speaking, *a TU is taken to represent a time slice of one of the terminal lineages whose relationships are being studied in a particular analysis.*

Why do I say "in a particular analysis?" Because the definition of TUs, even for the same group of organisms, may change in analyses at different scales. There unfortunately isn't one fundamental TU suitable for any and all analyses; for several different reasons. Epistemologically speaking, since TUs are dependent on character-state divisions in the characters being employed, they are discovered and defined in the course of character analysis (as discussed in detail below), and of course different characters are useful at different scales of analysis. There is thus a reciprocal relationship between character states and TUs as they are being discovered during character analysis at different levels. Ontologically speaking, larger-scale lineages are usually composed of smaller lineages nested inside them, and the choice of which lineage to represent in a particular analysis depends on the questions begin asked. Furthermore, the lineages at these different levels potentially have different histories; in other words the smaller lineages are not always proper subsets of the larger ones. This is sometimes called the gene tree/species tree distinction (Maddison and Maddison 1992), but that distinction is far too simplified; there are many nested levels of potentially incongruent lineages, not just two (more on this topic later).

Even if one wanted to try to avoid these problems by using only semaphoronts in a data matrix, one would still need to pay attention to the same issues of scale. One would still need to decide conceptually which lineages are being represented by what semaphoronts. It is nearly impossible in practice to use single semaphoronts as terminals rather than compositely coded TUs that have data taken from a number of semaphoronts. For one thing, not all semaphoronts bear all the characters; there may be juvenile specializations or sexual dimorphism present in a lineage. Some specimens will be missing reproductive organs or other key features. Different genes will often be sequenced from different individuals. Furthermore, data are often taken from the literature (e.g. a previously published ultrastructural analysis) or from a database (e.g. another laboratory's gene sequence), in cases where no reference can be made to an original semaphoront (e.g. if no voucher specimen was deposited in a museum). Thus, data are virtually always compiled from studies of different individual organisms considered to represent the same terminal lineage. TUs are nearly always composites in practice; their composition varying depending on the scale of analysis.

This topic obviously touches on the species debate, on which I have some opinions (Mishler and Brandon 1987; Mishler 1999; Mishler and Theriot 2000a, b, c), but which I am attempting to steer clear of in this essay to maintain focus. I am speaking here to how data matrices are made: classification (including naming species) is something that happens *much* later in the process. So, while this is not the place to debate species concepts, I do need to point out that the fractal scaling of nested lineages includes those well below the traditional species level. Thus, species are not somehow different from lineages at any other level; they are not 'privileged' TUs—they simply need to be justified like any other.

In summary, there is never a given, *a priori* set of TUs to begin a phylogenetic analysis with. Certainly, named taxa (including species) should not be taken as TUs without question. TUs need to be constructed during each analysis,

and re-checked each time a group is re-studied. They need to be carefully justified and re-justified using character evidence. This causes problems with easy comparison between analyses based on different data, but is an unavoidable fact of life in systematics and needs to be taken into account in such areas as database design (more below).

4.3 What exactly is a character (that is, a column in the data matrix)?

The fundamental activity in phylogenetic systematics is *character analysis* (Neff 1986) in which characters and states are hypothesized, tested, and refined in a reciprocal manner, in concert with the assembly of TUs, as part of the development of a data matrix. In addition to the logical primacy of data matrix construction, there is a temporal primacy as well. It is an established fact that a systematist spends 95% of his/her time gathering and analyzing character data and less than 5% time turning the assembled data matrix into a tree. Character analysis must be the all-important part of the phylogenetic reconstruction process if there is going to be a hope of discovering the history of a group. Fortunately, there have been some clear treatments of the elements of character analysis (Wiley 1981, Farris 1983, Neff 1986), but these were published some time ago and seem to be unknown to many recent workers. Younger systematists would do well to put more energy into investigations of the principles of character analysis and building better matrices, than into ever more complex model building for tree reconstruction, keeping firmly in mind the principle of 'garbage in, garbage out.' No model of the evolutionary process can be brought to bear successfully if the data matrix does not represent cogently argued character and character-state statements.

Before using a tool (characters in this case) it is wise to think carefully about what one is trying to do with the tool. What we are trying to do in phylogenetic analysis is to infer the existence of some past lineage by finding characters that changed state in that lineage and can thus serve as a potential marker for reconstructing that branch in the future (the Hennig Principle). The goal of

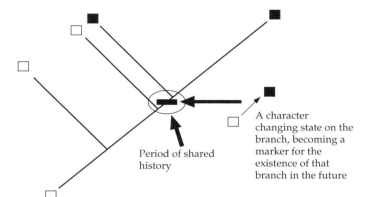

A character changing state on the branch, becoming a marker for the existence of that branch in the future

Period of shared history

Figure 4.1 Illustration of the concept of a phylogenetic marker.

character analysis is find as many potential markers as possible that can serve as evidence for the past existence of lineages shared by one or more of the TUs (see Fig. 4.1). These markers are the only tools a phylogeneticist has to reconstruct the branching history of life, but of course the kind of markers that are useful for branches at one level of depth in time won't necessarily be so for another level. Thus markers need to be searched for carefully in light of the particular branching events one is trying to reconstruct. Since semaphoronts are chosen to build TUs that are representative of the branching events under study, then we need to find 'good' characters that differentiate the chosen semaphoronts.

Much has been written about what constitutes a 'good' character. Ontologically speaking, potentially informative markers need to support a hypothesis of homology across the group under study; thus they need to be comparable in a convincing way across the study organisms. They need to be independent, so they can be taken as separate pieces of evidence for the existence of past lineages in the face of confounding effects such as convergence. They need to have discrete states so they can be inferred to contain a record of evolutionary events marking at least one specific past lineage. The epistemological rules of character analysis can thus be summarized as follows. Potential characters need to be evaluated by evidence for: (1) homology and heritability of a character across the TUs being studied, (2) independent evolution of different characters, and (3) presence in each character of a system of at

least two discrete states. I elaborate somewhat on each of these criteria in turn below:

(1) Homology is certainly one of the most important concepts in systematics, and therefore also one of the most controversial. Following on from the work of Hennig and later phylogenetic systematists, when we say that two semaphoronts share the same characteristic, we mean they share a profound historical continuity of information (Roth 1984, 1988). They are postulated to have shared a common ancestor that had that characteristic. Thus an important contribution of cladistics has been the explicit formulation of a phylogenetic criterion for homology: *a hypothesis of taxic homology (i.e. a potential synapomorphy) by necessity is also a hypothesis for the existence of a monophyletic group* (Patterson 1982; Stevens 1984). Each postulated homology (i.e. a column in the data matrix) is essentially a miniature phylogenetic hypothesis all by itself (especially as viewed in the context of its assigned states), and can be tested against other postulated homologies. Therefore, congruence among all postulated homologies provides a test of any single character in question; some characters initially thought to be homologous are later inferred not to be because they are in conflict with the majority of characters. The initial hypotheses of homology are based on detailed similarity in structure and development (see the discussion in Wiley 1981); these go into the matrix for eventual testing by congruence.

(2) For character changes to count as independent pieces of evidence in the congruence test (Patterson 1982), it is necessary that they not be

genetically, developmentally, or functionally correlated with other characters. There are many biological processes acting to distort the phylogenetic signal present in characters (e.g. reversal to primitive states caused by heterochrony, convergent evolution across different characters caused by natural selection, parallel changes to the same state within one character caused by functional constraints, etc.), along with random effects such as long branch attraction (caused by the accumulation of homoplastic matches on long, non-sister branches making them appear to be sister groups). The only weapon the phylogenetic systematist has against this inevitable distortion is many independent sources of information that are, as best as can possibly be determined, not impacted by the *same* biasing factors.

Note that there is another meaning of 'correlation', phylogenetic congruence, that does not disqualify characters from counting as independent! Congruence is what gives us supporting evidence for the existence of a monophyletic group. Thus the rules of character analysis need to be carefully drawn to encompass all the valid potential markers possible while rejecting those that are not suitable.

(3) Why is it necessary for a useful character to have at least two distinct states? Again, we need to think back to what we are trying to do: discrete states are needed because we are trying to reconstruct a discrete thing, an evolutionary event in which a prior state changed to some new posterior state, thus marking the existence of a shared ancestral lineage. The literature on the practice of how to define character states has had a checkered past. In most cases, people have simply made character state distinctions without any justification at all, and many methods proposed for 'gap coding' are flawed in various ways (Stevens 1991). The empirical details are beyond the scope of this chapter; see Mishler and De Luna (1991) for a discussion of this issue and a recommended approach using ANOVA and multiple range tests to seek statistically homogeneous groups of TUs representing character states.

To summarize, a 'good' character for phylogenetic analysis shows greater variation among TUs than within them. This variation must be heritable

and independent of other characters. This view of taxonomic characters also requires that each be a system of at least two discrete transformational homologs, or *character states*. Note that just as with TUs, there is never a given, *a priori* set of characters to begin a phylogenetic analysis with. Characters need to be discovered and evaluated during each analysis, and re-checked each time a group is studied.

4.4 What is the relationship between TUs and character states (that is, the individual entries in the data matrix)?

Neither the concept of TU nor the concept of character can be fully understood alone, without reference to each other and to the 'fractal' nature of the tree of life (as discussed earlier). The nature of both TUs and characters change as you go up and down this fractal scale.

As discussed earlier, the rows in a data matrix are virtually never based on data taken from a single individual, given that different labs are producing the data over time, and that different data-gathering techniques (ranging from DNA extraction through preparation for anatomical study) often require destructive sampling; thus data are often compiled from study of different organisms considered to represent the same TU. Thus TUs are nearly always composites in practice, their composition varying depending on the scale of analysis.

Likewise, what counts as a useful character changes depending on the scale of analysis. They have been selected based on their apparent utility for the task at hand, homologized (e.g. aligned) for the organisms under study, and pre-screened for their fit to the criteria of a good taxonomic character. Thus, the columns in a data matrix are already highly refined hypotheses of phylogenetic homology, defined with respect to the scale of the current study.

To make things more complicated, there is clearly a reciprocal relationship between TUs and character states. As detailed earlier, a TU can best be defined as a set of individual samples (semaphoronts in Hennig's terminology) that are homogeneous for character states currently known, while a character can best be defined as a potential marker for shared history of some subset of the known TUs. This means that TUs and characters emerge during a process of "reciprocal illumination" (Hennig 1966). To a large extent their definitions and discovery are interlinked. How do we proceed empirically in a way that avoids circularity? Before answering this question we need to consider the scaling problem in more detail.

4.5 Deep vs. shallow phylogenetics

The reconstruction of 'deep' relationships is fundamentally different than reconstruction of 'shallow' relationships (Mishler 2000). This is because the problems faced at these different temporal scales are quite distinct. In shallow reconstruction problems, the branching events at issue happened a relatively short time ago and the set of lineages resulting from these branching events is relatively complete (extinction has not had time to be a major effect). In these situations the relative lengths of internal and external branches are similar, giving less opportunity for long-branch attraction. However, the investigator working at this level has to deal with the potentially confounding effects of reticulation and lineage sorting. Character-state distinctions may be quite subtle, at least at the morphological level. At the nucleotide level it is necessary to look very carefully to find genes evolving rapidly enough; however, such genes may be relatively selectively neutral, and thus less subject to adaptive constraints which can lead to non-independence.

In deep reconstruction problems, the branching events at issue happened a relatively long time ago and the set of lineages resulting from these branching events is relatively incomplete (extinction has had a major effect). In these situations, the relative lengths of internal and external branches are often quite different; thus there is more opportunity for long branch attraction, even though there is little to no problem with reticulation and lineage sorting since most of the remaining branches are so old and widely separated in time. Due to all the time available on many branches, many potential morphological characters should be available, yet they may have changed so

much as to make homology assessments difficult; the same is true at the nucleotide level, where multiple substitutions in the same region may make alignment difficult. Thus very slowly evolving genes may be sought, but that very conservatism is caused by strong selective constraints which increases the danger of convergence leading to character dependence. Another approach is to increase sampling density—if TUs can be added that more evenly sample the true tree, thus reducing the asymmetry between internal and external branches, then faster-evolving genes may have better performance (Källersjö et al. 1998, 1999).

These considerations suggest that the problems being faced, and their best-justified solutions, will change as you go up and down this fractal scale. The nature of TUs and usable characters are going to change, and we need to have a way to scale phylogenetic results from one level to the next if we are going to have a hope of building a complete tree of life. There is effectively an infinite number of semaphoronts out there; there will never be a 'complete' data matrix for all of them for the practical reason that there are too many. But more importantly, it isn't at all clear that a single global analysis of all semaphoronts living on Earth would be desirable, even if we could do it. There is the fact discussed earlier that a given semaphoront doesn't bear all the relevant data, and thus composite TUs would need to be constructed in practice. There is also the fact that character homologies can be drawn much more easily when comparing only closely related TUs. Very few characters can be coded reliably across the whole tree of life. So we need to examine the scaling issue closely to see how we might combine or concatenate data matrices and phylogenetic results from more-shallow analyses into deeper and deeper ones until eventually a global tree of life can be produced.

4.6 How should we connect up analyses and data matrices that are 'nested' inside each other at various different time scales?

How will we ultimately connect up deep and shallow analyses, each with their own distinctively useful data and problems? Some hold out hope for eventual global analyses, once enough universally comparable data have been gained and computer programs get much more efficient, to deal with all extant organisms at once. Others would go to the opposite extreme, and use a *supertree* approach, where shallow analyses are grafted on to the tips of deeper analyses. An intermediate approach, called *compartmentalization* (Mishler 1994, 2000), uses shallow topologies (that are based on analyses of the characters useful locally) to constrain global deep analyses (that are based on analyses of characters useful globally). All of these issues surrounding how to use phylogenetic markers at their appropriate level to reconstruct the extremely deep tree of life are likely to be among the major concerns of phylogenetics in coming years, as reconstruction of the whole tree of life from twigs to trunk is attempted.

The different approaches to concatenating analyses at different scales can be best viewed as a spectrum (see Fig. 4.2). At the left-hand end of this spectrum, the approach is to include all possible TUs and potential characters in one matrix. Generally this is not actually done, because the sheer amount of data (millions of possible TUs) makes thorough phylogenetic analysis computationally impossible. The most-common approach in practice in global analyses is to select a few representatives of a large, clearly monophyletic group (the *exemplar method*). Care is sometimes taken to select representatives that are 'basal' TUs within the group to be represented (i.e. cladistically basal relative to the imaginary root defined by outgroups); however, this still does not avoid two important problems: (1) within-group variation is not fully represented in the analysis, and (2) an increase both in terminal branch lengths and in asymmetry between lengths of different

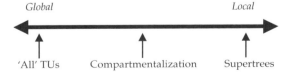

Global   *Local*

'All' TUs   Compartmentalization   Supertrees

Figure 4.2 How to concatenate different analyses to build the tree of life? Shown is a spectrum of approaches ranging from global to local. See text for explanation.

branches is introduced. These problems can lead to erroneous long branch attractions in global analyses.

At the right-hand end of the spectrum, local analyses are simply grafted together into supertrees at the place where shared taxa occur, without reference back to the original data. There are many ways to do this in detail (as reviewed by Sanderson *et al.* 1998), but the important thing is that the analyses on real character data are only done locally, and the concatenation is based on a combination of local topologies rather than an integration of local data sets into a global data set.

Both of these approaches may be problematic, one too global, the other too local. Thus the appeal of a promising intermediate approach called *compartmentalization* (by analogy to a water-tight compartment on a ship—homoplasy is not allowed in or out). This approach represents diverse yet clearly monophyletic clades by their inferred ancestral states in larger-scale cladistic analyses (Mishler 1994, 2000). A well-supported local topology is sought first, then an inferred "archetype" or *Hypothetical Ancestor* (HA) for the group is inserted into a more inclusive analysis. In more detail, the procedure is to: (1) perform global analyses, determine the best supported clades (these become the compartments); (2) perform local analyses within compartments, including more taxa and characters (more characters can be used within compartments due to improved homology assessments among closely related organisms—see below); (3) return to a global analyses, in one of two ways, either (a) with compartments represented by single HAs (the archetypes), or (b) with compartments constrained to the topology found in local analyses (for smaller data sets, this approach appears better because it allows flexible character state assignments to the base of the compartment based on optimizations to the local topology).

The compartmentalization approach differs from the exemplar approach in that the representative character-states coded for the archetype are based on all the TUs in the compartment, thus the reconstructed HA is likely to be quite different from any particular TU. As an estimate of the states of the most recent common ancestor of all the local

TUs, the HA is likely to have a much shorter terminal branch with respect to the global analysis, which in turn can have the beneficial global effect of reducing long branch attraction. In addition to these advantages of compartmentalization at the global level, the local analyses will be better because one can: (1) include all local TUs for which data are available; (2) incorporate more (and better justified) characters, by adding in those characters for which homology could not be determined globally (e.g. genes that can only be aligned locally); (3) avoid spurious homoplasy that can change the local topology due to long-branch attractions with distant outgroups. The effects of compartmentalization are thus to cut large data sets down to manageable size, suppress the impact of spurious homoplasy, and allow the use of more information in analyses. This approach is self-reinforcing; as better understanding of phylogeny is gained, the support for compartments will be improved, leading in turn to refined understanding of appropriate characters and TUs both within the compartments *and* between the compartments.

4.7 Structural vs. DNA sequence characters

The choice of data for use at different scales of analysis is the crux of the matter. One important issue to consider is how intrinsically useful are different categories of characters at these different scales? It is clear that, as general categories, structural data (i.e. anatomical, morphological, or genomic) and DNA sequence data have different and complementary strengths and weaknesses. DNA sequence characters are much more numerous than structural characters, thus increasing the chance that sufficient markers can be found for all branches of a tree. They are especially useful in organisms with simple morphology, such as fungi and bacteria, that may lack a sufficient number of structural characters. Objectively defining character states in structural comparisons can be difficult, particularly in shallow reconstructions, while the states are usually clear-cut in DNA sequence data. It has been argued that it is useful that DNA sequence data are independent from

morphological characters that are perhaps subject to adaptive convergence (although convergence of course cannot be ruled out in DNA sequences, particularly at deeper levels). Sequences of highly conserved genes can be homologized across very broad groups that share little morphologically, although these same highly conserved regions are probably highly subject to adaptive convergence. Finally, models of evolutionary change are easier to postulate for DNA sequence evolution, a perceived advantage for those who like to use maximum likelihood methods.

On the other hand, especially in deeper comparisons, structural characters (i.e. traditional morphological characters but also modern genomic characters such as rearrangements and intron insertions; see next section) often have much greater complexity, and may exhibit ontogeny, allowing a temporal axis of comparison not available with DNA sequence data. Structural characters often change in an episodic pattern, which is necessary for evidence of deep, short branches to remain detectable. Clock-like markers are the worst kind of data for those sorts of branches; the markers keep changing and thus erasing history. It is much better for discovering those deep, short branches to have a clock like those found frozen in place on the sunken ship *Titanic* (still showing the time the ship went down); a clock that stopped ticking when some major change occurred. Furthermore, the number of possible character states is usually much higher in morphological character systems (and in genomic rearrangements) than in DNA sequence data, which serves to make long branch attraction less of a problem (see Mishler 1994 for discussion). Morphological data are more easily gathered from large numbers of specimens, and from fossils, making it much easier to robustly sample the true phylogeny. For all these reasons, morphological data have remained among the characters of choice at deeper phylogenetic levels, and have been joined recently by an exciting new class of structural characters derived from genomic comparisons. The latter promise to be very useful in the future, particularly for those deep, relatively short internal branches that have proven resistant to phylogenetic reconstruction with DNA sequence data.

4.8 Genomic characters

This is the era of whole-genome sequencing; molecular data are becoming available at a rate unanticipated even a few years ago. Sequencing projects in a number of countries have produced a growing number of fully sequenced genomes, providing computational biologists with tremendous opportunities. However, comparative genomics has so far largely been restricted to pairwise comparisons of genomes; for instance, to identify syntenic regions, orthologous genes, and common regulatory elements between human and mouse. The importance of taking a phylogenetic approach to systematically relating larger sets of genomes has only recently been realized.

A recent synthesis of phylogenetic systematics and molecular biology/genomics—two fields once estranged—is beginning to form a new field that could be called phylogenomics (Eisen *et al.* 1998). Something can be learned about the function of genes by examining them in one organism. However, a much richer array of tools is available using a phylogenetic approach. Close sister-group comparisons between lineages differing in a critical phenotype (e.g. desiccation or freeze tolerance) can allow a quick narrowing of the search for genetic causes. Dissecting a complicated, evolutionarily advanced genotype/phenotype complex (e.g. development of the angiosperm flower) by tracing the components back through simpler ancestral reconstructions can lead to quicker understanding. Hence, phylogenomics allows one to go beyond the use of pairwise sequence similarities and use phylogenic comparative methods to confirm and/or to establish gene function and interactions.

Cross-genome phylogenetic approaches have the potential to provide insights into many open functional questions. A short list includes understanding the processes underlying genomic evolution, identifying key regulatory regions, understanding the complex relationship between phenotype and genomic changes, and understanding the evolution of complex physiological pathways in related organisms. Using such a comparative approach will aid in elucidating how these genes interact to perform specific biological processes. For example, Stuart *et al.* (2003) used microarray data from four completely sequenced genomes (yeast, nematode, insect, and human) to show coexpression relationships that have been conserved across a wide spectrum of animal evolution.

Most importantly for the systematist, the new comparative genomic data should also greatly increase the accuracy of reconstructions of the tree of life. Even though nucleotide sequence comparisons have become the workhorse of phylogenetic analysis at all levels, there are clearly phylogenetic problems for which nucleotide sequence data are poorly suited, because of their simple nature (having only four character states) and tendency to evolve in a regular, more-or-less clock-like fashion. In particular, as stated earlier, deep branching questions (with relatively short internodes of interest mixed with long terminal branches) are notoriously difficult to resolve with DNA sequence data. It is fortunate, therefore, that fundamentally new kinds of structural genomic characters such as inversions, translocations, losses, duplications, and insertion/deletion of introns will be increasingly available in the future.

These characters need to be evaluated using much the same principles of character analysis (discussed earlier) that were originally developed for morphological characters. They must be looked at carefully to establish likely homology (e.g. examining the ends of breakpoints across genomes to see whether a single rearrangement event is likely to have occurred), independence, and discreteness of character states. Given the close link between characters and TUs discussed above, it is also necessary to consider carefully the appropriate TUs for comparative genomic analysis, especially since different parts of one organism's genome may or may not have exactly the same history. Thus close collaboration between systematists and molecular biologists will be required to code these genomic characters properly, and to assemble them into matrices with other data types. Challenges resulting from combining different data sources, in light of the possibility of different histories for different parts of the same genome, are discussed in the next two sections.

4.9 Dealing with heterogeneous data types

There is every reason to search carefully for good potential markers in all kinds of data, particularly for the deep branching questions discussed earlier. Deep phylogenetic reconstructions are by their nature difficult, and all characters should be sought and used if they meet the criteria of good potential markers (Mishler 2000). However, it remains controversial how data from different sources are to be evaluated and compared (Swofford 1991). Some have argued that data sets derived from fundamentally different sources should be analyzed separately, and only common results taken as well supported (i.e. consensus tree approaches), or at least that only data sets that appear to be similar in the trees they favor should be combined (Huelsenbeck *et al.* 1996). Others have argued that all putative homologies should be combined into one matrix (i.e. 'total evidence'; Kluge 1989; Barrett *et al.* 1991; Donoghue and Sanderson 1992; Mishler 1994). Theoretical arguments at present favor the latter approach: if characters have been independently judged to be good candidates for phylogenetic markers, then they are equivalent and should be analyzed together.

There is one major exception to the preference for a 'total evidence' position: data should not be combined into a single matrix if there is evidence that some characters had a different branching history than the rest (Mishler 2000). However, this is not easy to detect. There are several sources of homoplasy other than different branching history, including evolutionary convergence. If several data partitions show different highly discordant trees due to convergence, the only way to see the 'true' tree topology is to combine them. The only weapon a systematist has against convergence is the likelihood that truly independent characters will be subject to different confusing factors and thus the true history may emerge when these independent characters are combined (Barrett *et al.* 1991). Probably all character systems are influenced by constraints that tend to bias phylogeny reconstruction one way or another, yet a combination of very different character data

may allow the noise to cancel out, and the historical signal to come through.

Therefore, observing a particular gene or other data partition exhibiting serious conflict with another is not sufficient reason to reject combining them. There must also be additional evidence, outside of the phylogenetic analysis, for reticulation or lineage sorting. The best current examples of such discordance are in shallow analyses, where organellar genomes may have different phylogenies than those of associated nuclear genomes and morphologies (Smith and Sytsma 1990; Rieseberg and Soltis 1991). Barring that sort of clearly explainable discordance via reticulation, all appropriate data should be used, especially in deep analyses because, as argued earlier, reticulation and lineage sorting are much less likely to be problems in deep analyses, while convergence is likely to be a greater problem. But even if its effects may be negligible in many deeper analyses, the problem of reticulation is a difficult one, worthy of a more detailed look.

4.10 Reticulation

As introduced earlier, the tree of life is essentially composed of nested sets of lineages. Look closely at one lineage, and it turns out to be composed of smaller lineages, all the way down to within the organism (e.g. cell lineages and gene genealogies). None of the levels of nested lineages can be considered fundamental (Mishler and Theriot 2000a, b,c)—it depends on the scale of the specific question being asked. To build the large-scale framework of the tree of life one can probably ignore the fine-scale lineages within organisms and between organisms within populations. But to study microevolutionary differentiation processes and design conservation plans at the population level, one needs to look at the fine-scale lineages, and to look at the spread of cancer cells in a body, one needs to look at finer levels still.

The major problem that arises is that these nested sets of lineages are not always proper subsets. Especially at the finer levels, sublineages of a larger lineage may not all have the same history, and/or may not have the same history as the larger lineage. For example, parts of the genome within one organism can have different

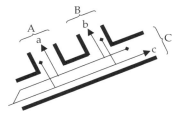

Figure 4.3 Illustration of lineage sorting. Three larger-scale lineages are outlined with dark lines and labeled with capital letters. Three extant smaller-scale lineages are included, together with extinct relatives, and shown with lighter lines and lower-case letters. Note that the relationships of the larger-scale lineages are A(B,C) while the relationships of the smaller-scale lineages are (a,b)c because of the particular pattern of extinction that occurred. This would result in apparent homoplasy at the level of the larger-scale lineages.

histories, for two main reasons. The first of these is *lineage sorting* (see Fig. 4.3), which occurs when genes exist in families within the genome due to past duplication events, and differential extinction has taken place in derived higher-level lineages such that the relationships of the genes appear not to match the relationships of the higher-level lineages (Avise 1989). The problem in this case is one of mistaken homology—paralogy is confused with orthology because not all the gene lineages are present in all higher-level lineages.

The second major reason for differential histories is *reticulation*, which occurs when once separate lineages blend back together. At the genome level, recombination can bring genes with different histories together into a single lineage. Of all the different sources of homoplasy, such as adaptive convergence, gene conversion, developmental constraints, mistaken coding, lineage sorting, and reticulation, the last is the most problematical. This is because reticulation violates a fundamental assumption underlying cladistic analysis, that of a branching model of history. The other factors are all cases of mistaken hypotheses of homology of one sort or another, whereas 'homoplastic' character distributions due to reticulate evolution involve true homologies whose mode of transmission was not tree-like. The possibility of reticulation further complicates the relationship between TUs and characters discussed earlier, since it ensures that some lineages nested inside of larger ones truly have different histories than others.

Because of this important violation of a fundamental cladistic assumption, Hennig (1966) and later Nixon and Wheeler (1990) were correct in focusing on reticulation and the problems it causes for cladistics. However, the problems posed by reticulation are more complicated than their proposed 'solution,' i.e. their suggestion that the species level can be used as a dividing line by supposing that reticulation only occurs below the species level. This assumption (made by many, but not all, cladists) of an abrupt cessation of interbreeding at the species level, separating rampant reticulation below from clean divergent evolution above, was wrong in two important respects. One is the implication that reticulation can be disregarded at higher levels, and the other is the implication that cladistic methods are not appropriate below the species level. Mishler and Theriot (2000a, b, c) refuted both implications; here are their arguments in summary:

(1) There is no consistent demarcation between reticulate and branching relationships at any particular level. Hybridization takes place between clades of various patristic/cladistic degrees of relatedness. Reticulate relationships range from intense (in panmictic, sexually reproducing groups where individual relationships are exclusively reticulate), to less intense (in spatially or temporally subdivided groups where both reticulate and divergent relationships exist among individuals), to none in clonally reproducing organisms. Rare, high-level hybridizations may occur among very divergent lineages, such as among genera of orchids; viral-mediated lateral transfer of genetic material is suspected at much higher levels.

(2) Just as barriers to reticulation are often not complete, reticulation is not a complete barrier to cladistic analysis. There is much phylogenetic structure within named species; indeed, a whole new field of phylogeography was founded to explore this (Avise 1989). We can reconstruct relationships in the face of *some* amount of reticulation (how much is not yet clear, but is amenable to study). For example, McDade (1992) showed that incorporating a few known hybrids in an analysis of 'good' species does *not* seriously affect the cladistic topology of the good species. There may be a self-correcting mechanism here as there

is with other sources of homoplasy: even major convergence (e.g. among cave animals) can be uncovered via cladistic analysis. As with convergence, where the application of cladistic analysis provides the only rigorous basis we have for identifying homoplasy and thus demonstrating non-parsimonious evolution, the only way we can identify reticulation on the basis of character analysis alone is through the application of cladistic parsimony, followed by the examination of homoplasy to attempt to discover its source (see discussion by Vrana and Wheeler 1992). As was pioneered by Slatkin and Maddison (1989), cladistic analysis of non-recombining genes can even be used to measure gene flow between populations. Thus, cladistic analysis can be used to study reticulation, at any level.

(3) Thus, just as there may be no largest cladistic unit for which reticulation is impossible, there may be no smallest 'irreducible' cladistic unit within which no further diverging phylogenetic patterns occur. Ontologically speaking, we are dealing with a fractal pattern again; if you look inside one lineage you see a pattern of divergence of lineages within (and some reticulation, perhaps increasingly greater as one looks at less-inclusive lineages). This fractal pattern of reticulation and branching presents a problem for simple phylogenetic inference. But, as argued above, phenomena such as lineage sorting and reticulation can be discovered as incongruence between organismal and gene phylogenies, or incongruence between different genes or different regions of the genome.

4.11 TUs, characters, and database design

One of the big challenges in modern biology is informatics. There are so many data available, and a number of projects are attempting to represent the information in databases. However existing databases (e.g. GenBank or Tropicos) are essentially a flat file with respect to phylogeny. Data are entered with whatever taxon name happens to be attached to them. The only sense of evolutionary relationships is given by a schema of higher-taxon names (say families and phyla) that can be used to group the basic entries. These higher taxa may or

may not be monophyletic, and essentially function as static sorting bins for pulling out the basic records—there is no way to access or display emergent properties of data at higher evolutionary levels or to discover finer-scale patterns at lower levels. In other words, databases are not yet sensitive to the fractal nature of phylogenies (with their many hierachically nested levels). As argued above, there are no basic comparable taxa (terminal or otherwise), or characters. Both TUs and characters are defined with respect to a certain level in the phylogeny.

As a new generation of phylogenetic databases are built (in part coordinated by a large NSF ITR grant supporting a national resource in phyloinformatics, Cyber Infrastructure for Phylogenetic Research (CIPRes); see www.phylo.org), there needs to be much more flexibility built in. The main themes of this chapter need to be explored to appropriately present the richness of phylogenetic data to users. Fundamental open questions that need to be addressed for databases include: (1) how can the elements of the data matrix (TUs, characters, and states) as defined and recognized in some particular study be stored and potentially retrieved for use in a future study at a different level? (2) How can heterogeneous data types (e.g. DNA sequences, genomic rearrangements, morphology) be compared/combined? (3) How can data sets and analyses at very different scales be concatenated (e.g. supertree, compartmentalization, or global approaches as discussed earlier)? (4) How can phylogenetic results at these different concatenated scales, where TUs are nested inside larger ones, and character definitions (e.g. alignments) change as you move up and down the scale, be presented to the community in comprehensible and useable ways?

The centerpiece of all future biological databases will need to be phylogenetic classification, a deeply nested hierarchy of named nodes linked to all available structural and functional data at each level dynamically, as new data enter the database. All biological data fall somewhere on the tree of life, which is the one thing that can unify them all. This new approach to biodiversity informatics will take advantage of the richness of the phylogenetic structure of biological data.

4.12 Tree building

This chapter has focused on the first phase of phylogenetic analysis, building the data matrix, rather than the second phase, building a tree from the matrix. Still, a few words on the latter are appropriate. The simplest model for evaluating congruence among characters (different hypotheses of homology) is equally weighted parsimony (Farris 1983), which remains the preferred method for comparing diverse sorts of characters. Each column in a data matrix can be regarded as an independently justified hypothesis about phylogenetic grouping (the criteria for justifying these individual character hypotheses is discussed above), an individual piece of evidence for the existence of a monophyletic group. Parsimony assumes that an apparent homology is more likely to be due to true homology than to homoplasy, unless evidence to the contrary exists, i.e. a plurality of apparent homologies showing a different pattern (Funk and Brooks 1990; Mishler 1994). Parsimony does involve some simplifying assumptions, i.e. that all character-state changes are similar in their probability of change, and thus they can all be equally weighted. This assumption, while robust, can lead to mistaken reconstructions under some extreme circumstances of asymmetric probabilities of change within and among characters, and in such cases simple parsimony can be modified using more complicated models of change by either character and character-state weighting (Albert *et al.* 1992, 1993; Albert and Mishler 1992) or maximum likelihood approaches (Felsenstein 1981; Yang 1994).

Debates will no doubt continue over how complicated an evolutionary model it is prudent to include in an analysis, but it is clear that all the parsimony and maximum likelihood methods, by using individual character data (specific hypotheses of homology), belong to a related Hennigian family of methods. Fortunately, one important empirical observation is that differential weighting and maximum likelihood have little effect on simple parsimony reconstructions. Weighted parsimony and maximum likelihood topologies are almost always a subset of the equally weighted parsimony topologies, especially when applied to

data with an appropriate rate of change for the problem at hand (more on this later). Thus, paradoxically, pursuit of well-supported weighting schemes has ended up convincing many of us of the broad applicability and robustness of equally weighted parsimony (Albert *et al.* 1993). Furthermore, all reconstruction methods work best with 'good data', i.e. characters chosen with respect to a particular level of phylogenetic question. It is with more problematic data (e.g. with a limited number of informative characters, a high rate of change, or strong constraints) that results of different methods begin to diverge. Weighting algorithms and maximum likelihood approaches may be able to extend the use of problematic data, but only if the evolutionary parameters that are biasing rates of change are known. As biases become greater, precise knowledge of them becomes ever more important for avoiding spurious reconstructions. Therefore, given the large number of potential characters made available by modern technology, it is desirable to be highly selective about the characters that are used to address any particular phylogenetic question; to the extent possible, the problematic data should be left out (possibly to be used at a different, more appropriate level: see discussion on compartmentalization in Mishler 1994, and elsewhere in this chapter).

What is the relationship between this chapter emphasizing the data matrix, and the general themes of this book on parsimony? Simple. A rigorously produced data matrix has already been evaluated carefully for potential homology of each feature when being assembled. Everything interesting has already been encoded in the matrix; what is needed is a simple transformation of that matrix into a tree without any pretended value added. Straight, evenly weighted parsimony is to be preferred, because it is a robust method (insensitive to variation over a broad range of possible biasing factors) and because it is based on a simple, interpretable, and generally applicable model. More-complicated models for tree building are fundamentally attempts to compensate for marginal data. Given the surfeit of data available these days, it would be wiser to avoid the use of marginal data!

These issues of how to use phylogenetic markers at their appropriate level to reconstruct the extremely fractal tree of life are likely to be one of the major concerns in the theory of phylogenetics in coming years. In the future, my prediction is that more-careful selection of characters for particular questions (i.e. more-careful and rigorous construction of the data matrix) will lead to less emphasis on the need for modifications to equally weighted parsimony. The future of phylogenetic analysis appears to be in careful selection of appropriate characters (discrete, heritable, independent, and with an appropriate rate of change) for use at a carefully defined phylogenetic level.

4.13 Acknowledgements

This chapter has benefited from analyses and collaborations supported by three NSF grants, and I acknowledge my co-principal investigators for their help in understanding these issues: the Deep Gene Research Coordination Network (DEB- 0090227; http://ucjeps.berkeley.edu/bryolab/deepgene/), the Green Tree of Life Project (EF-0228729; http://ucjeps.berkeley.edu/TreeofLife/), and the ITR grant entitled Cyber Infrastructure for Phylogenetic Research (CIPRes; EF-0331494; www.phylo.org/).

CHAPTER 5

Alignment, dynamic homology, and optimization

Ward C. Wheeler

5.1 Introduction

Systematics is the production of cladograms that link taxa through their observed variation. These cladograms must optimize an objective function such that they can participate in hypothesis testing on the basis of this function. The core activity of systematics is to assay the relative merits of a pair of competing scenarios and judge one superior. The repeated and transitive application of this elemental comparison results in a globally optimal solution that is the ultimate goal of systematics.

This depiction of systematics raises three points: the nature of the objective optimality criterion; the manner of determination of this value; and the assessment of the relative merits of cladograms. This chapter is concerned with the second of these three, the realm of character analysis, homology and optimization. The arguments here will be based on the optimality criterion of parsimony or minimum cost. Likelihood or other criteria could well be used, however, and most of the character-optimization discussions would remain largely unchanged other than the specifics of their implementation and numerical values. The comparison of cladograms is the province of cladogram or tree searching and is not discussed in any depth here.

Homology is the relationship between features that is derived from their shared, unique origin. Given a single cladogram, two features are homologous if their origin can be traced back to a specific transformation on a branch of *that* cladogram, but the same pair of features may not be homologous on alternate cladograms. Homology is entirely cladogram-dependent and the relative optimality of alternate cladograms determines whether or not features exhibit this relationship.

The dynamic homology framework (Wheeler 2001a) is an analytical concept that extends through optimization of transformations to the correspondence among features (often referred to as putative or primary homologies) themselves. The joint scenario of correspondence and transformation is chosen such that the overall cladogram cost is minimal. The correspondences among features (nucleotides in this context) are not predetermined, but a result of the analysis. In this framework, there is no distinction between putative or primary and secondary homology (de Pinna 1991)—all variation is optimized *de novo* for each cladogram.

5.2 Sequence data

There are two properties that have been used to differentiate sequence data from other sorts of information: simplicity of states and length variation. Unlike complex anatomical features (e.g. limb or wing) that can express themselves in a myriad of forms, nucleotides exhibit only four conditions. Complexity and difference imply that states (e.g. presence/absence, or conditions) are not comparable across characters. Nucleotide states, on the other hand are identical no matter where they occur. Nucleotide sequences may also differ in length. These two aspects of molecular sequence data remove the complexity and positional information so often used in establishing primary homologies in anatomical systems.

5.3 Alignment and optimization

Two approaches have been developed to deal with the absence of preordained homologies and analyze sequence data. On one hand, methods have been devised to create the missing primary homology statements that are then analyzed by standard techniques—broadly referred to as multiple alignment. Traditionally, sequence data have undergone this pre-phylogenetic analysis step to permit familiar procedures akin to those used with anatomical characters. A second approach is to directly optimize sequence variation during cladogram searching. This methodology requires no notions of primary character homology or any global (i.e. topology-independent) homology statements whatsoever (other than that the compared sequences themselves be homologous). Direct optimization is also applicable to simple, serial morphologies, for which characters and their states are similarly constrained. For the sake of discussion here, the terms alignment and optimization will refer to these alternate approaches.

5.4 The problem

In computational terms, the problem of determining the cost of a given cladogram reduces to the determination of the set of internal vertex (hypothetical ancestral) sequences such that the overall cost is minimized. Whether expressed in terms of alignment or optimization, the problem (known as the tree-alignment problem) is NP-hard (see Wang and Jiang 1994); hence, we are very unlikely to achieve exact solutions. NP-hard problems are members of a class of computational problems for which there is no known polynomial time solution. These problems are often combinatorially 'explosive', with the size of the solution space expanding factorially. That is, when the sequences can vary in length, even the determination of the cost of a single cladogram will be heuristic (for 10 sequences of length 5—an unrealistically small and well-behaved case—there are 1.35×10^{38} homology schemes; Slowinski 1998).

Given that determining cladogram cost is heuristic, the transformation and homology statements derived from the cladogram are heuristic as well.

When coupled with cladogram search, we are faced with a compound NP-complete problem and all of our statements will be based on approximate solutions.

Both alignment and optimization may be viewed as heuristic approaches to solving this problem. Alignment accomplishes this based on static, global, primary homology statements, whereas optimization techniques propose cladogram-specific homology scenarios.

5.5 Alignment methods

As a heuristic solution, alignment decomposes the nested homology/search procedure into two sequential problems. Length-variable sequences are converted into a series of column vectors (primary homology statements) through the insertion of gap characters (-) as placeholders that denote the results of insertion/deletion events. Alignments minimize a cost function (in the case of two sequences the cost to 'edit' one sequence into the other) that is based on the relative costs of transformation events (especially insertion/deletion—'indel'—costs), which may or may not be cladogram-based.

There are several components to the alignment process. These progress from pairwise alignment of two sequences to the exact solution for multiple sequences and then to the heuristic methods employed in real-world analyses.

5.5.1 Pairwise alignment

Alignment of sequence pairs is the foundation of all more elaborate procedures. The problem, simply stated, is to create the series of correspondences between the nucleotides in two sequences via the insertion of gaps, such that the edit cost (the weighted sum of all events—insertions, deletions, nucleotide substitutions—required to convert one sequence into another) between the sequences is minimized (or some other function optimized). Costs must be assigned to each type of event, or trivial, zero-cost alignments can result (e.g. indels costing zero and an alignment that places each nucleotide opposite a gap). The first algorithmic solution to this form of string-matching problem was proposed by Needleman and Wunsch (1970) and is used throughout most alignment procedures (see Gusfield 1997 for more extensive discussion).

Consider two sequences ACGT and AGCT and alignment parameters of nucleotide-substitution cost equal to 1 ($Cost_{Subs}$) and indel cost equal to 10 ($Cost_{InDel}$). The algorithm follows a dynamic programming approach by solving a series of small, dependent sub-problems that implicitly examine all possible alignments. There are two components to the procedure. The first determines the cost of the best alignment (or alignments—there may be multiple solutions). This is often referred to as the wavefront update. The second component is the traceback, which yields the alignment itself (more-complex examples can be found in Phillips *et al.* 2000). Needleman and Wunsch described a maximization-of-identity algorithm, where here a minimization of difference is presented. The underlying principles are unchanged.

$$Cost_{i,j} = \min\{Cost_{i-1,j} + Cost_{InDel}, Cost_{i,j-1} + Cost_{InDel}, Cost_{i-1,j-1} + Cost_{Subs\ i,j}\}$$

The first part of the algorithm fills a matrix M of size $(n+1) \times (m+1)$ to align a pair of sequences a and b of length n and m respectively. Each cell (i,j) is the cost of aligning the first i characters of a with the first j characters of b (i.e. aligning $a_1 \ldots a_i$ and $b_1 \ldots b_j$). Each value is calculated using the previously aligned subsequences: that is, the cost of cell (i,j) will be

$$\min\{(i-1,j) + indel, (i,j-1) + indel, (i-1,j-1) + align\ character\ a_i\ and\ b_i\}$$

or less formally, the minimum among

aligning $a_1 \ldots a_{i-1}$ and $b_1 \ldots b_j$ and
   aligning character a_i with a gap,
aligning $a_1 \ldots a_i$ and $b_1 \ldots b_{j-1}$ and
   aligning character b_j with a gap,
aligning $a_1 \ldots a_{i-1}$ and $b_1 \ldots b_{j-1}$ and
   aligning character a_i and b_j.

The additional first row and column (the reason for the $+1$ in the matrix dimensions) represents the alignment of a sequence with an empty string; that is, initial gaps. Each decision minimum is recorded, to follow the path that leads to the cost of aligning a and b; that is, the cost in cell (n,m) (Fig. 5.1).

In order to create the actual alignment between the sequences a traceback step is performed that proceeds back up and to the left of the matrix, keeping track of the optimal indels and substitutions performed in the matrix-update operations. The minimum cost path is followed back, where the best move is diagonal if the nucleotides of the sequences correspond, and the left and upwards moves signify indels (Fig. 5.1). The minimal cost alignment for these sequences (ACGT and AGCT) with the cost regime {indels = 10, substitutions = 1} is 2 with two base substitutions implied between the sequences (C $\leftrightarrow$ G, and G $\leftrightarrow$ C).

If a complementary cost scenario is specified, e.g. indels = 1 and substitutions = 10, a different optimal solution is found (Fig. 5.1, right). In this case as well, the minimum cost is two, but no substitutions are implied—only indels (2).

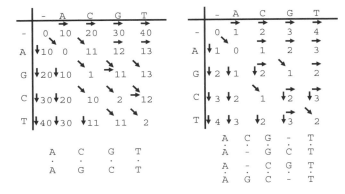

Figure 5.1 Needleman and Wunsch (1970) alignment matrix tables for two cost scenarios. On the left, indel events cost 10 steps and nucleotide changes 1, while these are reversed on the right. Both cost scenarios yield minimum cost alignments of cost 2, although minimizing indels in the former (left) and nucleotide substitutions in the latter (right).

Furthermore, there are two equally optimal solutions differing in the placement of the gaps. This ambiguity comes from the equally costly paths found at matrix element 3,3 (of 0,0 to 4,4). The non-unique nature of such solutions is a frequent property of alignments and can have dramatic effects on phylogenetic conclusions (Wheeler 1994).

5.5.2 Exact multiple alignment

The pairwise procedure can be generalized in a straightforward fashion to align more than two sequences. The matrix would have an axis for each sequence (l sequences would require l dimensions) and there would be $2^l - 1$ paths to each cell representing all the possible combinations of gaps and substitutions possible (seven in the case of three sequences). These two factors add enormously to the calculations, making true multidimensional alignments unattainable for real data sets.

An additional complexity arises in analyses of data sets with more than three sequences. The cost calculations at each cell may (as Sankoff and Cedergren 1983 suggested) be based on the cladogram of relationships of the sequences. If this is known, or at least specified *a priori*, the cell cost can be calculated directly. If, however, the cladogram is unspecified, a search would be performed for each cell, or the entire multidimensional alignment repeated for multiple (potentially all) cladograms.

The immense computational burden of exact multiple alignment ensures that heuristic solutions are used in nearly all real-world cases.

5.5.3 Heuristic multiple alignment

Current heuristic procedures are similar in that many attempt to render multiple alignment tractable by breaking down simultaneous *n*-dimensional alignments into a series of manageable pairwise alignments related by a "guide tree" (in the parlance of Feng and Doolittle 1987). These differ in the techniques used to generate the guide tree and conduct the pairwise alignments at the guide tree nodes. Furthermore, the procedures

may or may not be explicitly linked to optimality criteria (Fig. 5.2).

By far the most commonly used heuristic multiple-alignment implementation is CLUSTAL, mainly because it is fast and relatively easy to use. Many others are freely available, however, and take different approaches to the problem. Several of these approaches are illustrated in this sample. More-complete lists can be found at http://pbil.univ-lyon1.fr/alignment.html and more comparisons in Phillips *et al.* (2000).

CLUSTAL (Higgins and Sharp 1988 *et seq.*) creates a single multiple alignment based on a single guide tree. A neighbor joining tree (Saitou and Nei 1987) is calculated from the pairwise alignments via a 'corrected' distance formula. This tree is used as a guide tree for progressive pairwise alignment of terminal sequences and internal consensus sequences (a down-pass). A second (up) pass resolves the placement of gaps in internal and ultimately observed sequences. There is no optimality value associated with a CLUSTAL alignment.

TREEALIGN (Hein 1989a, b) Also produces a single multiple alignment based on a single guide tree, but that guide tree is constructed (with some tree refinement) as the alignment is created. A parsimony step is included as part of the tree-reconstruction procedure. Although alignments are not searched as such, the generation of the guide tree examines multiple alternatives. A final, single multiple alignment is generated with an attached parsimony score, but no comparisons to other complete alignments are made.

DALIGN (Morgenstern *et al.* 1996) differs from other methods in looking for alignments of contiguous gap-free fragments of DNA that may have mismatches. This contrasts with the approach that attempts to align each position in a sequence. No gap penalty is employed. The idea behind this method is to create complete alignments by stitching together locally similar sequences that may be separated by highly divergent regions. An optimal alignment is one that maximizes the weighted sum of the matches in the smaller segments. Alignments can be compared on this basis. This method makes no reference to cladograms or trees whatsoever.

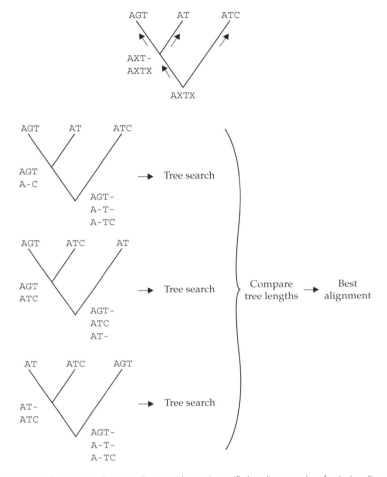

Figure 5.2 General heuristic multiple sequence alignment. Top, a guide tree is specified to direct a series of pairwise alignments which incrementally include sequences as the guide tree is traversed from tips to root (e.g. CLUSTAL). Usually, some form of consensus sequence is created at the internal nodes. Additional gaps are inserted as the tree is traversed a second time form root to tips. When an optimality criterion is employed (e.g. MALIGN) multiple guide trees are created and the derived alignments compared by some metric (such as phylogenetic tree cost).

COFFEE (Notredame *et al.* 1998, 2000) behaves as a 'wrapper,' using a genetic algorithm to optimize multiple alignments based on consistency with the pairwise alignments of the same sequences. Any pairwise alignment procedure can be used under the COFFEE optimality function.

The following alignment methods involve 'search' procedures. In MALIGN and the method of Hein *et al.* (2003), tree searches are conducted to produce multiple alignments, whereas POY searches for optimal cladograms directly and can generate alignments *post facto* for the optimal cladogram.

MALIGN (Wheeler and Gladstein 1994) uses multiple guide trees to generate a diversity of multiple sequence alignments, choosing the best on the basis of the parsimony score (indels included) of the most parsimonious cladogram derived from that alignment. Guide trees are searched and multiple alignments created for each candidate guide tree. Each alignment is used as the basis for a heuristic cladogram search (indels weighted and included). The cost of the most parsimonious cladogram is attached to the alignment as its optimality score. MALIGN will output multiple multiple-alignments if they are equally optimal.

Hein *et al.* (2003) employ the Thorne–Kishino–Felsenstein (TKF) model (Thorne *et al.* 1991) for likelihood-based multiple alignments related by a tree. The algorithm employed is based on Sankoff (1975) for likelihood. Currently, the implementation (designed for demonstration purposes) can manage a few sequences (ca. 7) but could well be extended to larger data sets.

POY Implied Alignment (Wheeler 2003a; Wheeler *et al.* 2003) is not an alignment program, but searches for parsimonious cladograms directly (see the next section). A multiple alignment can be generated, however, from the transformation series implied by the optimal cladogram. This is not a multiple alignment in the sense of other methods, but rather is inextricably linked to the cladogram from which it was derived (Wheeler 2003a).

5.6 Optimization methods

In contrast to alignment procedures, optimization methods skip the alignment step and proceed directly to the determination of cladogram cost. This is achieved by focusing on determining optimal hypothetical taxonomic unit (HTU) sequences at internal tree nodes. In doing so, homology schemes are created for each cladogram uniquely, and for cladogram costs based on them. Multiple-alignment methods create a single alignment upon which all cladograms are diagnosed. Optimization methods create individualized homology schemes for each cladogram.

5.6.1 Exact solutions

As mentioned earlier, the determination of the lowest cost for a single cladogram depends on the lowest cost assignment of HTU sequences, and this is an NP-hard problem (Wang and Jiang 1994). Exact solutions, therefore, will not be available generally.

Sankoff (1975) proposed a recursive procedure that would calculate the minimum-cost cladogram exactly. This method requires a number of steps proportional to $(2n)^m$ where n is the average length of the sequences and m the number of sequences for a given cladogram. An alternate, simple-minded exhaustive approach would be to simply

generate a list of all possible sequences, determine the edit cost between each pair (via some procedure akin to that of Needleman and Wunsch 1970), and try each possible sequence at each internal cladogram node by dynamic programming (Sankoff and Rousseau 1975). This type of explicit enumeration could be accomplished by extending the candidate set of sequences employed by search-based optimization (Wheeler 2003b) to include all possible sequences. Since this would entail an explosively increasing number of sequences this technique would become untenable rapidly. Some sort of branch-and-bound technique could be applied to this search given an initial upper-bound estimate, but it is unclear whether much additional headway can be made towards exact solutions.

5.6.2 Heuristic solutions

The operational goal of heuristic optimization procedures is to determine a set of HTU sequences that minimizes the overall cladogram length (= edge weight). Two general sorts of approach have been proposed based on attempts to estimate these internal vertex sequences using known sequences or on a search for them within the world of possible sequences.

The first-estimation heuristic was proposed by Sankoff *et al.* (1973) and Sankoff (1975). Given the high dimensionality of the exact recursive solution proposed by Sankoff (1975), a three-dimensional local-optimum heuristic was proposed. This would break the problem down into a series of single-point estimations surrounded by three known or previously estimated sequences (Fig. 5.3; as opposed to the two-point problem reduction in many heuristic alignment approaches). At the time, the method was too time-consuming for real data sets.

Wheeler (1996, 2002) proposed a two-dimensional heuristic (optimization alignment, later called direct optimization), which though more approximate that the three-dimensional approach, was more rapid (Fig. 5.4a). Later, Wheeler combined the Sankoff method with direct optimization and incremental character optimization (Gladstein 1997) in iterative-pass optimization,

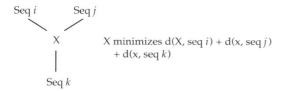

Figure 5.3 Median-state heuristic for *n*-dimensional optimization proposed by Sankoff (1975). The state of X (which could be an entire sequence) is that which minimizes the summed distances to the nodes which connect to it.

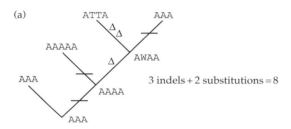

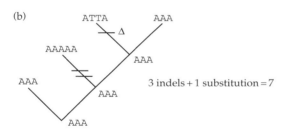

Figure 5.4 Estimation methods. (a) Direct optimization (Wheeler 1996) results in a cladogram of cost 8 for the input sequences AA, ATTA, AAAAA, and AAA when all events (indels and nucleotide substitutions) are equally costly. (b) Iterative-pass optimization (Wheeler 2003c) improves on this by 1 step. The horizontal bars signify indels and Δ represent nucleotide subsitutions whose location may be ambiguous.

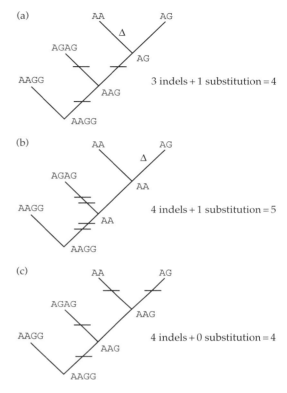

Figure 5.5 Search-based methods. (a) Direct optimization (Wheeler 1996) for the sequences AG, AA, AGAG, and AAGG. When all events are equally costly (indels and nucleotide substitutions) the cladogram has a cost of 4. (b) Fixed-state optimization (Wheeler 1999) limits the HTU sequences to those present in the terminals and results in a cost of 5. (c) Search-based optimization (Wheeler 2003b) allows the addition of HTU states—here AAG—and thereby reducing the cladogram cost to 4.

which improved cladogram-length calculations and can be used for larger numbers of sequences (Fig. 5.4b; Wheeler 2003c).

Sequence-search heuristics first appeared with fixed-state optimization (Wheeler 1999). The Fixed-state method limited the possible set of HTU sequences to those observed in terminal taxa, which are then diagnosed via dynamic programming based on a matrix of edit costs between the sequences. Given this constraint, less-satisfactory lower bounds on cladogram length are usually found (Fig. 5.5a; when sequences differ

greatly in length this may not be true). Since the method is not calculating ancestral sequence states but simply optimizing states, cladogram optimization time, after initialization, is independent of sequence length. As the number of sequences increases, the number of potential sequence states rises as well, both improving the cladogram cost estimation and increasing the cost of computation of a given cladogram (roughly m^3 for m sequences) (Fig. 5.5b).

Search-based optimization (Wheeler 2003b) relaxes the strict limit on sequence states by the addition of heuristically chosen sequences (Fig. 5.5c). Through the increase of the state set at the cost of execution time, progressively lower bounds can be found until further enlargement of the set is unproductive. The set could be made all-inclusive with an exact solution the result (but at great time cost).

5.7 Comparison of alignment and optimization

Although the goals of alignment (at least in phylogenetics) and optimization are the same—to find minimum-cost cladograms—the approaches are quite different. Alignment methods seek to find a single putative homology scheme upon which all cladograms are evaluated. Optimization methods perform this operation for each evaluated cladogram. As such, cladogram searches based on alignment methods are likely to be consistently faster than optimization approaches since the steps involved in determining the cost of a cladogram from a fixed alignment are much less burdensome. Optimization methods, however, are likely to find lower-cost cladograms (Wheeler 1996; Giribet *et al.* 2002; T. Grant. pers, comm.) and execution time comparisons should include the time consumed by alignment.

This can be illustrated by examining a simple set of three sequences (Fig. 5.6). There is not necessarily one globally optimal alignment. An alignment may be optimal for a particular cladogram (*a la* Sankoff and Cedergren) but any cladogram search based on such an alignment may well overlook other equal or lower-cost solutions. Optimization

procedures, by examining the cladograms themselves, do not suffer this shortcoming (direct optimization as implemented in POY (Wheeler *et al.* 2003) finds both cladograms).

5.7.1 Evaluation

Given the identical goals of alignment and optimization, how can these somewhat competing methods be evaluated? Speed and effectiveness are two obvious criteria. Speed would be measured straight-forwardly as the time required to complete the combined alignment/cladogram-search operation versus that for the optimization-based cladogram search. The determination of cladogram cost for fixed alignments can be accomplished extremely efficiently (Goloboff 1994; 1998b) even for fairly general dynamic-programming characters (Goloboff 1995, 1996a). Implementations such as TNT (Goloboff *et al.* 2002) are able to evaluate many tens of thousands of cladograms (containing hundreds of taxa) per second. Multiple alignment implementations that generate a single multiple alignment (such as CLUSTAL) can create an alignment of a thousand nucleotides for a hundred taxa in a few minutes. Multiple-alignment procedures that evaluate many candidate multiple alignments (such as MALIGN) will absorb much more time. Such a search using dynamic homology optimization (at least under present implementations) could take yet longer.

An example (for illustrative purposes, not exhaustive by any means) is provided by the analysis of 100 mollusk 18S rRNA sequences (G. Giribet, personal communication). The alignment programs CLUSTAL and MALIGN were used and compared to optimization-based POY. CLUSTAL produced alignments most quickly and MALIGN most slowly. When comparing the approach of CLUSTAL and POY, CLUSTAL was faster by 20% (without cladogram search and when minimal POY options were specified), but the multiple alignment (really an implied alignment) produced by POY was 30% less costly in terms of parsimony (see Table 5.1). Cladogram searching would add time to the total solution of the alignment methods, but this would be a small premium

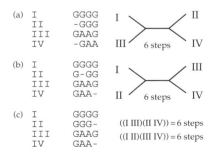

Figure 5.6 Simple alignments of four sequences GGGG, GGG, GAAG, and GAA. The alignments in (a) and (b) result in minimal cost, but different cladograms. The alignment in (c) yields both.

Table 5.1 Performance of CLUSTAL, MALIGN. and POY on mollusc test set. The analyses were performed with all transformations (indels included) costing 1. The data set consisted of 100 mullusk 18S DNA sequences of approximately 1000 bp (G. Giribet, personal communication). All runs were on a Pentium M computer at 1.7 Ghz under LINUX. Runs for MALIGN and POY specified indel cost as 1, CLUSTAL was run twice; once under the default values (Default) and a second time specifying all gaps and transformations as 1 (1 : 1 : 1 : 1). The POY run with TBR branch swapping yields two equally costly cladograms. The arrows denotes the cost of the implied alignments when analyzed using NONA (Goloboff 1993b). NONA diagnosed the POY cladograms as the cost found by POY, but was able to find more-parsimonious solutions using the implied alignments

Method	Options	Execution time(s)	Cost
CLUSTALW	1 : 1 : 1 : 1	688	11 999
CLUSTALW	Default	722	10 642
MALIGN	'Build' only	26 270	11 790
POY + Implied Alignment	'Build' only	920	7 989 → 7 970
POY + Implied Alignment	TBR	134 470	2 @ 7 690 → 7 684

in comparison with the alignment time. This could, however, narrow the CLUSTAL vs. POY execution-time difference.

The second criterion, effectiveness, favors optimization methods. Given that optimization methods are creating homology schemes specifically tailored to be optimal for each cladogram examined, it is only logical that this approach should result in less-costly (or higher-likelihood, for that matter) cladograms. This has been shown several times (Wheeler and Hiyashi 1998; Giribet *et al.* 2002; T. Grant. pers, comm.) and in the example above. Decreases in cladogram cost of 10% over alignment methods are not unexpected. The more length variation present in the sequences, the more opportunity there is for dynamic homology to find more-effective topology-specific solutions.

5.7.2 Interrelationship

There is a connection between multiple alignment and cladogram optimization. Transformation series inherent in a cladogram can be extracted and represented as an implied alignment (Wheeler 2003a). Such an implied alignment contains all the synapomorphy statements and transformation events required by the topology under an optimization approach. As such, they resemble standard multiple alignments, but are actually derived from the analysis of a specific cladograms as opposed to the basis for a search. Given the dependence of this sort of alignment on a specific cladogram, the

object created is not necessarily fair to topologies other than its basis cladogram. Each of those topologies would be tested best by their own implied alignments. Such a unique-alignment procedure is the approach optimization methods bring to phylogenetic analysis.

An effect of this is seen in the calculations of Bremer (1994) support. Support values calculated on the basis of a global alignment can overestimate support compared to those based on dynamic homology. Given the specific homology schemes created by optimization methods, alternate cladogram lengths should be lower (or at worst equal) to those based on an alignment that is optimal for some other cladogram. This will tend to inflate the differences in cladogram lengths, hence Bremer values.

Alignment and optimization can be used in tandem to reduce execution time in optimization-based searches. In essence, an implied alignment (or any alignment for that matter), represents a static approximation of dynamic homology. Given that an implied alignment is generated for a specific cladogram, it can be used as the basis for rapid cladogram cost evaluations among similar topologies. The implied alignment is used to identify candidate cladograms quickly for further, more time consuming, analysis. If a cladogram is found to be superior to previously identified solutions, a new implied alignment is created based on the new topology and the process continued. This approach can accelerate searches by a factor of four or more depending on the problem at hand (Wheeler 2003a).

5.8 Conclusion

Traditionally, alignment has been used to convert data without inherent putative homology statements into those that do. This step is operationally logical, but, given the ultimate goal of optimal cladograms, unnecessary. The criticism of optimization-based methods as lacking primary homology is largely based on this historical exercise. Clearly, *a priori* notions of homology (at least at the nucleotide level) are not logical or computational requisites of phylogenetic analysis. Criticisms of the optimization approach need to be based in effectiveness and logic—not on appeals to tradition.

As such, multiple alignment does not have separate standing in phylogenetic analysis. It is one approach to solving a complex, NP-complete problem. In comparison to optimization-based procedures, it may be fast, but it is approximate. In essence, alignment is a heuristic—and not a very effective one.

5.9 Acknowledgments

I thank Vic Albert for initiating and managing this effort, Gonzalo Giribet for use of unpublished sequences, Vic Albert, Lorenzo Prendini, Andres Varon, and an anonymous reviewer for helpful critique of the manuscript, NSF and NASA for support, and Steve Farris for decades of guidance.

CHAPTER 6

Parsimony and the problem of inapplicables in sequence data

Jan E. De Laet

''I don't know what you mean by 'glory,'' Alice said. Humpty Dumpty smiled contemptuously. 'Of course you don't–till I tell you. I meant 'there's a nice knock-down argument for you!'' 'But 'glory' doesn't mean 'a nice knock-down argument,'' Alice objected. 'When I use a word,' Humpty Dumpty said in rather a scornful tone, 'it means just what I choose it to mean–neither more nor less.''

(Caroll 1872, chapter VI)

6.1 Introduction

About 10 years ago, Maddison (1993; see also Platnick *et al.* 1991) drew attention to problems that can arise in parsimony analyses when data sets contain characters that are not applicable across all terminals. Examples of such characters are tail color when some terminals lack tails, or positions in DNA sequences in which gaps are present. Maddison (1993) examined various ways of coding such characters for various parsimony algorithms and concluded that no general solution was available. Since then, the problem of inapplicables has been rediscussed repeatedly (e.g. Lee and Bryant 1999; Strong and Lipscomb 1999; Seitz *et al.* 2000), but Maddison's conclusion still holds.

Farris (1983), focusing on regular single-column characters as classically used in phylogenetic analysis, characterized parsimony as a method that maximizes explanatory power in the sense that most-parsimonious trees are best able to explain observed similarities among organisms by inheritance and common ancestry. This led De Laet (1997; see also De Laet and Smets 1998) to formulate parsimony analysis as two-item analysis. In this view, parsimony maximizes the number of observed pairwise similarities that can be explained as identical by virtue of common descent,

subject to two methodological constraints: the same evidence should not be taken into account multiple times, and the overall explanation must be free of internal contradictions.

Here, I examine how this formulation can be used to deal with the problem of inapplicables. More specifically, I deal with the problem of inapplicables in sequence data, a harder and more general problem than most cases of inapplicability that Maddison (1993) had in mind. The review of parsimony analysis in the first section provides the basis for discussing the analysis of sequence data in the second section. The basic idea of the whole chapter is to explore the ramifications of the conceptual framework of Farris (1983) beyond the realm of single-column characters. This was in part prompted by the double observation that several authors seem to be using isolated elements of that paradigm when discussing methods for sequence analysis (see, e.g., Frost *et al.* 2001; Simmons 2004), while, at the same time, no coherent discussion of those ideas as applied to sequence data is available.

6.2 Parsimony analysis as two-item analysis

Some notes on terminology are appropriate first. Take a simple term such as 'autapomorphy'.

Originally, autapomorphies were defined as 'apomorphous features characteristic for a particular monophyletic group (present only in it)' (Hennig 1966, p. 90). In addition to this original meaning, a

Characters

	c1	c2	c3	c4	c5	c6	c7	c8	c9	c10
Terminals										
out1	0	0	0	1	0	0	0	1	0	0
out2	0	0	0	1	0	0	0	0	0	0
A	1	0	0	1	0	2	0	0	1	1
B	1	1	1	0	0	1	1	0	1	0
C	1	1	1	0	0	1	0	0	1	0
D	1	1	1	1	1	2	0	0	1	1
E	1	1	1	1	1	2	0	0	0	1
F	1	1	1	1	1	2	0	1	0	1

Figure 6.1 A data set with 10 unordered characters for eight terminals. Terminals *out1* and *out2* are interpreted as outgroups.

more restrictive usage that reserves the term for 'novelties that are coded as unique in a data set' (Kluge 1989, p. 9) is widespread.

Consider the data set of Fig. 6.1 and its most-parsimonious tree (*out1 out2* (*A* ((*B C*) (*D* (*E F*))))) (see Fig. 6.2). Under Hennig's original definition, the first seven characters all provide autapomorphies. As an example, character *c4* has apomorphous state *0* for monophyletic group (*B C*), and that state does not occur outside that clade. Under the more restrictive definition only character *c7* is autapomorphic. Obviously, questions as to whether autapomorphies should be taken into account or not when calculating the consistency index of a data set on a tree (e.g. Yeates 1992) take an entirely different meaning depending on the way in which the term 'autapomorphy' is used.

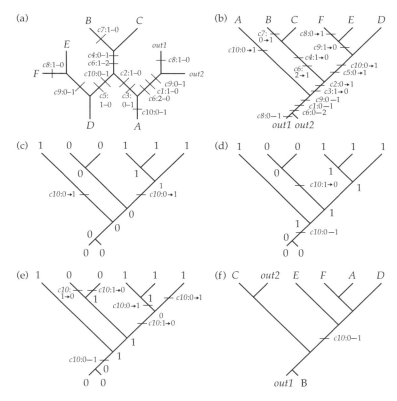

Figure 6.2 Parsimony analysis of the data of Fig. 6.1. (a) The most-parsimonious explanation of the data requires 14 steps. (b) To come to hypotheses of synapomorphy and monophyly in the ingroup, the ingroup is rooted using the branch that leads to the outgroups (note that this procedure does not imply such hypotheses outside the ingroup). (c, d) Two alternative optimal explanations of character *c10* on the most-parsimonious tree. (e) A suboptimal explanation of character *c10* on the most parsimonious tree. (f) An optimal explanation of character *c10* on a suboptimal tree.

Paraphrasing Farris (1983, p. 8), I share Humpty Dumpty's disdain for arguing definitions as such. Therefore I shall not discuss and evaluate the pros and cons of various possible meanings of the terms that I employ, nor indicate alternative terms with identical or similar meanings. But as the above example shows, it is important to make intended meanings clear, so in this section I shall explicitly point out my usages of terms.

At the same time, this process will provide an interlocked set of concepts that will allow a clear discussion of parsimony and inapplicables in the next section, and help to distinguish terminological issues from more substantial argument. To preempt any objection as should the conclusions hinge on major redefinitions of familiar terms, I shall indicate how my usages are rooted in existing literature. This, however, should not be taken to imply that these usages are always strictly in line with those references: whenever some existing, term is close enough, in spirit, to intended use (as would, e.g. Kluge's use of Hennig's autapomorphy above) I shall adopt existing terminology rather than propose a new term.

6.2.1 Characters and character analysis

Conceptually, a cladistic analysis consists of two main activities (see, e.g., Rieppel 1988; de Pinna 1991; Rieppel and Kearney 2002). The first comprises empirical observation, leading to delimitation of characters and character states, and to a data set in which those characters are scored for the terminals in the analysis. This is the activity of perceiving similarity and coding it into characters and data sets, to which I shall refer as *character analysis* (Kluge and Farris 1969, p. 9–10; see also Rieppel and Kearney 2002, p. 60). The second activity takes data sets as input, identifies their most-parsimonious hierarchic arrangment(s), and uses the resulting cladogram(s) as a basis for phylogenetic inference. I shall refer to this as *parsimony analysis* (Farris 1983, p. 10–12; see also later).

Character analysis and parsimony analysis stand in a continuous relationship of reciprocal illumination, at different levels (e.g. Rieppel 2003, p. 182; see also Hennig 1950, p. 26). As an example, the selection of terminals that will be included in a data set is in part guided by existing phylogenetic

hypotheses. Likewise, empirical work that results in new characters that are added to data sets can lead to cladograms with new or refined hypotheses of phylogenetic relationships. These, in turn, can point to characters that are highly incongruent with the general pattern and that may therefore be worth additional scrutiny. If an empirical basis can be found for a reinterpretation of such characters or their states, the data set can be adapted accordingly (see, e.g., Farris 1983, p.10).

At a given point in this process of continuous refinement, consider an individual character such as $c4$ in the data set of Fig. 6.1. From the point of view of character analysis this *character* is a statement about a feature that comes in two states, coded *0* and *1*, such that state *0* is observed in terminals *B* and *C* and state 1 in all other terminals. Theoretically, such a character expresses the hypothesis that the observed feature carries evidence on the genealogical relationships among the taxa that are involved. This directly limits characters and character states for phylogenetic analysis to features that are inheritable. A thought-provoking discussion of this seemingly trivial observation can be found in Freudenstein *et al.* (2003).

Beyond this, however, little more specific can be said other than that a character state as observed in different terminals 'must be sufficiently similar to be called the same [. . .] at some level of taxonomic generality' (Kluge 1997a, p. 89; the quote refers to derived states but the statement is valid in general), an observation that also holds for the character as a whole (see, e.g., Platnick 1979, p. 542; Jenner 2004, p. 301). For morphological and anatomical features, the criteria of composition, conjunction, ontogeny, and topography provide perspectives that can serve to evaluate if such sufficiency holds in particular cases (Kluge 1997a). Of those, topography or topological relationships are often considered to be the fundamental criterion (e.g. Rieppel 1988; de Pinna 1991, Hennig 1966, pp. 93–94; see also Remane 1952, pp. 31–66).

As discussed extensively by Rieppel and Kearney (2002, in the context of anatomy; see also Jenner 2004), care must be taken to give similarity statements as expressed in characters an observational basis. In order to do so one has to rely, however, unavoidably on background knowledge, and there is in principle no limit to the degree of background knowledge that can be incorporated in a character (Rieppel and Kearney 2002, p. 265). So even in this specific and restricted context of erecting character hypotheses for cladistic analysis, the concept of similarity unavoidably retains some elusiveness. This notwithstanding, similarity assessments as expressed in characters and their states, in the theoretical framework as just dicussed, are the empirical basis on which further phylogenetic inference is built.

6.2.2 Single-character phylogenetic inference

If no other comparative data were available for the terminals that are involved, a character such as c4 would constitute a data set on its own. It is a useful exercise to subject such a minimal data set to parsimony analysis. Within the constraint of terminal sampling, this leads to the following inferences: (1) the feature arose in a common ancestor of these terminals, from which they inherited it; (2) differentiation into two states occurred at a later stage; (3) for each state, the terminals with that state are only connected through ancestors that have that same state. These inferences do not yet include a polarity statement for which state is considered apomorphic and which plesiomorphic.

The apomorphy/plesiomorphy pair of terms is defined as follows: for a given evolutionary transformation, the condition or state from which the transformation started is *plesiomorphic* or *primitive* and the condition after the transformation *apomorphic* or *derived* (Hennig 1966, p. 89). As discussed by Hennig (1966, p. 93), coming to an hypothesis of features that are involved in such a transformation on the one hand and deciding on the evolutionary direction of such a transformation on the other are entirely different questions. The inclusion of outgroups in data sets is arguably the most general and least assumption-laden way to address the latter question.

Roots and outgroups
In general, when studying the phylogenetic relationships among a group of terminals, one assumes that these are part of a *monophyletic group*

at some level of inclusiveness, meaning that they share a common ancestor that they do not share with terminals outside that group (Hennig 1966, 73–74; see Farris 1991 for a review of this and related terms). The terminals that are assumed to be part of the monophyletic group are called *ingroup terminals* and are collectively referred to as the *ingroup*. Terminals outside that group are called *outgroup terminals* or *outgroups* for short.

When outgroups are included in a data set, they can be used to root the ingroup after the globally most-parsimonious arrangements of the data have been identified (Farris 1972, p. 657; see Figs 6.2a and 6.2b for an example). In the ingroup, hypotheses of relative apomorphy and plesiomorphy and of the direction of transformations then directly follow (Farris 1982a; see Figs 6.2c and 6.2d for some examples). This is the procedure that is now almost universally used to root ingroups and polarize characters, and it is mostly referred to as the *outgroup method* or the *outgroup criterion* (see, e.g., Farris 1979, p. 511). Confusingly, these and similar labels were also used in a series of papers in the 1980s for a series of methods of prior character polarization that are fundamentally different and mostly no longer in use. A historical account and a discussion of these methods can be found in Nixon and Carpenter (1993). The precise way in which hypotheses on character polarity come about does not affect the argumentation in this paper, so without loss of generality the discussion is restricted to outgroups.

In a data set that has only one character, as above, the general use of outgroups as just described becomes simplified because the best tree for the data set coincides with the structure of its single character. In the above example, the outgroup hypothesis could be the assumption that terminals A through F (the ingroup) share a most recent common ancestor that is not shared with terminals *out1* and *out2* (the outgroups). Observing that state 1 of character c4 is present in the outgroups as well as in the ingroup, it follows that state 1 is plesiomorphic in the ingroup; that state 0 is apomorphic in that same group; and that (B C) is a monophyletic subgroup of the ingroup.

Outgroups do not always lead to such unambiguous single-character inferences. An example is character *c6*, where (*A D E F*) and (*B C*) could both be monophyletic; or, alternatively, either could be paraphyletic with the other monophyletically nested in it. In addition, contradictions can arise between a character hypothesis and the outgroup hypothesis, even with binary characters. An example is character *c8*: the two following statements, derived from the character, contradict the outgroup hypothesis: terminals *out1* and *F* are only connected through ancestors that have state 1; the other terminals are only connected through ancestors that have state 0. Such cases are mostly but not necessarily interpreted to mean that the hypothesis of ingroup monophyly is incorrect. In general, nothing more can be said other than that the data do not support the prior assumption of ingroup monophyly (Farris 1972, p. 657), an observation that is also consistent with the alternative interpretation that the data are wrong. Neither issue addressed in this paragraph affects the argumentation of this paper.

Premises
Obviously, the above conclusion of monophyly for (*B C*) is conditional: it depends on the correctness of the outgroup hypothesis, on the correctness of the similarity assessments that led to character *c4* and its coded states, and on the correctness of several other, hidden, assumptions that remained unexpressed (such as absence of reticulate evolution). So, it would be more precise to say that (*B C*) is a putative monophyletic group, or a presumed monophyletic group, or that *B* and *C* are hypothesized to be monophyletic, each time conditional on the premises stated above (see Farris 1983, p. 13 for a similar use of the term 'putative'). Below, I shall use such verbose formulations only when confusion could arise otherwise, or when I wish to stress the difference between hypothesis or inference on the one hand and true historical account on the other. For the latter I shall then use the convenient adjective 'true', following existing practice (see, e.g., Farris 1983, p. 12), while observing that the philosophical problems that surround the notion of truth (see, e.g., Boyd 1991) do not affect this usage. The same applies to some other terms that I already have used: outgroup, apomorphy, and plesiomorphy are defined in terms of phylogenetic history but are often used to refer to just a hypothesis about that history.

Hennig (1966, p. 89) introduced the terms *symplesiomorphy* and *synapomorphy* to decribe the presence of plesiomorphies and apomorphies among terminals. As above, these terms are defined with respect to true evolutionary history, but are often used to refer to inferences as well. Such context-dependent shifts in meaning of these and similar terms are widespread in the literature, Hennig (1966) being a prime example. Related to this, when considering a transformation series such as $a \rightarrow a'$, Hennig (1966, pp. 88–89) sometimes referred to *a* and *a'* as 'character conditions,' sometimes as 'special characters' and sometimes even just as 'characters.' Combined with context-dependent meanings of terms, such use of different terms for the same thing, with meanings that often differ from current usage, can make it hard to understand Hennig's writings. This is even more problematic because Hennig used an argumentation scheme to order and polarize characters that is very different from current practice. In the above example, Hennig referred to *a* and *a'* as characters 'in the sense that they distinguish their bearers from one another' (Hennig 1966, p. 89). At the level of character analysis they are, in current usage, just character states.

When used conditionally, the precise meaning of terms such as synapomorphy and plesiomorphy in particular cases can drastically change according to the exact conditionals that are used or implied. Consider, for example, isolated character *c9* and the outgroup hypothesis. In that case the presence of state 1 in terminals *A*, *B*, *C*, and *D* is a (putative) synapomorphy compared to the presence of state 0 in terminals *out1*, *out2*, *E*, and *F*, which is a (putative) plesiomorphy. On the other hand, when considering the whole data set of Fig. 6.1 and its most-parsimonous tree (Fig. 6.2b), the presence of the same character state 1 in the same terminals *A*, *B*, *C*, and *D* is now a (putative) symplesiomorphy compared to the presence of state 0 in terminals *E* and *F*, which has become a (putative) synapomorphy. The presence of state 0 in the outgroups remains a (putative) symplesiomorphy. More interestingly, the presence of apomorphic state 1 in its original form (terminals *A*, *B*, *C*, and *D*) and in its more derived form (terminals *E* and *F*) is now a putative synapomorphy for terminals *A*–*F*.

6.2.3 Homology, the Hennig–Farris auxiliary principle, and parsimony analysis

A crucial assumption in the above interpretation of a single character is *Hennig's auxiliary principle*, stating 'that the presence of apomorphous characters in different species…is always reason for suspecting kinship [i.e. that the species belong to a monophyletic group], and that their origin by convergence should not be assumed a priori' (Hennig 1966, p. 121; square brackets present in original). In this quote, the term 'character' refers to a 'special character' (Hennig 1966, p. 89), which is a character state as used in this chapter, whereas an apomorphous (special) character refers to a special character that 'can certainly or with reasonable probability be interpreted as apomorphous' (Hennig 1966, p.121), i.e. an hypothesis of apomorphy or a putative apomorphy; monophyly is used in its true historical meaning.

Without this principle, one could equally well assume that, for example, state 1 of character *c5* of Fig. 6.1 arose multiple times. As an example, on the most-parsimonous tree (Fig. 6.2b) state 1 could have arisen a first time in the branch that leads up to terminal *D*, and a second time in a common ancestor of *E* and *F* that is not a common ancestor of *D*. Under this interpretation, the shared presence of 1 in *E* and *F* would be interpreted as evidence for monophyly of clade (*E F*), to the specific exclusion of terminal *D*, even if *D* has the same state.

However, given that the delimitation of character *c5* is grounded in empirical observation, this is not a very plausible interpretation of the character. Indeed, if any empirical evidence were available that state 1 as present in terminal *D* is not sufficiently similar to state 1 as found in terminals *E* and *F* to be called the same at some level of generality, these terminals would not have been assigned the same numeric state code to begin with. Since this was not the case, preferring the second interpretation over the first amounts to

discarding some of the evidence that bears on the problem at hand (viz. the perceived similarity between terminal *D* on the one hand and terminals *E* and *F* on the other. The remaining evidence (viz. the perceived similarity between *E* and *F*) then supports monophyly of *E* and *F* to the exclusion of *D*.

Homology should be presumed in the absence of evidence to the contrary
Hennig's formulation of his auxiliary principle, quoted earlier, is logically inconsistent because it can lead to internal contradictions: if the presence of presumed apomorphies is always to be a reason for suspecting true monophyly (first part of the principle), then it is not simply sufficient that multiple, convergent, origins of that state should not be assumed a priori (second part). This would still leave open the possibility that some terminals with the presumed plesiomorphic state obtained that state through a reversal. In that case, the group of all terminals with the presumed apomorphic state would no longer be truely monophyletic, which contradicts the first part. So that first part by logical necessity requires an additional statement that the origin of presumed plesiomorphies should not a priori be interpreted as reversals (for characters with more than two states, a similar statement is required for each state). As an example, without this addition a character such as *c5* could be taken as evidence for, e.g., a monophyletic group (*A D E F*) because it is not precluded that state 0 in terminal *A* arose as a reversal within that clade. In this interpretation, state 0 as present in terminal *A* would be derived relative to state 1 as present in terminals *D*, *E*, and *F*.

Such additional statements are implicit in Farris' (1983, p. 8) formulation of Hennig's auxiliary principle: 'homology should be presumed in absence of evidence to the contrary', where *homology* refers to similarities among organisms that have arisen historically through inheritance from a common ancestor, irrespective of these similarities being apomorphic or plesiomorphic. More explicit discussions of the necessity, in parsimony analysis, of explaining plesiomorphic similarities as due to common descent can be found in Farris *et al.* (1995, p. 215) and

Farris (1997, pp. 132–133). I shall therefore refer to the auxiliary criterion in its logically consistent form as the *Hennig–Farris auxiliary principle*.

When, as above, the Hennig–Farris auxiliary principle is applied to single–character data sets, it can be interpreted as a condition that makes the apomorphic state by necessity mark a true monophyletic group: the state arose only once and never reverted. That group will be present on any tree that requires only a single origin for that state, which is in line with Farris' (1983, p. 12) observation that grouping by true synapomorphy would have to behave exactly as parsimony, in the sense that it would lead to preference for the tree(s) on which no homoplasy is present (*homoplasy* being a point of similarity among organsims that cannot be explained by inheritance and common descent on a particular tree; Farris 1983, p. 18; see also below). These are, by definition, the shortest trees possible, so they are also most parsimonious trees.

Parsimony and the Hennig–Farris auxiliary principle
In practice, however, one is constrained to work with actual observable traits of organisms rather than with true historical synapomorphies. Character codings of such traits seldom if ever capture all true evolutionary transformations, let alone their order, as exemplified by the presence of homoplasy in all but the smallest and simplest data sets (note that absence of homoplasy in such data sets would hardly justify the conclusion that all relevant transformations have been captured— absence of evidence is not evidence of absence). This led Farris (1983, p. 17–19; see also Farris and Kluge 1986, p. 300; Farris 1986, pp. 15–16) to a general characterization of parsimony analysis in terms of a methodological principle that is fundamental to science in general: maximization of explanatory power or conformity between observation and theory. More specifically, the observations are the similarity statements as coded in characters, and the theory is that these similarities have arisen through inheritance and common descent. Most-parsimonious cladograms are then preferred because they are the trees on which the greatest amount of such observed points of similarity among organisms can be explained by inheritance and common descent (contra Grant

and Kluge 2004, p. 29). As such they provide the best explanation of the observations on account of the theory.

Note that, at this level of analysis, characters and their states can indeed be treated as simple observations, even if, as discussed above, they are complex theories or hypotheses on their own. Likewise, little confusion arises if the presence of the same character state of a given character in two terminals is simply called an *observed point of similarity* between those two terminals. Such usages of these terms can be found, for example, throughout Farris (1983).

Similarities as coded in characters can very well be true homoplasies rather than true homologies. Likewise, it cannot be ruled out that character similarities that can be explained as homologies on most-parsimonious cladograms are true homoplasies instead, even when using single-character data sets as above. Combined with the observation that parsimony minimizes putative homoplasy, such observations are sometimes taken to mean that it is an assumption of parsimony analysis that homoplasy is rare in evolutionary history. However, even if rarity of homoplasy may be a sufficient condition to prefer most-parsimonious trees (see, e.g., Felsenstein 1981), it is definitely not a necessary condition.

Consider a data set for terminals *out*, *A*, *B*, and *C* where 10 characters support clade (*B C*) and just one character supports clade (*A C*) (this example and discussion is based on Farris 1983, pp. 13–14, see also p. 12, pp. 18–19). If clade (*A C*) is genealogically correct, then the 10 characters that support (*B C*) are (true) homoplasies; if, on the other hand, clade (*B C*) is genealogically correct, then the single character that supports (*A C*) is a (true) homoplasy. These simple observations point out an interesting asymmetry in the relationship between characters and genealogies: a given genealogy implies that characters that contradict this genealogy are homoplasious but requires nothing concerning characters that do not contradict the genealogy. Now assume that true homoplasy is so abundant that only one out of those 11 characters has escaped its effects. Under the assumption that this one character can equally well be any character in the data set, a simple statistical argument leads to preference for clade (*B C*): the probability that this single historically correct character supports this clade is 10 times higher than the probability that it supports (*A C*). Thus it is seen that even under extremely high levels of homoplasy most-parsimonious trees can still be the best phylogenetic hypotheses one can make on the basis of the available data, even if some of the putative homologies may be true homoplasies instead.

The underlying assumption of the above conclusion is best stated in the negative: absence of any assumption about the distribution of homoplasies in data sets. In a statistical framework, this can be understood as the use of an uninformative prior. Obviously, one can postulate distributions of homoplasy such that the most-parsimonious trees will no longer be the best bets. Such distributions are typically derived from stochastic models of sequence evolution (see, e.g., Felsenstein 1978a; Huelsenbeck and Lander 2003). The mere fact, however, that such distributions can be postulated does not by itself invalidate parsimony analysis as a method to analyze empirical data. Indeed, such a conclusion would crucially hinge on the realism or plausibility of the underlying stochastic models (and not on their simplicity, as Huelsenbeck and Lander 2003 seem to suggest). Farris (1983, pp. 14–17, p. 12; see also Farris 1999) amply discussed these issues and found the models that were in use at that time greatly lacking in realism. Stochastic models of sequence evolution have dramatically increased in complexity since then (see Felsenstein 2004 for a review), but they still seem mostly inadequate to model even small-sized real data sets (D. Pol, personal communication). Therefore, Farris' discussion and conclusions remain as valid and to the point as they were more than 20 years ago.

Considering all this, the Hennig–Farris auxiliary principle can be phrased as the following rule for erecting character hypotheses and interpreting their optimizations on trees: 'features that on the basis of empirical evidence are deemed sufficiently similar to be called the same at some level of generality should be treated as putative homologues in phylogenetic analysis (even if they may be true homoplasies instead).' In combination with the principle of maximizing explanatory power,

this makes similarity-based statements of putative homology the centerpiece of phylogenetic inference: most parsimonious trees are trees on which the greatest amount of putative homology statements that return from character analysis *can* be explained as due to inheritance and common descent, and such trees are the best available phylogenetic hypotheses for the terminals at hand, whether or not the individual similarity statements or their explanations are historically correct.

As just discussed, the premises under which this holds are best stated in the negative: complete non-reliance on specific premises regarding correlations of evolutionary rates within and across characters and lineages. As such, parsimony analysis can be considered the most general method for phylogenetic analysis that is available. Tuffley and Steel (1997; see also Steel and Penny 2000) and Goloboff (2003) have examined similar but less extreme positions of agnosticism with respect to the details of evolutionary processes, using stochastic modeling. In both cases the most-parsimonious tree(s) are the best phylogenetic hypotheses, reinforcing the above conclusion.

6.2.4 Quantifying and maximizing homology

Given a tree and a data set such as in Fig. 6.1, Farris (1983) did not directly quantify the amount of points of similarity that can be explained by common descent and inheritance on that tree. Instead he used, as a relative measure, the minimum number of independent statements of homoplasy that are required on that tree. This works because an instance of homoplasy is present on a tree whenever a point of similarity as expressed in a character cannot be explained as homology on that tree (Farris 1983, p. 18).

So, when comparing two trees, the tree with the lower level of homoplasy will have the greater amount of similarity that can be explained as homology, and hence the greater power to explain the data on account of the theory. In practice, most parsimony programs calculate the minimum number of steps that are required, which, for a given character, differs from the minimum number of independent statements of homoplasy by a constant factor. As a result, the same ranking of trees is obtained. Several points are worth elaborating here.

Inner-node state assignments and the requirement of internal consistency

First, whether or not a particular pairwise similarity as coded in a character can be explained as a homology on a particular tree does not just depend on the structure of the tree and on the state distribution of the character that is involved, but also on assumptions that are made about the character states that are present at the internal nodes of the tree.

Take character *c10* of the data set of Fig. 6.1 and the most-parsimonious tree for that data set (Fig. 6.2b). Representing a pairwise similarity that is expressed as the presence of a same state *i* of a character in two terminals *X* and *Y* as $S_i(X \ Y)$, or, equivalently, $S_i(Y \ X)$, the similarity among terminals *A* and *D* as coded in *c10* is $S_1(A \ D)$. With inner node state assignments as in Figs. 6.2c or 6.2e, this pairwise similarity cannot be explained as a homology because independent derivations of state 1 from state 0 are involved. On the other hand, with state assignments as in Fig. 6.2d, that same similarity can be explained as a homology. Similarly, $S_0(out1 \ B)$ can be explained as a homology in Fig. 6.2c but not in Figs. 6.2d and 6.2e. In general, a pairwise similarity $S_i(X \ Y)$ can be explained as a homology on a tree when all nodes that connect *X* and *Y* have been assigned that same state *i*; in that case, the statement is said to be *accomodated* on the tree. In all other cases, it is a homoplasy, and the statement is not accomodated (only cases in which unique states are assigned to inner nodes are considered in this paper; polymorphic inner nodes, as in Farris (1978a) or in Felsenstein (1979), are left undiscussed).

The connection between the explanation of a character and assignments of states to inner nodes can be seen as a methodological constraint that ensures that the set of all homology statements that can be derived from a tree and a character state distribution is free from internal contradictions (De Laet and Smets 1998, pp. 374–376). Or, put positively, it ensures that the overall explanation is logically possible or consistent. This, in turn,

makes the explanation of the character on the tree logically capable of phylogenetic interpretation (Farris *et al.* 2001b). For example, on this tree one can explain either the similarity between *A* and *D* (e.g. Fig. 6.2d) or the similarity between *out1* and *B* as a homology (e.g. Fig. 6.2c); one cannot possibly, however, simultaneously explain both similarities as homologies because they are mutually exclusive. This logical requirement of non-contradiction is also met in maximum likelihood methods that integrate over all possible sets of inner-node state assignments, such as that of Felsenstein (1981). It is not met in quartet and triplet methods (De Laet and Smets 1998). Pairwise similarity statements that can simultaneously be explained as homology on a given tree will be referred to as (mutually) *compatible statements*.

When the terminals of a tree are labeled with the observed states of a particular character and the inner nodes have been assigned character states as well, the tree can be cut into a number of parts in which all nodes have the same state, and such that neighboring parts have different states. I shall refer to such parts as *regions*. There is a straightforward connection between number of regions and number of steps: any boundary between two regions implies a step, so the number of steps is one less than the number of regions. By definition, all similarities within a region can be explained as homologies, while similarities across regions are homoplastic. Because these regions are non-overlapping and because homologies do not cross the borders of such regions, the problem of quantifying the amount of similarity of the character that can be explained as homology on the tree can be broken down easily into the smaller problem of determining the amount of homology in such a region. For the same reason, the different states of a character can be treated independently under those conditions.

Independence and the units of empirical content of comparative data sets

A second issue is logical independence of pairwise homology (and homoplasy) statements within characters (Farris 1983, pp. 19–20, 21–22; De Laet and Smets 1998, pp. 369–374; this is different from logical dependence *between* characters, as discussed, e.g., in Wilkinson 1995, pp. 297–298). Consider state 1 of character *c10* as it returns from character analysis. At that point, all its six pairwise similarity statements can be interpreted as homologies: $S_1(A\ D)$, $S_1(A\ E)$, $S_1(A\ F)$, $S_1(D\ E)$, $S_1(D\ F)$, and $S_1(E\ F)$. Not all of these are independent though: if, e.g., $S_1(A\ D)$ and $S_1(A\ E)$ can be interpreted as homologies, then, by necessity, $S_1(D\ E)$ can be interpreted as a homology as well. In general, if n_i terminals have the same character state for a given character, there are $n_i * (n_i - 1)/2$ different pairwise similarity statements that can be made, but no more than $n_i - 1$ of those can be independent. Adding statements beyond this number will introduce redundancy in the description of the data. This maximum number of independent pairwise similarity statements is at the same time the minimum number of statements that must be considered to deduce the complete set: when removing statements from a largest set of independent statements, there is no longer sufficient information to generate all data.

Non-redundant descriptions. I shall call such maximal sets of independent pairwise similarity statements *smallest generating sets*. The exact identity of the members of such sets does not matter, the important points are completeness and absence of logical dependencies. As an example, {$S_1(A\ D)$, $S_1(A\ E)$, $S_1(A\ F)$} and {$S_1(A\ D)$, $S_1(D\ E)$, $S_1(E\ F)$} are two different smallest generating sets for state 1 of character *c10*; {$S_1(A\ D)$, $S_1(A\ E)$, $S_1(A\ F)$, $S_1(E\ D)$} is a generating set, but not a smallest one because not all of its elements are independent. Next consider how the pairwise similarities in a character state can be explained on a particular tree with a particular set of inner-node state assignments, such as, for example, in Fig. 6.2c. There are two regions that have character state 1: isolated node *A* and subtree (*D* (*E F*)). All similarities within a region are homologies and all similarities across regions homoplasies, so $S_1(D\ E)$, $S_1(D\ F)$, and $S_1(E\ F)$ are homologies, while $S_1(A\ D)$, $S_1(A\ E)$, and $S_1(A\ F)$ are homoplastic.

A non-redundant description of this can be determined as follows. For each region that is involved, establish a smallest generating set (in general, a region with *j* terminals will have smallest

generating sets of cardinality *j* − 1). These sets non-redundantly describe the homologies of the character state on the tree, and the total number of independent statements that are accomodated is the total number of statements in these sets. Then pool these generating sets and augment the resulting set to obtain a smallest generating set for all similarities in the character state, without reference to a tree. The added statements form a maximal set of independent pairwise similarity statements that are not accomodated. This procedure establishes that the number of independent accomodated homologies and homoplasies for a given state add up to a number that is tree-independent. As a result, minimizing the number of independent statements of pairwise homoplasy in a character state and maximizing the number of independent statements of pairwise homology in that same state are equivalent problems indeed. Because independent homologies can be counted one region at a time, this remains true when summing over all states in a character, and/or over all characters in a data set.

In this example, the first region (isolated node *A*) has no similarities and therefore an empty smallest generating set; {$S_1(D\ E)$, $S_1(E\ F)$} is a smallest generating set for the second region. Adding, for example, homoplastic statement $S_1(A\ E)$ is sufficient to fully describe the character state and its explanation on the given tree. As an example, given that $S_1(D\ E)$ is accomodated and that $S_1(A\ E)$ is not accomodated, it follows that $S_1(A\ D)$ is not accomodated either.

Explanation. When assessing how well a tree with inner-node state assignments can explain a character state as due to inheritance and common descent, the correct measure is the number of independent accomodated pairwise similarities, not the total number of accomodated pairwise similarities. Consider a character in which 100 terminals have state 0 and another 100 state 1, and two trees on which the first 100 terminals occur in one region and the other 100 in two regions. Assume that in the first tree, the first region with state 1 has one terminal and the second 99; and that, in the second tree, both regions with state 1 have 50 terminals. The total number of pairwise similarities in this character state is $99 \times 100/2 = 4\,950$, of which at most 99 are independent. Summing over regions, in the first case a total of $0 + 4\,851 = 4\,851$ similarities are accomodated, in the second case only $1\,225 + 1\,225 = 2\,450$.

Yet in both cases, the same number of 98 independent pairwise similarities are required for a non-redundant description of the situation. Or, conversely, in both cases only a single independent pairwise similarity cannot be explained as a homology. This is in direct agreement with the observation that both cases can equally well explain the observations on account of the theory, which in this restricted case is possible historical identity of state 1 through inheritance and common descent on the given trees with the given sets of inner-node state assignments for the given character. The total number of pairwise homologies gives a different answer (the first tree is considered about twice as good: score $4\,851$ vs. $2\,450$) because that number also depends on the numbers of terminals that are present in each region of a tree in which the state is homologous. As these numbers do not feature in the theory on account of which the data are explained, the total number of accomodated similarities is not suited to measure agreement between theory and observation.

Weighting. An alternative way of viewing the difference between all and independent pairwise similarity statements is in terms of dynamic weighting of similarity statements (see De Laet and Smets 1998 for a similar discussion in the context of triplet and quartet methods). More particularly, if the weight that is assigned to an independent accomodated similarity statement in a given region is calculated dynamically as the total number of statements in that region divided by the number of independent statements in that region, then the total number of unweighed accomodated statements equals the number of weighted independent accomodated statements. This weighting scheme is highly unnatural and hard if not impossible to defend, which just reinforces the conclusion of the previous paragraph. But it also raises the general question of weighting.

I have been assuming equal weighting of similarity statements throughout, but the principle of parsimony as discussed here does in itself not prescribe that all parts of the data be equally weighted. Farris (1983, p. 11) discussed this issue at the level of differential weighting of entire characters and characterized his preference for equal weighting as a stance of ignorance: in the absence of any convincing reason for doing otherwise, all characters in a data set are treated as if they provide equally cogent evidence on phylogenetic relationship. The same reasoning applies at the level of the independent similarity statements that make up characters.

Algorithms such as Farris (1970; additive characters) or Sankoff and Rousseau (1975; step matrices) can be seen as methods that apply differential weighting within characters. Such differential weighting is defined in terms of transformations, not in terms of similarities: transformations between different pairs of character states can receive different weights. This may seem problematic for the current approach because the simple equivalence of minimizing homoplasy and maximizing homology, as discussed above, in general only holds when all transformations and all unit homologies are weighted equally. However, differential weighting as in Farris (1970) and Sankoff (1975) can also be characterized in terms of similarities that are hierarchically nested. A full discussion of this issue is beyond the scope of this review.

A methodological requirement. The unit of evidential value of a data set on a tree that arises from this discussion is an independent accomodated pairwise similarity statement. Likewise, independent pairwise similarity statements are the currency in which the empirical content of a data set is measured. This ultimately permits to interpret the preference for independent accomodated statements (versus all accomodated statements) as a methological requirement when maximizing the number of pairwise similarity statements that can be explained as homology: it enforces that each unit or quantum of empirical content of a data set is considered precisely once. Note that, in itself, this does not amount to equal weighting: whether or not all quanta of comparative empirical content should receive the same weight is an entirely different question.

Again, this methodological constraint is not met in quartet and triplet methods (De Laet and Smets 1998). Likewise, it is not met in methods that base the inference on a square matrix of pairwise distances among terminals, such as neighbor joining (Saitou and Nei 1987), for the simple reason that the required information to do so is not present in such matrices. To be sure, neighbor joining can in principle operate directly on character state data (Saitou and Nei 1987, p. 410), but such data sets are mostly reduced to square distance matrices first. In maximum likelihood methods such as Felsenstein (1981), the constraint is met. The difference with parsimony analysis is that in such methods the explanation of a similarity statement on a tree is based on integration over all possible inner-node state assignments, using stochastic models of character evolution and best-scenario branch lengths (see, e.g., Steel and Penny 2000 and Goloboff 2003 for a discussion). As seen above, when looking for best trees, parsimony analysis evades uncertainty as to the true historical status of a similarity statement that can be explained as a homology on a tree at an entirely different level, thus enabling it to remain largely agnostic about details of the processes of character evolution.

Maximizing the amount of homology
Given a data set of characters, one has to identify the tree or trees on which the highest number of independent compatible pairwise similarity statements can be explained as homology. This involves an optimization at two different levels. First, which is the highest number of such homology statements on a given tree? Second, given a procedure to solve the first problem, which is (are) the tree(s) on which this number is maximal?

The first problem can be tackled one character at a time because there are no logical interactions among the explanations of different characters (this is a fundamental assumption that is not met when inapplicables are present). Within a character, though, it cannot be tackled one state at a time because the explanation of any given state imposes methodological constraints on allowed explanations of the other states. As discussed above, such constraints are met when inner-node state assignments are taken into account, in addition to the observed states at the terminal nodes. Therefore, a crude solution for optimizing a character on a tree is to generate all possible sets of inner-node state assignments and to count the number of independent accomodated statements for each (three different possibilities, on the same tree, are illustrated in Figs. 6.2c–6.2e, with scores 5, 5, and 2). If the sets of inner-node state assignments are generated in a clever enough order, this can be improved using a branch-and-bound mechanism.

However, a much more efficient approach is possible, starting from the above observation that the number of independent compatible homologies and homoplasies for a character add up to a number that is tree-independent. As a result, a set of inner node state assignments that minimizes independent homoplasies also maximizes independent homologies. Next, the minimum number of independent homoplasies for a given character and a given optimal set of inner-node state assignments equals, up to a tree-independent constant, the number of regions as imposed by the inner-node state assignments, which in turn is one more than the minimum number of steps in the character. Therefore, algorithms that minimize the number of steps in such characters can be used to maximize homology. Examples are the algorithm of Farris (1970) for binary characters and additive multistate characters, or the algorithm of Fitch (1971; see also Hartigan 1973) for unordered characters.

The second problem is illustrated in the two trees of Figs 6.2b and 6.2f: even if the second tree can explain some characters better than the first tree (e.g. *c10*), the first tree is preferred because it provides a better explanation of the data as a whole. The problem of deciding whether a given tree is an optimal tree for the data at hand is NP-complete (Foulds and Graham 1982). Practically, this means that in general the only way to find the best tree(s) is the hard approach of examining all possible trees that exist for the given terminals, either explicitly or implicitly, by using a branch-and-bound approach (for which see Hendy and

Penny 1982). Unfortunately, the number of trees grows so extremely fast as the number of terminals grows (see, e.g., Felsenstein 1978b) that this approach is only feasible for relatively small numbers of terminals. Exactly how many terminals can be analysed in this way depends on the structure of the data set and on the computing power and time that is available, but as a rule of thumb it is somewhere between 15 and 25. So, when dealing with increasingly larger numbers of terminals, one is practically forced to restrict the tree search to increasingly smaller subsets of all possible trees, proportionwise. In doing so, heuristics such as branch swapping are used to make sure that no or little computing effort is wasted on trees that are manifestly not optimal (for a broader discussion and some developments beyond simple branch swapping see, e.g., Goloboff 1999; Moilanen 1999; Nixon 1999; Moilanen 2001).

Both levels of optimization are logically independent, even if they are in practice often tightly integrated in heuristic approaches (see, e.g., Goloboff 1996b for examples). One could do a tree search using any imaginable function that computes a number from a tree and a data set, and, heuristic uncertainty aside, the resulting trees would be optimal according to that function. Therefore, when comparing and evaluating different methods, it is sufficient to examine the meaning of the function used to evaluate any single tree.

6.2.5 Characters revisited

Summarizing this long introductory section, observation-based pairwise similarity statements are the fundamental statements of comparative research. When searching for trees on which the highest number of such similarities can be explained as homologies, two methodological requirements must be met: (1) the overall explanation of the data must be free of internal contradictions, which can be enforced by assigning, for each character, states to inner nodes of the tree; (2) the same piece of empirical content should not be used multiple times, which translates into counting only homologies that are logically independent.

From this point of view, a character that describes the distribution of a number of states in a

number of terminals is just a convenient non-redundant summary of elementary putative homology decisions that are made, during character analysis, in all possible pairwise comparisons of some observable characteristic in those terminals (see De Laet and Smets 1998, pp. 378–380; the unhappy informal use of the term 'essence' does not invalidate their discussion). In each such pairwise comparison, the mere fact that the characteristic is being compared entails the hypothesis that at some level of generality it is historically the same. At a lower level, the different states of the character are hypotheses of alternative expressions of the characteristic, each of which is also hypothesized to be historically the same. As discussed above, all such hypotheses are to be seen through the lens of the Hennig–Farris auxiliary principle.

To clarify, consider some angiosperms and a character that codes a floral structure that comes in two forms, rounded (state 0) and square (1). The fact that these two forms are coded as states of the same character reflects the hypothesis that the structures, despite the observed difference in form, are homologous at a more general level. Mostly, such an hypothesis is based on a combination of criteria. As an example, when the development of floral buds in different terminals is compared, the meristem that gives rise to the structure could originate in almost identical topological relationships relative to other meristems. In addition, the adult structures, whether round or square, could share many anatomical and morphological similarities. As a whole, the character then reflects the higher-level prior hypothesis that the structure in all these terminals is identical through common descent and inheritance. Within the character, the difference in general form (round vs. square) is considered important enough to warrant recognition of two different states, reflecting the lower-level prior hypotheses that the roundness and the squareness of these structures can be explained as identity through common descent and inheritance as well.

The different roles of characters and character states
It has often been observed that there is a large discrepancy between the formalized nature of phylogenetic analysis once a data set has been

constructed and the much more subjective deci-sions that are involved in character analysis, when it comes to deciding if observed features in two terminals should be coded as the same state of a character, two alternative states of a character, or part of different characters altogether (see e.g., de Pinna 1991, p. 380). Pleijel (1995) argued that this is especially relevant for the assumptions regarding homology of states within a character (are two such floral structures homologous, irre-spective of their general form?). Contrary to hypotheses of homology within states (is the roundness of two round structures homologous, is the squareness of square structures homologous?), such higher-level hypotheses are never questioned during subsequent phylogenetic analysis (Pleijel 1995, p. 312). As an example, consider character $c9$ of the data set of Fig. 6.1, and assume that state 0 codes the square and state 1 the round structure of the above character. On the most-parsimonious tree for these data (Fig. 6.2b), the squareness of the structure that is observed in terminals *out1* and *out2* is not homologous to the squareness of the same structure that is observed in terminals E and F, and the initial lower-level hypothesis has to be revised.

Similar posterior revisions of the higher-level hypothesis cannot be made because the homology of round versus square structures has been hard-coded in the analysis, precisely because they have

been coded as states of the same character. To remove such hard-coded higher-level assump-tions, Pleijel (1995) proposed to use absence/pre-sence coding of character states, which is formally identical to non-additive binary coding, a tech-nique that stems from phenetics (see, e.g., Sokal 1986). Whether it is feasible or desirable to exclude such assumptions from the analysis will be examined below.

But whatever the answer, the use of absence/presence coding as a means of doing so can lead to internal inconsistencies in the phylogenetic expla-nation of data, a result that is particularly relevant for this paper because Pleijel (1995) advanced absence/presence coding as a promising way to deal with inapplicables. Consider the data set of Fig. 6.3a and assume, without loss of generality, that none of the character states codes for absence. In the recoded version of Fig. 6.3b each column stands for one character state of a character of Fig. 6.3a, with 0 coding for absence of that state and 1 for presence. When analyzing Fig. 6.3a, the three trees of Fig. 6.3c are obtained (nine steps; loss of two independent pairwise similarities). With the recoded data, only one shortest tree is found, the middle tree of Fig. 6.3c; the two other trees are suboptimal by one step (18 vs. 17).

Pleijel (1995, p. 313) pointed out that, with absence/presence coding, hypotheses concerning

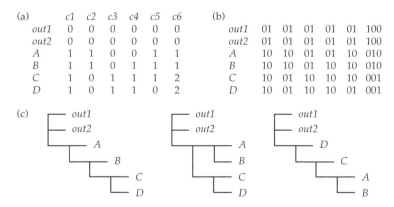

Figure 6.3 Absence/presence coding of character states aims to remove prior hypotheses of homology among states (Pleijel 1995) but can lead to internal inconsistencies. (a) A dataset with characters that reflect nested hypotheses of homology as determined during character analysis (characters unordered). (b) The characters of (a) with absence/presence recoding of character states. (c) The three most-parsimonious trees for (a). With the data coded as in (b) only the middle tree is considered optimal. The two other trees are rejected even if they explain the data equally well under acceptable hypotheses of homology that they imply.

transformation series between the analysed states will emerge as part of the results, but he remained somewhat vague about the logical and technical implications of this observation. As an example, take the three recoded states of the original character $c6$, each with a perfect fit on the single most-parsimonious tree for the recoded data. Because 0 stands for absence of the corresponding state, an inner node that is optimized as 0 can be hypothesized to have one of the two other states (other possibilities exist but are not relevant for the argument). Combining and summarizing all possible such optimizations of the three recoded states of $c6$, and using the outgroup hypothesis, three possible *implied transformation series* emerge from the tree: $1 \leftarrow 0 \rightarrow 2$, $0 \rightarrow 1 \rightarrow 2$, and $0 \rightarrow 2 \rightarrow 1$. Each of these has a perfect fit on the tree as well, and in each case only two steps are required to explain the state distribution. When doing the same excercise for the groups of states as defined by the other characters of Fig. 6.3a, all these other states can be explained by postulating a total of only seven steps (note that some of the implied transformation series incorporate non-homology of states as defined *a priori*; an example is character $c4$).

The middle tree of Fig. 6.3c is considered the best tree for the recoded states because it has the shortest length for the recoded data. But on the basis of possible transformation series that emerge as part of the analysis, one can construct a phylogenetic explanation of the data on that tree that requires fewer steps. So, whatever the length of an absence/presence recoded matrix on a tree means, it definitely does not measure how well that tree can explain the data phylogenetically under the assumption that character states can transform into one another, and maximization of phylogenetic explanatory power under that assumption cannot be the rationale for preferring trees that minimize this recoded length. Indeed, analyzing the two other trees in the same manner, they can also be explained by postulating only nine steps (which should not come as a surprise, as it was already clear from the analysis of the data set of Fig. 6.3a that the states could be grouped such that only nine steps are required on those trees). Yet they are rejected if the length of the recoded matrix is used as an optimality criterion.

(a)	$c1$	$c2$	$c3$	$c4$	$c5$	$c6$	(b)	$c5'$	$c6'$
out	0	2	4	6	8	10		8	10
A	1	2	4	6	9	11		11	9
B	1	3	5	7	8	12		8	12
C	1	3	5	7	9	13		13	9

Figure 6.4 Absence/presence coding of character states, to remove prior hypotheses of homology among states, can lead to surprising optimal implied transformation series. (a) A dataset with six unordered characters as they return from character analysis; the groupings of character states in columns (characters) reflect nested hypotheses of putative homology; the most-parsimonious tree is (*out* (*A* (*B* *C*))), which is also the best tree when the data are recoded to remove prior assumptions of homologies among states. (b) Alternative grouping of the states of characters $c5$ and $c6$ that cannot be rejected on the basis of the optimized recoded states. For this grouping, the transformation series as implied by the optimized recoded characters provides a better explanation of the data than the original characters.

One step further, posterior groupings of states may exist that reduce the total number of steps below the number required by the groupings as they come out of character analysis. An example is presented in Fig. 6.4. As above, it can be assumed without loss of generality that none of the states in Fig. 6.4a codes for absence. When states 8–13 are grouped as in characters $c5$ and $c6$ of Fig. 6.4a, the transformation series that are implied by the optimizations of the recoded states on the best tree require a total of five steps on the best tree. But the alternative grouping as in Fig. 6.4b, implying $11 \leftarrow 8 \rightarrow 13$ and $10 \rightarrow 9 \rightarrow 12$, can explain the observed distributions of states 8–13 at only four steps. This optimal implied grouping of states obviously contradicts the empirical evidence on the basis of which the original characters were proposed. But then it is the aim of this approach to remove such untestable assumptions (Pleijel 1995, p. 312), and posterior acceptance of groups of states as in characters $c5'$ and $c6'$ is just a logical consequence. More precisely, recognition of such transformation series follows from the notion that hypotheses concerning transformation series among the analysed states should emerge as part of the results and from the general requirements that the analysis should be logically capable of phylogenetic interpretation and internally consistent.

It does not require much imagination to see that in practice this could easily lead to situations where square floral structures of one angiosperm would *a posteriori* be considered homologous with, for example, a type of root system as present in another angiosperm, and the round floral structures of this other angiosperm to the root system of the first. Most systematists would not hesitate to reconsider homology within states on the basis of a well-supported most-parsimonious tree (the squareness of the floral structures in these terminals is not the same as the squareness of such structures in those other terminals after all, despite my prior assessment to the contrary), but in general such reinterpretations across characters are much more difficult to accept (darn, these flowers are actually not flowers but modified root systems!).

So, even if statements of homology among states are untestable in the sense of Pleijel (1995), they put bounds on the degree of reinterpretation of character states one is willing to accept in the light of incongruence in the data, and these bounds reflect empirical evidence as obtained during character analysis. Outright removal of such bounds, as would seem to be a logical consequence of using absence/presence coding as advocated by Pleijel (1995), therefore amounts to throwing away important relevant empirical data. As a workaround, one could limit implied transformation series to include only groupings of states that are compatible with the results of character analysis. But that actually amounts to giving up the premise that prior statements regarding homology among states should be removed from the analysis. And as discussed above, absence/presence coding then results in the same trees as obtained with regularly coded characters, at least if the aim of the analysis is to maximize explanatory power in a phylogenetic context.

Beyond single-column characters

On the other hand, it is not uncommon in character analysis to find multiple possible interpretations for features, which is not surprising given the role of background knowledge as discussed earlier. As an example, depending on the view one takes, the vegetative region in some species of the angiosperm genus *Utricularia* (bladderworts) can be interpreted morphologically as a shoot-like leaf, a branched stem system without leaves, or a shoot

with stems and leaves (Rutishauser and Sattler 1989; a fourth, more complex, interpretation is also provided). Similar problems abound when dealing with fossils or when making comparisons across very divergent groups. In both cases one often has to deal with structures that cannot be easily homologized across the terminals being compared, which in turn often results in competing and conflicting prior interpretations. In studies of sequence data, this problem can come in the form of different prior hypotheses about orthology and paralogy of sequences (Fitch 1970) or in different alignments for the same set of putative orthologs (several examples of the latter case are discussed in the second section).

In each such case, when characters are coded according to just one of the competing interpretations, chances are that the chosen view will be favored by the resulting trees simply because the data have been exclusively interpreted as such to begin with. As observed by Endress (1994, p. 401–402), circular reasoning when dealing with such ambiguously interpretable features can be overcome by repeatedly testing all different possibilities. Only this approach amounts to a sincere attempt at falsification. Unfortunately, in formal analyses and with current algorithms this is not easy to achieve because the technical framework of independent single-column characters does not lend itself to simultaneous analysis of such alternative interpretations of the data in a logically consistent and correct way.

A hard work-around would be to manually construct and analyse as many data sets as there are different combinations of different interpretations in different characters, which may be practically feasible when the number of such combinations is not too large. The best phylogenetic hypotheses would then be the shortest trees across all those data sets, and optimal homologizations and details of transformation series would emerge from those trees as part of the analysis. The difference with absence/presence coding of states is that, as above, the level of reinterpretation of states that one is willing to accept in the light of incongruence is still bounded by the results of character analysis. The difference with an analysis of just one set of classic single-column characters is of a purely technical nature: these are cases in which the *a priori* acceptable hypotheses of homology among states cannot be expressed as a simple series of independent single-column characters. But the purpose remains maximization of the number of independent pairwise similarities that can be interpreted as identical through common descent and inheritance. From this point of view, the next section can be seen as an attempt to develop a formal and logically consistent method to deal with the problem of multiple *a priori* acceptable hypotheses of homology among states in the case of putative homology statements within putative orthologous sequences.

6.3 Parsimony analysis of sequence data

When dealing with sequence data, it is not unusual to find that putative homologous sequences have different lengths in different terminals. Such length differences are explained as the result of indel events, insertion and/or deletions that occurred in the course of evolutionary history. As a consequence of indel events, two sequences that are homologous as a whole will nevertheless contain subsequences that are not homologous: with a

deletion, the resulting sequence misses a part of the original sequence; with an insertion the resulting sequence has a subsequence that was not present before. In both cases, characters that describe the subsequences that are involved will be inapplicable in the other sequence.

For the purpose of phylogenetic analysis, it is common practice to establish the positions and sizes of indels by creating a multiple alignment prior to tree evaluation and tree search, thus turning the putative homologous sequences into a sequence of single-column positional characters that subsequently can be treated as a regular data set (see Fig. 6.5a for an example). Each such positional character describes the state distribution of the base that is found at that position of the alignment, with gaps (coded as dashes in this chapter) indicating inapplicability. As discussed by Maddison (1993, p. 578), this makes sequence data susceptible to the general problems that come with inapplicables.

However, the approach of generating multiple alignments prior to tree evaluation and tree search is fundamentally insufficient as a general method for analysis of sequence data, as will be discussed below. As a consequence, the question of inapplicables in sequence data cannot be discussed in

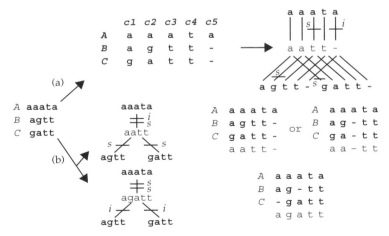

Figure 6.5 Three putative homologous sequences and two different approaches to evaluating them on the single unrooted tree for three terminals. (a) First a multiple alignment is constructed to establish base-level positional correspondences (dashes indicate gaps); the resulting positional characters are optimized using the algorithm of Fitch (1971), resulting in three substitutions (*s*) and one indel (*i*). (b) The unaligned sequences are optimized directly on the tree using the algorithm of Sankoff (1975); in this example, two optimal reconstructions of the sequence at the inner node exist, each at four steps; in each case, the optimal length imposes one or more optimal sets of positional correspondences.

general at that level. It is argued that a general method by necessity requires that unaligned sequences be directly optimized on trees, using algorithms such as Sankoff (1975) or Altschul (1989, pp. 307–308). Such algorithms treat the unaligned putative homologous sequences as one single complex character, to which I shall refer as a *sequence character*. It is widely believed that the various parameters that these algorithms employ to set up a cost regime, such as base substitution and gap costs, can only be specified or interpreted with reference to detailed models of the evolutionary processes that generated the data. However, the cost regime can also be set according to the principle of parsimony as discussed above, leading to a maximization of the amount of independent sequence similarity that can be interpreted as due to inheritance and common descent (De Laet 2004).

Throughout this section I use DNA sequences, but the discussion is general and applies to any kind of data that can be conceptualized to be hierarchically related through substitutions and indels, including, for example, serial homologs in morphology or different versions of manuscripts in stemmatology. Examples are constructed such that optimalities can be verified by hand.

6.3.1 Some background

Some additional notes on terminology are appropriate first. Gap and gap cost terminology can be confusing because the same terms are sometimes used for different things and the other way around. As an example, in a sequence like *a t t - - - t t a c* the term gap is sometimes used for each of the three consecutive missing positions in the middle (three gaps), or alternatively for the whole stretch of three missing positions (one gap). In this paper, a *gap* always refers to a maximum stretch of missing positions, not to smaller composing parts. The *length* of a gap is the number of positions over which it extends. The smallest composing part of a gap is referred to as a *unit gap*. The character that is used to indicate a unit gap, a dash in this chapter, is sometimes called the *gap character*, a term that has also been used for characters in data sets that describe the distribution of a putative indel events (e.g. Simmons and Ochoterena 2000).

All gap costs in this paper are of the form $a + (n-1) * b$, in which n is the *length* of the gap, a the *(gap) opening cost*, and b the *(gap) extension cost*. If gap opening cost and gap extension cost are equal, the term unit gap cost refers to either, and the cost for a gap of length n is n times the unit gap cost. Such a cost regime can be expressed as a 5×5 step matrix (see Sankoff and Rousseau 1975) in which the unit gap is included as a fifth state, in addition to a, c, g, and t.

The *minimal mutation algorithm* of Sankoff (1975) is illustrated in the example of Fig. 6.5b. It reconstructs *inner node sequences* and *positional correspondences* among observed sequences such that the total number of mutations is minimized under the assumption that a gap of length n constitutes n mutation events. This corresponds to a cost regime in which all base substitution costs, the gap opening cost, and the gap extension cost are equal. Sankoff and Cedergren (1983) generalized the approach to a step matrix with arbitrary metric distances, still treating a gap of length n as n events. A further extension to include gap costs of the form $a' + n * b$, in which n is the length of the gap, $a' + b$ the gap opening cost, and b the gap extension cost, was examined by Altschul (1989, pp. 307–308). With such gap costs, the first unit gap of a gap incurs a cost $(a' + b)$, each next unit a cost of b.

Sankoff (1975) used the concept of optimal *frame sequences* to specify reconstructed sequences and positional correspondences that lead to minimal costs. Sankoff and Cedergren (1983) framed their discussion in terms of the slightly less general concept of *tree alignments*. A tree alignment always refers to a particular tree with the given sequences at the tips and hypothetical or reconstructed sequences at the inner nodes. It consists of (1) that tree; (2) a matrix in which both observed and reconstructed sequences are aligned; and (3) correspondences between nodes of the tree and rows of the matrix. It is conveniently represented as a tree in which the nodes are labeled with the rows of the matrix, as, for example, in Fig. 6.10 (see below). In this way it is easy to see that, in a tree alignment, each branch of the tree defines a pairwise alignment between the sequences at the two nodes that the branch connects. The *cost of the tree alignment* is then defined as the sum of the costs of these pairwise alignments along all branches of the tree, always with reference to the cost regime in use. A 'classic' multiple alignment of the terminal sequences is obtained by deleting the rows with inner-node sequences from the matrix of a tree alignment. Multiple alignments that are obtained in this way have been called *implied alignments* (e.g. Schwikowski and Vingron 1997; Wheeler 2003a). Some examples of optimal implied alignments can be found in Fig. 6.5b.

With cost regimes that make no difference between gap opening cost and gap extension cost, the cost at any position in a pairwise alignment of a tree alignment is independent from the costs at its other positions. By extension, this also applies to the costs of complete colums of a tree alignment. As a result, each such column can be interpreted as a single-column character with a set of inner-node state assignments. In this way the algorithm of Sankoff (1975; all substitution costs and unit gap cost equal) can be seen as a generalization of the minimum mutation algorithm of Fitch (1971). Indeed, under the conditions of Sankoff (1975), each column of an optimal tree alignment specifies a character and set of inner-node state assignments that are also optimal under the conditions of Fitch (1971). The generalization lies in the fact that different optimal tree alignments for the same data on the same tree can imply different sets of Fitch characters (see Fig. 6.5b for examples). The algorithm of Sankoff and Cedergren (1983; tree alignments with step matrices) is a similar generalization of the algorithm of Sankoff and Rousseau (1975), which, in turn, generalized Fitch (1971) to accomodate differential weighting within characters. Under the conditions of Altschul (1989; different gap opening and gap extension costs), the costs of the different columns of a tree alignment are no longer independent. As a result, such tree alignments cannot be understood in terms of independent single-column positional characters.

As was the case with inner-node state assignments for simple single-column characters (compare, e.g., Figs. 6.2c and 6.2e), tree alignments on a given tree can be optimal or suboptimal. Sankoff and Cedergren (1983) called the cost of an *optimal* tree alignment for a set of observed sequences on a given tree the *tree distance* of those sequences on that tree. Their and similar algorithms (Sankoff 1975; Altschul 1989) can be used to calculate such tree distances and the reconstructions that come with them. In terms of the current approach, the tree distance as defined by Sankoff and Cedergren (1983) is the *length* of the sequence character on that tree. As such, the algorithms of, for example, Fitch (1971) and Sankoff (1975) are comparable in the sense that they both calculate the cost of an optimal reconstruction of a character on a tree. As will be discussed below, they are vastly different when it comes to computational complexity. For tree alignments, the second level of optimization—the problem of finding, among all possible trees, trees of minimal length or tree distance—is often called *generalized tree alignment* (e.g. Jiang and Lawler 1994; Vingron 1999) but other terms are used as well; Hein (1989a), for example, refers to it as the *general parsimony problem*.

6.3.2 Putative homologous sequences: a sequence of characters or a sequence character?

It has been argued that all substitution costs and the unit gap cost should be set equal in Sankoff (1975) style analyses of sequence data (Frost *et al.* 2001), a position that will be examined more closely later. However, first it is argued, in this subsection, that a general method of sequence alignment must by necessity move beyond prior multiple alignments (contra Simmons and Ochoterena 2000; Simmons 2004). The argumentation does not depend on the particular settings of the cost regime, but for clarity I tentatively accept the position of Frost *et al.* (2001) and contrast (equally weighted) Fitch (1971) analysis of prior alignments with Sankoff (1975) analysis of unaligned sequences.

When optimizing a sequence character on a tree, base-level correspondences among the observed sequences are not determined and fixed *a priori* but calculated as part of the optimization process, as already illustrated for three terminals in Fig. 6.5. The full implication of this can be seen when analyzing more than three sequences, such that alternative trees exist and have to be examined. Consider the data set of Fig. 6.6a. For four taxa

(a) A gc   (b) A gc   (c) A gc   (d) A gc-
    B cg   B cg   B cg   B cg-
    C c   C -c   C c-   C --c
    D gg   D gg   D gg   D gg-

Figure 6.6 A simple dataset (a) and three different multiple alignments (b, c, d).

A–D, three unrooted trees exist: (A B)(C D), (A C) (B D), and (A D)(B C). Using Sankoff (1975), the latter two are both diagnosed at cost 3 (each time two substitutions and one indel) while (A B)(C D) comes at cost 4 (three substitutions and one indel). Looking at the two optimal trees, (A C)(B D) comes with the implied alignment of Fig. 6.6b, (A D)(B C) with the different implied alignment of Fig. 6.6c. So it is not just that base correspondences are not fixed prior to analysis, *a posteriori* they can be different in different optimal trees.

A simple case of symmetry
The data set of Fig. 6.6a has a peculiar symmetry: when the labels of A and B are switched and the directions of all sequences reversed, the original data set is recovered. As such it provides a perfect example where mutually exclusive sets of putative homology statements cannot be distinguished at the level of character analysis. The higher-level hypothesis in this data set is that the sequences are orthologs. Within the orthologs, however, the symmetry makes it logically impossible to decide *a priori* if the single c of terminal C is to be considered homologous to the c in the second position of A or to the c in the first position of B. Conceptually, this is like the situation in bladderworts, discussed above, where it cannot be determined *a priori* if the vegetative system should be considered a shoot-like leaf or a leaf-like shoot system (even if the situation with bladderworts is more complex because there are still other homologizations that are considered acceptable on *a priori* grounds).

Turning to trees, the symmetry has, as a consequence, that these data cannot possibly distinguish between (A C)(B D) and (B C)(A D), two unrooted trees in which the labels of A and B have been exchanged. This conclusion follows directly and solely from the internal structure of the data set. As such it can be used to establish the following strong test for candidate phylogenetic methods: (A C)(B D) and (B C)(A D) should get the

same score. Any method that does not meet this test is in serious trouble.

As discussed, Sankoff (1975) optimization diagnoses (A C)(B D) and (B C)(A D) at the same cost and thus meets the test. Turning to prior alignments, the first question is which prior alignments to consider. With data as simple as this it is easily established that alignments in Figs 6.6b and 6.6c are the only valid candidates. All other alternatives, such as, for example, Fig. 6.6d would need some special argumentation as to why, in this case, the c that is observed in terminal C should not *a priori* be considered homologous to the c that is observed in A or to the c that is observed in B. Given that it is accepted, *a priori*, that the sequences as a whole are homologous (they are putative orthologs), this seems hard to do. A Fitch (1971) analysis of alignment 6b yields tree (A C)(B D) at cost 3, with (B C)(A D) one step more costly; alignment 6c yields (B C)(A D), also at cost 3, and with (A C)(B D) one step more costly (in both cases, (A B)(C D) has a cost of 4). So, when looking at just one alignment, the two trees get a different score and the method fails the above test. As a result, depending on the prior alignment that is used, positive support is found for either (B C)(A D) or (A C)(B D), whereas in fact relationships are ambiguous.

Similar symmetry observations can be made with respect to alignments 6b and 6c: they can be turned into one another by exchanging the labels of A and B and reversing the direction of each sequence. Therefore, if either is considered optimal according to some criterion, the other should be as well. So a way out of the problem of finding spurious relationships with single prior alignments suggests itself: rather than to construct and analyse just one prior alignment, identify and analyse all different prior multiple alignments that are considered optimal, and accept only groups that are common to all. This may sound trivial but it raises the non-trivial question of how to calculate the relevant prior optimal multiple alignments. For this particular example, that question comes down to finding a criterion that gives an optimal score to alignments on Figs 6.6b and 6.6c and a worse score to all other alignments.

Optimal alignments of two sequences can be calculated using dynamic programming algorithms as pioneered, in biology, by Needleman and Wunsch (1970) and Sellers (1974). A description of the basic algorithm and some historic notes can be found in Kruskal (1983); extensions are reviewed in, for example, Gusfield (1997). For the current purpose, approaches that generalize such algorithms to more than two sequences can be grouped according to whether or not they use the tree-alignment approach.

In optimal tree alignments, the kind of data symmetry in Fig. 6.6a is reflected directly in symmetry of calculations when comparing trees $(A\ C)(B\ D)$ and $(B\ C)(A\ D)$. So it was not just coincidence that the above Sankoff (1975) optimization of the data of Fig. 6.6a gave identical scores for those trees, with implied alignments that display among themselves the same symmetry as the data. Theoretically then, one could use a tree-alignment analysis to generate implied alignments that are next used as prior alignments. There would be no need to analyze the implied alignments, though, because their best trees would already have been identified in the preliminary tree alignment analysis. In fact, while the approach provides a solution to the problem discussed here, it actually comes down to giving up the notion that sequences should be aligned prior to tree evaluation and tree search.

Among the multiple alignments methods that do not use tree alignments, *SP alignments* or *sums-of-pairs alignments* (Murata *et al.* 1985; Carillo and Lipman 1988) and especially *progressive alignment* methods (e.g. Feng and Doolittle 1987; Thompson *et al.* 1994; Notredame *et al.* 2000) are probably most widely used. First consider SP alignments. An SP alignment of a set of sequences is an alignment for which the sum of pairwise alignment scores between all possible pairs of sequences is minimal. Setting all substitution costs and the unit gap cost to 1, it is easily verified that the alignments of Figs 6.6b and 6.6c have identical SP scores of 9, leaving the SP criterion as a potential solution to the problem.

Another case of symmetry
However, consider the data of Fig. 6.7a. Reading each sequence in reverse, nothing changes for B and E, but the sequence of A is turned into the

	(a)		(b)		(c)	
	A	ct	A	ct	A	ct
	B	c	B	-c	B	c-
	C	tc	C	tc	C	tc
	D	tc	D	tc	D	tc
	E	tt	E	tt	E	tt

Figure 6.7 A simple data set (a) and two different multiple alignments (b, c). According to the SP criterion, alignment (b) is better than alignment (c) (SP scores 13 and 14).

sequence of C and D, and the sequences of C and D are turned into the sequence of A. Therefore, the structure of the data set is such that these data cannot distinguish between trees that differ only in the positions of A vs. $(C\ D)$, as, for example, the pair $(B\ (C\ D)\ (A\ E))$ and $(A\ B\ (E\ (C\ D)))$. Using Sankoff (1975), these trees both have a cost of 3, which is the optimal cost over all trees as well. Tree $(B\ (C\ D)\ (A\ E))$, or any other tree that has an AE–BCD partition, comes with optimal implied alignment 7b; tree $(A\ B\ (E\ (C\ D)))$, or any other tree that has an AB–CDE and an ABE–CD partition, comes with alignment 7c. As above, these implied alignments have among themselves the same symmetry as the unaligned data. So Sankoff (1975) optimization does not tell these trees apart, and correctly so.

This is necessarily so as long as the ancestor of C and D has a reconstructed sequence that is identical to and perfectly aligned with the sequences of C and D in optimal tree alignments. If this is the case, the data symmetry is directly reflected in the Sankoff (1975) calculations that are performed on the two trees that are involved, and an identical cost on both trees follows. The assumption about the reconstructed sequence for the ancestor of C and D is easily proved by showing that its negation leads to a contradiction. Assume that an optimal tree alignment exists in which the ancestor of C and D has a sequence that is different or differently aligned. In that case, the tree alignment can be improved—contradicting the premise—by changing that ancestor and its alignment as indicated above. That this is an improvement can be seen as follows: for any position in the ancestor of C and D with an entry (base or unit gap) that is different from the base at the corresponding position in C and D, changing that entry into the corresponding entry of C and D will improve the cost by two mutations; at the same time, that change can incur at most one additional mutation, between the ancestor of *C* and *D* and the third node to which this ancestor is connected. So, in conclusion, optimal tree alignments are not tricked by data symmetries such as in Fig. 6.7.

This does not hold for SP alignments: alignment 7b has a better SP score than alignment 7c (13 vs. 14; the score for 7b is optimal), proving the case by counter example. As a result, if the SP criterion were used to construct and select prior alignments, alignment 7b would be selected and trees with *AB–CDE* and *ABE–CD* partitions considered suboptimal in the subsequent phylogenetic analysis. To salvage the approach, one could consider to examine suboptimal SP alignments, like 7c, up to the degree that all prior alignments have been accepted that are involved in symmetries such as in Figs. 6.6a and 6.7a. But this would not work, for two reasons. First, there is no general way to tell how far one has to descend into suboptimality before all relevant alignments have been taken into account. Second, many additional and unwanted alignments might pass as well. So accepting suboptimal SP alignments cannot be a general solution to this problem of data symmetry.

Similar problems can arise with progressive alignments using *guide trees* (e.g. Thompson *et al.* 1994; see also Feng and Doolittle 1987). Such trees are usually constructed on the basis of a square overall distance matrix that is derived from pairwise alignment scores. Multiple alignment then proceeds by traversing this tree from terminals to the root. At each node that is visited, a partial multiple alignment is created that includes and combines the partial alignments that are found at the daughter nodes (terminal nodes are initially assigned a trivial partial alignment that includes just the observed sequence of that node). In this way, all sequences are included in the alignment after the root node has been visited. At any node, the alignment of partial alignments mostly proceeds by using some modification of the SP criterion, considering only those pairwise alignments across the node being considered. Moreover, this criterion is mostly applied only locally: gaps that have been inserted before will never be removed. In general, this group of methods cannot guarantee

that symmetries as discussed here are properly taken into account.

A case of local symmetry

Based on the premise that multiple alignments should be constructed prior to tree search on the basis of a similarity criterion, Simmons (2004, p. 876; see also Ochoterena 2004) recently proposed the following tree-independent procedure for constructing optimal prior alignments. In a first step, construct one or more multiple alignments using, for example, programs that try to maximize (an unspecified measure of) similarity, or information from secondary structure. Next, evaluate these alignments using the number of 'differences' that are implied, and try to lower that score by adjusting those alignments. Such adjustments can be done manually or, ideally, using optimization programs. The rationale is to further increase the amount of similarity that is present in the alignment. The best alignments that are obtained are then subjected to parsimony analysis.

In the above, the number of differences is best explained by first looking at a regular data set such as in Fig. 6.1. For each character in the data set, the observed variation *m* (Farris 1989a, p. 417) is one less than the number of states in the character, and that number is the minimum of steps that the character can have on any tree. The observed variation for the data set as a whole, *M*, is the sum of the observed variation in all its characters, and can be interpreted as the number of steps that the best tree for the data set would have if all characters were congruent. If indel events would not occur, the number of differences in the sense of Simmons (2004) would be equal to *M*. But indel events do occur and complicate matters because single indel events can affect multiple columns of an alignment. However, as will be clear below, further details of the calculations that are involved in such cases (see, for example, Simmons and Ochoterena 2000) are not required for the current argument. Simmons (2004) observed that minimization of differences in this sense can lead to trivial alignments that require only as many indels as there are sequences in the data set, irrespective of the tree being considered (see Fig. 6.13c, below, for an example). To circumvent that problem,

(a) *out*	tttttttttttggggtttt	tcca	(b) tcca	(c) tcca	(d)	
A	aattttttttggggtttt	c	-c--	--c-		
B	aaaattttttggggtttt	c	-c--	--c-		
C	aaaaaatttggggtttt	c	-c--	--c-		
D	aaaaaaaattggggtttt	c	-c--	--c-	I	c
E	aaaaaaaaaaggggaaaa	cg	-cg-	-cg-	E	cg
F	aaaaaaaaaacccctttt	gc	-gc-	-gc-	F	gc
G	aaaaaaaaaaccccaaaa	aca	-aca	aca-	G	aca
H	aaaaaaaaaaccggaatt	gg	-gg-	-gg-	H	gg

Figure 6.8 An example of localized data symmetry. (a) A data set consisting of two sets of putative homologous sequences. (b, c) Two multiple alignments for the second set. (d) Reduced data set that exhibits the same kind of symmetry as discussed for Fig. 6.6.

Simmons (2004, p. 876) suggested not to add positions to alignments as obtained in the first step during possible adjustments in the second step.

This optimality criterion assigns the same scores to the symmetric alignments of Figs. 6.6 and 6.7, and in each case all other alignments have a worse score. Therefore this approach could correctly identify the relevant prior alignments for these problematic data sets. However, consider the data set of Fig. 6.8a, a case where two different sets of putative homologous sequences are analysed simultaneously (the example uses two sets of sequences for reasons of clarity only; similar examples can be constructed that use only one set of putative homologs). The structure of the first set of sequences jumps out so clearly that it is easily seen that the best trees for that part of the data are (*out* (*A* (*B* (*C* (*D* (*E* (*H* (*F G*)))))))) and (*out* (*A* (*B* (*C* (*D* (*F* (*H* (*E G*)))))))). Moreover, it is easily established that the first set of sequences is so strongly structured that the problem of finding the best trees for the data set as a whole reduces to evaluating the second set of sequences on those two trees.

In both trees, consider the ancestor of terminals *D–H* and this second set of sequences. In each case, that node will be optimized as *c* for the alignments of Figs. 6.8b and 6.8c, or indeed for any alignment in which the *c*'s of terminals *A–D* are aligned (it is easily seen that such must be the case for optimal explanations). Next consider the data set of Fig. 6.8d, where terminals *out*, *A*, *B*, *C*, and *D* have been replaced by a single hypothetical terminal *I* that is assigned that reconstructed sequence *c*. This reduced data set exhibits the same kind of data symmetry as discussed above: change the

labels of *E* and *F*, reverse the direction in which the sequences are read, and the original data set is recovered. Considering all this, the second set of sequences of Fig. 6.8a cannot be used to distinguish between the two candidate trees, as these only differ in their relative positions of *E* and *F*. Therefore, any method that assigns different scores to these trees for these data is in serious trouble.

The algorithm of Sankoff (1975) properly takes into account data symmetries such as in Fig. 6.8d. It also treats the whole data set of Fig. 6.8a correctly, which can be shown, as above, by observing that optimal tree alignments on optimal trees have to reconstruct the ancestral sequence for terminals *D–H* as *c*, and such that this *c* is aligned with the *c*'s of terminals *A–D*. The score for the complete data set of Fig. 6.8a on both trees is 30, and this is also the optimal score. Two corresponding implied alignments are shown in Figs. 6.8b and 6.8c. As above, these display the same symmetry as the raw data (other optimal tree alignments exist, but that does not affect the argumentation).

Evaluating these implied alignments using the criterion of Simmons (2004) cannot be done by simply summing over isolated columns because some gaps affect more than one column, and more elaborate calculations are required. However, these are not really required in this case because reversing the sequences in both alignments establishes mutual symmetry of gap positions for such calculations. So, whatever the contribution of the gaps in the first alignment, it will be the same in the second and their unit gaps can therefore be treated as missing entries for the purpose of assessing the relative scores of the alignments. This results in relative score three for Fig. 6.8b but four for Fig. 6.8c, and the procedure of Simmons (2004) therefore would lead to prior rejection of the alignment of Fig. 6.8c. The net result is that this procedure leads to rejection of a tree that the data cannot distinguish from a tree that it accepts.

Comparing the alignments of Figs 6.8b and 6.8c, the preference of the optimality criterion of Simmons (2004) for the first one boils down to the fact that it puts the last *a* of terminal *G* in the same column as the last *a* of the outgroup. But on the best tree for this alignment, the *a* that *G* and the outgroup share cannot be explained as identical by common descent and inheritance. Consider the consequences of this observation in the light of the overall analysis, where tree (*out* (A (*B* (*C* (*D* (*E* (*H* (*F G*)))))))) is accepted but (*out* (A (*B* (*C* (*D* (*F* (*H* (*E G*)))))))) rejected. Given the local symmetry in the second sequence character, both trees explain the data equally well, albeit with different posterior homologizations of positions and base identities. But they are different in their amounts of homoplasy: overall, the first tree has a homoplastic pairwise base similarity (the last *a* of terminal *G* and the outgroup) that the second tree lacks. Moreover, the preference for the first tree when using the procedure of Simmons (2004) is based solely on this difference: of the two trees with equal amount of similarity that can be explained as homology, it selects the tree that has the higher amount of homoplasious similarity. In more complex cases, this effect can ultimately lead to rejection of trees with higher amounts of homologous similarity in favour of trees with lower amounts of homologous similarity. The same problem can also occur with the related tree-independent optimality criteria for multiple alignments that have recently been discussed by Carpenter (2003, pp. 6–7) and Nixon and Little (2004).

General conclusions
None of this is accidental. Data symmetries such as in Figs 6.6a, 6.7a, and 6.8a have a consequence that no distinction can be made between particular trees or groups of trees. As a result, methods of analysis that do not directly take into account the structure of trees (e.g. SP alignment or the procedure of Simmons 2004), or do so in a way that violates the symmetry (e.g. progressive alignment,

or even just the use of suboptimal tree alignments), will not in general be able to deal with such situations. This leaves, by definition, optimal tree alignment methods. As a corollary, unless one is willing to defend methods that in some cases can give different scores to trees that cannot be distinguished by the data at hand, alignment and tree search cannot be properly separated in phylogenetic analysis of sequence data. Note that this conclusion is argued and reached in logical space. Whether or not it results in a practically feasible method will be discussed below.

The examples of Figs 6.6a, 6.7a, and 6.8a are unusual in that some terminals have sequences that are the exact reverse of other sequences, a situation that will hardly if ever arise in real data sets. But such perfect crab canons are not necessary for the phenomenon to occur. Sequences such as those can be embedded as short motifs in longer sequences that as a whole are not identical when read in reverse, and similar distortions could result. For simple examples as above, one could argue that the problem can easily be spotted and solved by carefully inspecting the data and the alignments by eye, but this approach would no longer work in such more complex cases.

In addition, the motifs that are involved do not have to be identical when read in reverse, only their alignment scores with the other sequences must remain unchanged. Lastly, even when the symmetry in the motifs is not perfect, by deviations in motif sequence and/or substitution costs that are involved, systematic distortions, though less well defined, would still arise. So situations where short subsequences can have alternative optimal alignments, with different local costs on different trees, may well be relatively common in empirical data. Moreover, when such data sets are aligned progressively according to a guide tree (using, for example, CLUSTAL; Thompson *et al.* 1994), such ambiguities that include groups of the guide tree may systematically be resolved in favor of the guide tree.

Summarizing, alignment and tree evaluation cannot be *properly* separated in phylogenetic analyses of sequence data. As a consequence, the view that a set of sequences that are deemed putative homologues should be turned into a sequence of positional characters prior to tree search and eva-luation is erroneous or at best incomplete. Instead, such sequences constitute a single complex char-acter, a sequence character, that can be optimized on trees using optimal tree alignment algorithms such as that of Sankoff (1975). These conclusions follow from very general considerations of data symmetry and do not depend on details of the cost regime that is used.

6.3.3 Quantifying and maximizing homology in sequence characters

Frost *et al*. (2001, pp. 354–355; they use the term 'indel' for a unit gap as used here) discussed the method of direct optimization (Wheeler 1996), and argued for setting all substitution costs and the unit gap cost equal because this amounts to equal weighting of all hypothesized transformations, which in turn 'renders the highest degree of des-criptive efficiency and maximizes the explanatory power of all lines of evidence (i.e. characters).' Direct optimization has been proposed and is still often discussed as a sequence optimization method that is qualitatively different from optimal tree alignment methods, but the method is best seen as a heuristic approximation for optimal tree alignments (De Laet and Wheeler 2003; see also below), and the claimed novelty of the approach rests on a lack of familiarity with or misunder-standing or misrepresentation of the work of Sankoff (1975) and Sankoff and Cedergren (1983) (see, e.g., Wheeler 1996, 1998; Giribet and Wheeler 1999; Phillips *et al*. 2000; Wheeler 2001b, 2002, 2003a). Therefore, the argumentation of Frost *et al*. (2001) amounts to a preference for the minimum mutation algorithm of Sankoff (1975).

Consider the sequence character *aaa*, *gat*, and *agt* and two alternative tree alignments on the single tree for three terminals as presented in Fig. 6.9. With the above cost regime, tree alignment 9a is better than 9b (three steps versus four). On the other hand, when looking at independent accom-modated pairwise similarities, as a measure of the amount of similarity that can be explained as homology, 9b performs better than 9a: it accomod-ates one more independent pairwise base match. This should not come as a surprise. For pairwise

Figure 6.9 Two different tree alignments of the putative homologues *aaa*, *agt*, and *gat* on the single tree for three sequences. (a) This reconstruction requires three steps (three substitutions, no indels) and retains three independent pairwise base similarities among observed sequences. (b) At four steps (one substitution, three indels) this reconstruction requires one more transformation, even if it retains one more independent pairwise similarity among observed sequences.

alignments, Smith *et al*. (1981; their equation 4b with $w_k = 0$) showed that maximization of base-to-base matches is equivalent to minimization of cost when all base substitution costs are set at twice the unit gap cost, a different regime than advocated by Frost *et al*. (2001). This result of Smith *et al*. (1981) cannot directly be extended to comparisons of more than two sequences, but a generalization to tree alignments (see below) still yields a cost regime that is different from the one favored by Frost *et al*. (2001). With more than three sequences, this difference can lead to a preference for different trees.

On a general level, this example merely reflects the well-known fact that the choice of substitution, gap opening, and gap extension costs affects the result of alignment and tree-building procedures. When examining the logical basis of sequence analysis, however, the paradoxical situation arises that the objectives of maximizing explanatory power and maximizing independent homologous similarity seem to be at odds. As discussed below, this contradiction is only apparent because the pre-mises at either side of the comparison are faulty: setting all costs equal does not maximize expla-natory power, and independent base-to-base homologous similarity is not all there is to sequence homology.

Subsequence homology and compositional homology The latter is easily seen when considering a data set, such as in Fig. 6.10, where sequences differ only in length. The two tree alignments that are shown do not differ in the number of independent

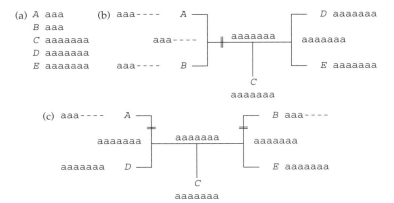

Figure 6.10 A data set in which the sequences only differ in their lengths (a) and two trees with optimal inner-node reconstructions and positional correspondences under the assumption that insertion/deletion of a stretch of contiguous bases is counted as one transformation (b, c). Double bars indicate indel events. Note that on each tree alternative sets of optimal positional correspondences exist.

base-to-base matches among observed sequences that they accommodate: in both cases there are 20 independent base-to-base comparisons, and all these are matches. Yet, the first tree alignment can be considered a better explanation of the data at hand because it captures an element of homologous similarity between the sequences of A and B that is not retained in the second one. However the tree of the first tree alignment is rooted, A and B share the absence of bases 4–7 with their direct ancestor. Depending on the position of the root, these three contiguous nodes lack the insertion of that subsequence, or they share its deletion; in both cases, this comes down to one unit similarity that can be explained as a homology. On the second tree alignment, the shared absence of bases 4–7 in A and B must be explained as a homoplasy. The main conclusion that can be drawn from this simple example is that sequence homology has a component that cannot be reduced to mere base-to-base composition. This component I shall refer to as *homology of subsequences*, as opposed to *base-to-base* or *compositional homology* within homologous subsequences.

The two components of sequence homology can be optimized separately but there would be little use in doing so. When just optimizing base-to-base similarities, gaps will be inserted 'at will' to maximize matches (Smith *et al.* 1981, p. 42). On the other hand, maximizing subsequence homology

without regard for the composition of those subsequences comes down to optimizing the length of the observed sequences as a regular unordered character, irrespective of the amount of substitutions that are implied. Optimized in isolation, neither will in general result in a globally optimal explanation of the data.

Instead, what is needed is an optimal balance between subsequence and compositional homology. This optimal balance can be found by using a cost regime that is the sum of the two cost regimes that are involved, provided that there is a mechanism to avoid or deal with logical contradictions between optimizations of both components. Such a mechanism is implicit in tree alignments because tree alignments are internally consistent explanations of the data. Therefore, expressions to describe the amount of subsequence homology and the amount of compositional homology in tree alignments can be derived independently and then simply summed to get an expression for the total amount of sequence homology. This expression, finally, can be used for purposes of optimization.

Quantifying the amount of subsequence homology of a tree alignment

The amount of subsequence similarity in a tree alignment that can be interpreted as homology can be measured indirectly and in a relative way by counting n_{indels}, the number of independent indel events, provided that the insertion/deletion of a series of contiguous bases is counted as a single event. This is so because each such indel event effectively marks a subsequence that is not homologous across a branch. Therefore, an independent indel event can be seen as an independent unit of non-homology in subsequence homology.

As discussed above, the cost of a tree alignment is obtained as the sum of the costs of the pairwise alignments across the branches of the tree. Technically, counting independent indel events in such a pairwise alignment is achieved by setting substitution costs to 0, gap opening cost to 1, and gap extension cost to 0. In addition, when evaluating such a pairwise alignment, paired gaps have to be removed first, a procedure that Altschul (1989) called *projection*. Projection is required because paired gaps just indicate that both sequences miss something that is present elsewhere on the tree and because the indel events that caused such a shared absence are accounted for along other branches. As an example, going from *-gaat---ccct-* to *-gaat--ccccc-* in, for example, the second tree alignment of Fig. 6.14, (see below) means going from *gaat-ccct* to *gaatccccc*. As far as subsequence homology is concerned, this comes at cost 1 (1 times the gap opening cost of 1 plus 0 times the extension cost of 0).

Quantifying the amount of compositional homology of a tree alignment
Specifying an expression for compositional similarity that can be explained as homology is more elaborate. A tree alignment can be seen as a regular multiple alignment with, for each position, reconstructions at the inner nodes. If, in a single column, the tree path between two observed bases passes through an inner node that is optimized as a unit gap character, these bases are not comparable because they are part of non-homologous subsequences; if, on the other hand, the connecting path has no nodes with unit gaps, they belong to homologous subsequences; more specifically, they occur at the same position within those homologues. I refer to such bases as *comparable bases*.

The observed bases in a single column of a tree alignment can be sorted into a number of groups such that two bases from the same group are comparable but two bases from different groups are not comparable. I shall refer to these groups of comparable bases as *subcharacters*, a concept that is closely related to the concept of regions as defined above, and denote the number of subcharacters in a column of a tree alignment as nsc_c. This number is related but not identical to the number of indel events in which this column of the alignment is involved.

Within a subcharacter, denote the number of observed bases as nob_{sc}. If two such bases are identical and all nodes in the path that connects them are labeled with that same base, then the two bases match and their shared presence can be explained as a homology. If any node in the path that connects two such identical bases has a base that is different, then they don't match and their shared presence cannot be explained as a homology. Two non-identical bases of a subcharacter or two bases that belong to different subcharacters, finally, do not contribute to base-to-base homology. The minimum number of pairwise comparisons that have to be made to classify the bases of a subcharacter into subgroups of such matching bases is $nob_{sc} - 1$. The number of mismatches nmm_{sc} in any such set of $nob_{sc} - 1$ independent pairwise comparisons can be thought of as the number of base substitutions or steps within the subcharacter.

With these definitions, the amount of compositional homology in a subcharacter is obtained just as the amount of homology in a regular character: the maximum number of independent pairwise comparisons minus its number of steps, or $nob_{sc} - 1 - nmm_{sc}$. With nob_c the total number of observed bases and nmm_c the total number of substitutions in a column of a tree alignment, the amount of compositional homology in a column is $nob_c - nsc_c - nmm_c$. The amount of compositional homology in the whole tree alignment is the sum of this value over all columns. Switching signs, $nsc_c + nmm_c - nob_c$ describes a cost function that varies directly with compositional homology in a column. In this expression, nsc_c can be considered a cost factor that accounts for local loss of compositional homology due to indel events (that may encompass multiple neigbouring columns), and nmm_c a regular substitution cost factor.

Maximizing homology in sequence characters

Adding it up, the total amount of homology of different tree alignments for a given set of sequences can be compared using cost function $n_{indels} + \Sigma(nsc_c + nmm_c - nob_c)$, where the summation is over all columns of the tree alignment: the lower the cost, the higher the amount of homology. In this expression, losses in subsequence homology and compositional homology are weighted equally. Differential weighting, for example to downweight subsequence homology, can be done by applying different weights to the two terms that are involved. As Σnob_c is identical for different tree alignments for the same data, the cost function for a tree alignment can be reduced to $n_{indels} + \Sigma(nsc_c + nmm_c)$. Using n_{subc} for Σnsc_c and n_{subst} for Σnmm_c, the relative amounts of total homology of two different tree alignments can be compared using $n_{indels} + n_{subc} + n_{subst}$, the sum of indel events, subcharacters, and substitutions.

Alternatively, the problem can be presented as a maximization of a similarity measure; this similarity measure would count independent homologous base-to-base matches but assign a penalty to indel events, much as the original algorithm of Needleman and Wunsch (1970). More specifically, the penalty would be -1 for each indel event in the tree alignment, irrespective of the length of the indel. In comparisons of two and three sequences, such similarity measures with length independent gap penalties have been studied by Fredman (1984) (*fide* Hein 1989a, p. 650).

In Figs. 6.11–6.15, the positions of all inferred indel events are indicated throughout the tree alignments, using vertical bars. The subsequences that are defined in that way can be considered *logical subsequences*. In simple cases, such logical subsequences are identical to the subsequences that effectively take part in the inferred indel events (e.g. Figs 6.11–6.14), but in more complex cases a single inferred indel event along a particular branch can affect a series of contiguous logical subsequences (see Fig. 6.15 for examples). The total number of subcharacters in a tree alignment can be easily determined as the sum of the lengths of its different homologous logical subsequences.

For any given tree alignment, $n_{indels} + n_{subc} + n_{subst}$ is a straightforward expression that is easily checked, but finding the tree alignment(s) for which this expression is minimal is quite something else. Even for a single given tree, the problem of deciding if a tree alignment is optimal has been shown to be NP-complete (Wang and Jiang 1994). Algorithmically, as the subsequence homology component requires use of variable gap costs (gap opening cost 1, gap extension cost 0), the algorithms of Sankoff (1975) and Sankoff and Cedergren (1983) are not adequate. Altschul (1989) does accomodate variable gap costs but still this is not sufficient because his algorithm does not keep track of the number of subcharacters in a column. This directly implies that the current cost function cannot be expressed just in terms of substitution, gap opening, and gap extension costs. To optimize this function, the dynamic programming recurrences of Altschul (1989) would have to be adapted and extended to keep track of observed bases and subcharacters in columns as well.

6.3.4 Discussion

So, when applied to sequence data, the simple principle of maximizing similarity that can interpreted as homology, in a logically correct way, leads to a preference for those trees on which the sum of indel events, base substututions, and subcharacters is minimal. In this final subsection, some properties and wider connections of this parsimony criterion are discussed.

Heuristics

Even with simple Hamming distances, as when using Fitch (1971) optimization of prior alignments, the problem of deciding if a tree is optimal is NP-complete (Foulds and Graham 1982). So, when combining tree search and tree alignment, one NP-complete problem is nested within another. As pointed out by Hein (1989a, p. 651), the computational complexity of this problem makes the use of heuristic approximations unavoidable. Examples of algorithms for heuristic approximations of optimal tree alignment costs, or algorithms that can be interpreted as such, can be found in, for example, Sankoff *et al.* (1973, 1976), Hein (1989a, b), Jiang and Lawler (1994), Wang *et al.* (1996), Wheeler (1996, 1999, 2003c; all available in Wheeler *et al.* 2003, where they are tightly integrated with a wide range of tree search heuristics; see De Laet and Wheeler 2003), and Schwikowski and Vingron (1997, 2003). Still other approaches can be found in the reviews of Vingron (1999) and Notredame (2002).

It currently remains largely an open question how well these various approaches perform in practice. In the end, even the use of an *a priori* alignment can be seen as a quick and dirty heuristic for the analysis of a sequence character. Even if any single such analysis is too shallow to be satisfactory, analyses of many different prior alignments may be effectively combined into a more elaborate search strategy, following the heuristic logic as developed in Farris *et al.* (1996) (see also Goloboff and Farris 2001).

Most heuristic tree alignment methods attack the optimal tree alignment problem by approximate decomposition into a set of simpler problems that can easily be solved exactly using pairwise alignments (e.g. Hein 1989a; Wang *et al.* 1996; Wheeler 1996, 1999) or threewise alignments on a star tree (e.g. Sankoff *et al.* 1973; Wheeler 2003c). Interestingly, compositional homology in a pairwise alignment amounts to the number of base matches, a number that can be maximized by setting the unit gap cost to half the substitution cost (Smith *et al.* 1981). To maximize total sequence homology in a pairwise alignment, an additional penalty has to be added for losses in subsequence homology, which, as discussed above, can be done

using the gap opening cost. With equal weighting of both components of homology, this penalty equals the substitution cost. As an example, using a substitution cost of 2, the corresponding gap-opening cost is $2+1$, and the corresponding gap extension cost 1. The same result holds for three-wise comparisons on a star tree.

Beyond three sequences this simpler cost regime is no longer equivalent to the criterion developed here, as can be seen from the following counter-example. The tree alignment of Fig. 6.11b explains the sequence character of Fig. 6.11a better than Fig. 6.11c because it can explain an additional independent pairwise base match: the *a* that terminates the sequences of *B* and *D*. This difference is correctly measured by the sum of indels, subcharacters, and substitutions, but with the simpler cost regime, both tree alignments come at the same cost of 12. In more complex examples, such situations can lead to a preference for different trees alltogether. The simpler cost regime may nevertheless be a good choice when using heuristic tree alignment methods that are based on pairwise or threewise comparisons of sequences.

For some approximation methods an upper bound can be established for their deviation of optimality. As an example, consider lifted alignments (Jiang and Lawler 1994; Wang *et al.* 1996; see also Wheeler 1999; Lutzoni *et al.* 2000), in which possible inner-node sequences are chosen from and restricted to the set of observed sequences. Under these restricted conditions, an efficient algorithm exists to find the optimal assignments of sequences to inner nodes of a given tree, and the resulting tree alignment can be shown to have a cost that is at most twice the cost of the unrestricted

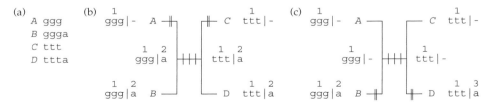

Figure 6.11 An example of the parsimony criterion for sequence characters. (a) A sequence character. (b) An optimal tree alignment on the optimal tree. (c) A suboptimal tree alignment on the optimal tree (same number of indel events and substitutions, but one more subcharacter). Single bars across branches indicate substitutions, double bars indel events. Logical subsequences are indicated using vertical bars, and numbered for clarity.

optimum for that tree (Wang *et al.* 1996) As discussed by Gusfield (1997, p. 358), such bounded-error approximation methods can help to understand the behaviour of difficult optimization problems; from a practical point of view, they may be combined with other methods, such as local improvement methods, to obtain more elaborate heuristic search strategies.

Inapplicables

The example of Fig. 6.12 illustrates that indel events divide the sequences of the tree alignment into subsequences that can be considered independently: the two optimal alignments that are shown have identical subsequences and only differ in the way that those subsequences (and their sub-characters) are presented. Incidentally, this example also shows that postulated indel events may improve the explanation of the data even in cases where all observed sequences have the same length.

This independence is a direct consequence of the fact that, in the current approach, base-to-base comparisons are only made within subsequences that can be explained as homologs. As a consequence, comparisons of sequences and their bases automatically occur at the correct levels of generality, and the problems with inapplicables that Maddison (1993) described simply dissolve. Indeed, Maddison (1993, p. 580) observed that all solutions that he considered to deal with inapplicables were in the end problematic because they did not properly restrict counting of steps to parts of trees where comparisons were valid, and he correctly surmised that an eventual solution would lie in the development of new algorithms. Most cases of inapplicability, however, would not require an algorithm as complex as the one discussed here, because there are fewer degrees of freedom in *a priori* acceptable hypotheses of homology.

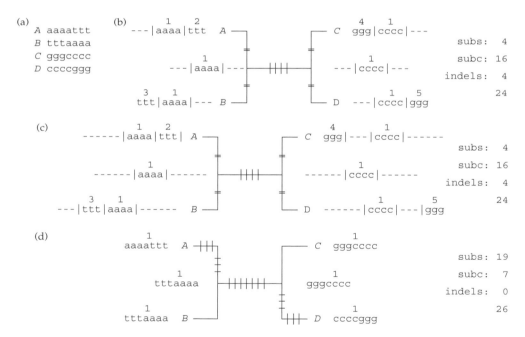

Figure 6.12 An example of the parsimony criterion for sequence characters. (a) A sequence character in which all sequences have equal length. (b, c, d) Three tree alignments of the character on the optimal tree (*A B*)(*C D*). The first two, requiring four indel events, are optimal; the third, not requiring indel events, is suboptimal by two units. The two optimal alignments that are shown imply the same five subsequences that take part in indel events and differ only in the way that these subsequences are presented (still other possibilities exist). Subs, subc, and indels are numbers of substitutions, subcharacters, and indel events. Single bars across branches indicate substitutions, double bars indel events. Logical subsequences are indicated using vertical bars, and numbered for clarity.

Consider again the multiple alignment of Fig. 6.8b, but now assume that the four columns are regular independent single-column characters, with a dash indicating inapplicability. Obviously, in this case there is no need to examine alternative groupings of states, such as in Fig. 6.8c, during tree search and optimization. Permitting such shifts would lead to the same problems as when using absence/presence coding of individual states. As the computational complexity of the current approach mostly derives from the need to examine alternative groupings of bases when optimizing sequences on a tree, this restriction has as a fortunate consequence that the general algorithm for dealing with this kind of inapplicability is much simpler and faster (De Laet 2003).

Maximizing homologous similarity vs. mimimizing transformations

The parsimony criterion as discussed here relies on the notion that one indel event counts as one unit loss of subsequence homology, irrespective of the number of bases that are involved. But this does not mean that it would in general produce trivial alignments that are obtained by simply juxtaposing all observed sequences, which requires only as many insertion events as there are sequences. An example is presented in Fig. 6.13. In the optimal tree alignment of Fig. 6.13b, two independent

pairwise base matches can be explained as homology. The trivial alignment that is obtained by juxtaposing all observed sequences (Fig. 6.13c) has no such base matches. In addition, compared to the first tree alignment, it has has four independent instances of subsequence non-homology. The total difference in explanatory power thus equals six, which is reflected in the relative tree scores.

This shows that the current criterion is not a minimum evolution method: the second tree alignment of Fig. 6.13 requires only four mutations (four insertions of subsequences of length four) but it is considered a much worse explanation of the data than the first one, which requires 10 mutations (10 substitutions). Given that one of the terms in the minimization for sequence character homology is the number of subcharacters, a quantity that has no direct relationship with evolutionary transformations, the non-equivalence of both approaches when dealing with sequence characters should come as no surprise. But this non-equivalence with minimization of evolutionary transformations does not imply that the current method is not logically capable of phylogenetic interpretation. Such an interpretation, however, is in terms of unit statements of similarity that can be explained in a logically consistent way as identity through

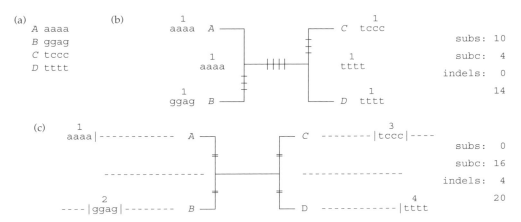

Figure 6.13 An example of the parsimony criterion for sequence characters. (a) A sequence character. (b, c) Two tree alignments on the optimal tree (*A B*)(*C D*). The first is optimal. The second, obtained by simply juxtaposing all observed sequences, is suboptimal by six units. Subs, subc, and indels are numbers of substitutions, subcharacters, and indel events. Single bars across branches indicate substitutions, double bars indel events. Logical subsequences are indicated using vertical bars, and numbered for clarity.

(a) A gaatcgct
     B gaatccgt
     C ataaaaacccac
     D ataaaaaccccgg
     E gaatccccc

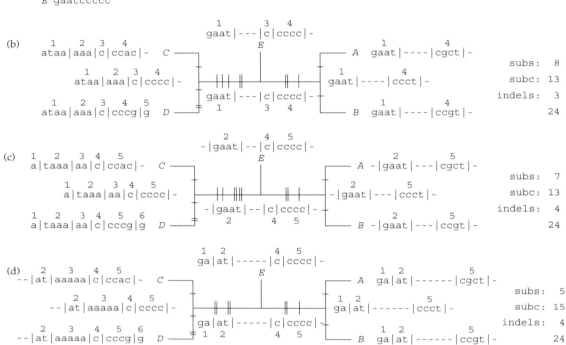

Figure 6.14 An example of the parsimony criterion for sequence characters. (a) A sequence character. (b, c, d) Three optimal tree alignments on its optimal tree. Subs, subc, and indels are numbers of substitutions, subcharacters, and indel events. Single bars across branches indicate substitutions, double bars indel events. Logical subsequences are indicated using vertical bars, and numbered for clarity.

common descent and inheritance, and not in terms of numbers of transformations that are required to that effect.

An example where different optimal tree alignments on the best tree have different numbers of indels plus substitutions is presented in Fig. 6.14. The two first tree alignments have more indel events plus substitutions than the third one (11 versus 9), but despite this higher total number of mutations, they provide an equally good overall explanation of the data in terms of the amount of total sequence similarity that can be explained as homology. More precisely, the first alignment accomodates 29 independent pairwise matches among observed bases, the second 30, and the third one 30 as well, as easily verified by examining

the tree alignments column by column. So just considering compositional homology, the first explanation is suboptimal. The difference, however, is exactly offset by its lower loss in subsequence homology (three indels versus four and four). With the cost regime that is advocated by Frost *et al.* (2001) (all costs equal), the optimization of Fig. 6.14c is preferred (cost 12 vs. costs 13 for 14b and 14 for 14d).

The difference between both cost regimes is further illustrated in Fig. 6.15. Maximizing the amount of sequence similarity that can be interpreted as homology, the tree of Fig. 6.15b is optimal, and an optimal tree alignment is shown. The tree of Fig. 6.15c is suboptimal by two units, as can be seen from the optimal alignment that

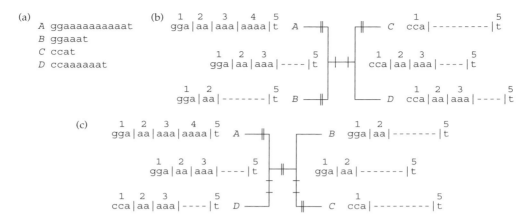

Figure 6.15 An example of the parsimony criterion for sequence characters. (a) A sequence character. (b) An optimal tree alignment on the optimal tree. (c) An optimal tree alignment on a suboptimal tree. Single bars across indicate substitutions, double bars indel events. Logical subsequences are indicated using vertical bars, and numbered for clarity. The number of subcharacters in both optimizations is the same.

is shown. Under the costs of Frost *et al.* (2001) the tree alignments of Figs 6.15b and 6.15c are also optimal for their respective trees, but the ranking of the trees reverses: the second tree is now preferred (costs 14 vs. 13). This shift in preference is a consequence of counting an indel event of length k as k events, as implicitly advocated by Frost *et al.* (2001). In this example, this amounts operationally to treating the lengths of the gaps that are involved as an ordered character.

A more extreme example of the same phenomenon occurs with a sequence character such as *ttaatt*, *ttaaatt*, *ttaaaatt*, and *ttaaaaatt* for terminals A, B, C, and D. With the cost regime of Frost *et al.* (2001), unrooted tree (A B)(C D) is preferred because, operationally, it best groups the series of *a*'s in the middle of the observed sequences according to their length. With the cost regime that maximizes homology, the three different unrooted trees for four terminals are considered equally good explanations of the character.

The preference of Frost *et al.* (2001, pp. 354–355) for equal substitution and unit gap costs follows from their position that all hypothesized evolutionary transformations should be weighted equally. However, this cost regime only accomplishes such equal weighting under the very restrictive assumption that indels only affect

single bases, which constitutes a severe knowledge claim about the processes that shape sequence evolution. It is hard to see then how this approach 'maximizes the explanatory power of all lines of evidence' (Frost *et al.* 2001, p. 354) even more so if one considers their apparent position that methods that make severe knowledge claims can be safely ignored (Frost *et al.* 2001, p. 354). No comparable claim is present in the current method, in which the lengths and positions of subsequences that take part in indel events are left open to optimization.

A similar methodological asymmetry exists between methods that impose irreversibility of inferred character evolution and methods that leave the possibility of reversal open during phylogenetic analysis. An extensive discussion of the issues that are involved can be found in Farris (1983, pp. 24–27). Frost *et al.* (2001) did not discuss such issues. In fact, they did not not even provide arguments why equal weighting of all evolutionary transformations should lead to equal substitution and unit gap costs. It can reasonably be argued that the principle of equal weighting of all transformations is instead better implemented by using equal substitution and gap costs, irrespective of the length of the gaps that are involved. However, for most sequence characters this cost regime would lead to trivial alignments such as in Fig. 6.13c, requiring only as many transformations as there are terminals, irrespective of the tree that is considered. Again, it is hard to see how such optimizations can be considered to maximize explanatory power. Yet they are optimal under the notion of minimizing equally weighted transformations.

Sequence characters and branch support
The example of Fig. 6.13 illustrates an interesting consequence for the concept of branch support. Consider the tree alignment of Fig. 6.13b. In that alignment, the (A B)(C D) branch is supported, not because of the four substitutions on that branch, but because collapse of the branch—resulting in an unresolved tree—would remove either the *a–a* base match between *A* and *B* or the *t–t* base match between *C* and *D*. This is in line with the observation of Farris *et al.* (2001a) that branch lengths do not measure support. Instead, support for any single branch is measured as the degree to which removal of the branch worsens the explanation of the data, which holds for sequence and non-sequence data alike. This, by definition, is Bremer (1988) support.

Alternatively, one could measure robustness of a branch using the jackknife (Farris *et al.* 1996) or related methods. However, as sequence characters have no predefined single-column characters, pseudoreplicates cannot be constructed in the usual way. This problem can be solved by resampling at the level of individual bases in the sequences to be compared, such that unsampled bases are made uninformative with a probability equal to the character removal probability of regular jackknifing (operationally, this can be done by replacing a base with a polymorphism code for '*a* or *c* or *g* or *t* or -'; or, a bit more conservative, for '*a* or *c* or *g* or *t*'). With a removal probability of 0.37, the (A B)(C D) branch in the above example would not survive, as it depends on the simultaneous presence of the four bases mentioned above. With the conservative approach, the probability that all four are retained in a pseudoreplicate is only $(1 - 0.37)^4$.

A likelihood conjecture
Miklós *et al.* (2004) recently described a probabilistic model of sequence evolution that allows insertions and deletions of arbitrary length, a more general approach than Thorne *et al.* (1992), the first probabilistic method that incorporated indels that affect multiple residues at once. In their model, substitutions are described using a regular time-reversible rate matrix; indels are modelled such that the rates for insertions as a function of their length *k* are a geometric function of *k*, and such that the ratio between the rates of insertions and deletions of length *k* is a constant.

Miklós *et al.* (2004) only dealt with comparisons of two sequences, but the model can in principle be extended to simultaneous comparison of more than two sequences that are related by a binary tree, similarly as Hein (2001) extended the two-sequence model of Thorne *et al.* (1991), the first stochastic model to include insertions and deletions (single residue indels only). In the approach of Hein (2001), rate parameters are assumed to be constant throughout the sequences. Removal of assumptions of that kind would turn the model into a no-common-mechanism model akin to the model of Tuffley and Steel (1997, pp. 584, 597) for regular *r*-state characters.

Envisioning such a double extension of the model of Miklós *et al.* (2004) it can be conjectured that, under a wide range of possible non-fixed rates, the trees that are found with a parsimony criterion along the lines as described here are also trees of maximum likelihood. As with single-column characters (see above), this does not imply that such a probabilistic process model would exhaustively describe and capture the current method.

Beyond sequence characters: the genome
Most examples above consist of data sets with just a single sequence character, but data sets can have several such characters, and in addition any number of single-column characters. Exactly which observations are coded as characters, the subject of character analysis, is ultimately outside the realm of the technical aspects of further analysis that have been discussed in this section. For sequence characters, a widely used criterion for

establishing hypotheses of putative sequence homology is almost identical to the technology to obtain those sequences in the first place: whatever is amplified using a particular primer pair. In addition, various other criteria can be used to identify biologically relevant structures, such as exons and introns in protein coding sequences, or stems and loops in rRNA sequences (see, e.g., Kjer 1995; Giribet 2002).

On the basis of such criteria, even contiguous stretches of the genome can be subdivided into sequences of sequence characters that can be optimized separately. When doing so, it may be a legitimate concern that the subsequent analysis might be constrained and even biased by preconceived ideas about the evolution of such structures. However, given that the complexity of the calculations when dealing with sequence characters makes the use of heuristics and approximations unavoidable, the procedure of breaking up long sequences in smaller components prior to analysis may very well be part of a heuristic search strategy. This approach could be especially powerful when combined with heuristic multiple alignment methods that try to assemble global alignments from alignments of fragments that are dynamically identified (e.g. Morgenstern *et al.* 1996; Morgenstern 2004).

On a more fundamental level, sequence characters as discussed here are thought to be hierarchically related through indels and substitutions only. This may be a biologically plausible assumption for shorter parts of the genome, but it definitely breaks down for complete genomes, where other processes such as inversions, duplications, and translocations play a role as well. Over the past few years, many combinatorial algorithms have been developed to study such phenomena (see, e.g., Sankoff and Nadeau 2000), and heuristic multiple-alignment methods that incorporate such rearrangment events are becoming available (see, e.g., Brudno *et al.* 2003, 2004). It remains an open question how such methods can be interpreted or generalized to accomodate a parsimony criterion as developed here.

Such extensions may well lead to revisions or further elaborations of the current framework.

Consider, for example, a process such as lateral transfer, which may well play an important role in the evolution of genomes (see, e.g., Kunin and Ouzounis 2003), or speciation through allopolyploidization (see, e.g., Vander Stappen *et al.* 2002). For any data set, positing sufficient such events in any phylogenetic tree will permit to explain all observed similarities as historically identical, whether through regular ancestor–descendant relationships of organisms or through non-hierarchic processes such as lateral transfer. It may be sufficient to restrict the current criterion to the former case, but, alternatively or additionally, a more general criterion might be conceived that maximizes the difference between similarity that can be explained as historical identity, whatever the underlying processes, and the minimum number of hypothesized historical events required to that effect.

This second approach would need careful elaboration of a broader theoretical concept of explanation than used here, which is beyond the scope of this chapter. However, one way to go would be to couple the principle of maximizing conformity between observation and theory to the principle of choosing the simplest theory or theories that can explain the data, which would lead to a true synthesis of two different but interwoven lines of argument that can be found in the work of Farris (see, e.g., Farris 1982b, 1983). As discussed extensively in this paper, the first principle leads to maximization of similarity that can be explained as homology. The second principle requires a measure of the simplicity of a phylogenetic explanation, which may well be the minimum number of logically distinct historical events that have to be postulated. The rationale for a combined optimality function as above would then be to find an optimal balance between both principles.

For single-column character data and under the above restriction, that approach would operationally be equivalent to the current parsimony criterion, because in such cases it amounts to minimizing twice the amount of homoplasy. For sequence characters as defined here (only indels and substitutions), it would amount to minimizing

$2n_{indels} + n_{subc} + 2n_{subst}$, which would obviously change details of several examples discussed in this section. For example, both trees of Fig. 6.13 are then considered equally good explanations; or the two first trees of Fig. 6.14 become suboptimal by two units. But the main conclusions, and especially those based on data symmetries, would remain valid.

6.4 Acknowledgments

This chapter was prepared on a gnu/linux system, mostly using vim and LaTeX. Thanks to Victor Albert, Norberto Giannini, Pablo Goloboff, Mark Simmons, Hubert Turner, John Wenzel, and two anonymous reviewers for constructive criticisms. Thanks to Kevin Nixon for bandwidth and disk space to make the computer program *goechel* available at www.cladistics.com. Goechel is a java shell around POY (Wheeler *et al.* 2003) to perform jackknife analyses as discussed in 6.3.4 (subsection *Sequence characters and branch support*). The author holds a return grant from The Belgian Federal Science Policy Office. Lastly I wish to mention Steve Farris. It's been my privilege to have him as a member of my PhD committee, late last century. Ever since, I consider myself a student of his, as should be obvious by even just a cursory glance at this paper. Thanks for it all, Steve!

III

Computational limits of parsimony analysis: from historical aspects to competition with fast model-based approaches

CHAPTER 7

The limits of conventional cladistic analysis

Jerrold I. Davis, Kevin C. Nixon, and Damon P. Little

7.1 Introduction

Software for cladistic analysis has been widely available for more than 20 years, and a series of advances made during this time have facilitated the analysis of matrices of ever-increasing size. A milestone was reached in 1993 with the assembly of a 500-terminal *rbcL* matrix (known as *zilla*), but optimal trees for this matrix were not discovered until years later. A range of analytical methods developed since that time have found what appear to be most-parsimonious trees for the *zilla* matrix, but there continue to be perceptions that shortest trees for this matrix cannot be discovered by conventional search methods with a single personal computer in a reasonable period of time. We provide an overview of the development of parsimony methods for cladistic analysis, describe strategies that have allowed the *zilla* matrix to be analyzed by conventional methods, and demonstrate that *zilla* was amenable to analysis by these methods as early as the mid-1990s, using then-available hardware and software. Preliminary analyses, even when unsuccessful at discovering most-parsimonious trees, can be used to identify appropriate software settings for use during thorough analyses. A useful indicator of the settings that yield the most efficient searches is the excess branch-swapping ratio, which is the ratio between the number of tree rearrangements conducted during a particular phase of branch swapping in which shorter trees are being discovered, and the minimum possible number of rearrangements during this phase. Two-stage search strategies, with intensive branch swapping conducted on a small percentage of the most optimal sets of trees obtained by a large number of relatively short searches, are more efficient than one-stage searches. Although data sets with substantially greater numbers of terminals than the *zilla* matrix are beyond the current limits of conventional cladistic analysis with a solitary personal computer, these techniques are likely to continue to be of importance when employed in association with more recently developed methods such as tree fusion, sectorial searches, tree drifting, and the parsimony ratchet.

7.2 A brief history of parsimony methods for phylogenetic analysis

Willi Hennig, the German dipterist, is widely considered to be the father of modern phylogenetics, and his book *Phylogenetic Systematics* (Hennig 1966) had a broad-reaching influence in the early development of the field. Hennig's greatest contributions are observed in his clear definitions of monophyly, in his discussion of the evidence used to determine monophyly (i.e. synapomorphy), and in his strict adherence to phylogenetic classifications. However, Hennig's explication of the methods by which one might determine synapomorphies, and thus monophyletic groups, and ultimately phylogenetic trees, were less precise. On the other hand, some methods that at least superficially embodied Hennig's proposals had been published several years earlier (e.g. Wagner 1952, 1961). Following the publication of *Phylogenetic Systematics* in the late 1960s, efforts were begun to reconcile Hennigian phylogenetics with quantitative methods that were becoming practicable through the availability of digital computers. Very quickly, the concept of parsimony as the overriding criterion in constructing phylogenetic trees began to predominate, although there was great resistance among the numerical taxonomists (pheneticists) of the day. Most of the advances in the development and application of parsimony to phylogenetics were due to the work of J. S. Farris, and much of the early seminal work, both by Farris and others, was published in the journal *Systematic Zoology*.

Among the first computational approaches for the production of phylogenetic trees using Hennigian concepts of synapomorphy was the Wagner tree method (Wagner 1961) as developed and implemented by Farris (1970). Initially, Wagner trees were computed by hand, which could be tedious with more than a few taxa. Such trees were then utilized as 'final' results—in other words, a Wagner tree was computed and this cladogram alone was the basis for interpreting the phylogeny.

Once computer programs became available for 'quickly' computing Wagner trees (e.g. Farris 1978b[1]), it became apparent that there were two major problems: first, Wagner trees were often suboptimal (i.e. there were shorter trees) when the data were noisy, or had significant levels of homoplasy, and second, there were often multiple equally parsimonious solutions. This was evident when the taxon order in an analysis was changed, which often resulted in the discovery of different trees of different lengths, often with more than one of the shortest length found. In the early days when the computer programs were mostly executed by stacks of cards, the easiest way to generate these extra trees was to 'shuffle' the taxon deck (each taxon, along with its character scores, was on a single card in the Fortran deck) and resubmit the job.

The first real breakthroughs in calculating parsimony trees came in the program PHYSYS

(Mickevich and Farris 1981). PHYSYS was only available for mainframe computers (PCs had not yet become widely available) and it was installed only at a few universities in North America. In addition to numerous other numerical techniques, PHYSYS included routines that performed branch swapping, including branch breaking (BB; later called tree bisection and reconnection (TBR) by Swofford 1990; branch-breaking actually had been implemented by Farris in 1970, in a little-known and undistributed program called Clad/OS (J. S. Farris, personal communication). It quickly became obvious that these 'heuristic' branch-swapping methods were more effective at finding shorter trees than those methods that merely shuffled the taxon order in a Wagner tree analysis. With this recognition also came the discouraging realization that such methods were not effective enough to guarantee the discovery of shortest trees for data sets of any realistic size, because of the massive numbers of trees that would need to be examined (see Felsenstein 1978b).

In the mid-1980s PCs became widely available, and MS-DOS machines with 64–640KB of RAM, running at 5–8 MHz, were the platforms for the cladistic parsimony program Hennig86 (Farris 1988; the 86 refers to the Intel 8086 chip family, not to the date of release, a common misconception). Many of the parsimony features of PHYSYS were implemented in Hennig86. The most important features included a 'branch and bound' command (*ie*, which stood for implicit enumeration) that could guarantee shortest trees on data sets with as many as 20 or more terminals, and the *mh** and *bb** commands that performed branch breaking (again, BB or TBR) on input trees (the starting trees usually were calculated with the Wagner algorithm). Although not obvious to users, the *mh** command did a series of quick analyses using different taxon-addition sequences, followed by branch breaking while holding few trees. The user would typically take the results of an *mh** analysis and perform branch breaking on the shortest of the trees from the initial sets while holding as many trees as would fit in available memory, with the command *bb** (which students sometimes confused with branch and bound). The *mh** + *bb** sequence in Hennig86 thus was the first common implementation of

---

[1] Note: we have attempted to provide accurate citations for the release dates of the computer programs cited in this chapter. However, we have encountered conflicting records concerning the release dates of certain versions of these programs, and some of the dates given may not be accurate.

a two-stage analysis (see below), with a series of 'quick' runs initially conducted, using multiple starting trees, followed by more exhaustive swapping on the best trees found during the first stage.

The first release of PAUP (Swofford 1984) initially ran on mainframes, then on MS-DOS PCs (Swofford 1985), and later the program was ported to the principal platform on which it is now used, the Macintosh (Swofford 1990). Although available for desktop computers before Hennig86 was released, PAUP had limitations on tree output, search options, and overall speed. Before the release of version 3.0 in 1990, PAUP did not implement branch breaking, and indeed the same method as branch breaking was named as TBR in that release (Swofford 1990). Because of the ease of use on the Macintosh platform, PAUP became and appears to remain (as PAUP*) the most widely used cladistic program. Unfortunately, the default settings and culture that developed around PAUP resulted in many analyses being conducted as single one-stage analyses (see below) with the maximum number of trees held for branch swapping being 100 (or in many cases the maximum that would fit in the available RAM).

The use of these methods, in association with what became known by many PAUP users as "swapping to completion" (conducting a complete round of branch swapping on an entire set of trees held in memory), came to be regarded as a thorough analysis by many investigators. Even these analyses often were not completed on moderately sized data sets, and on large data sets, such as the 500-terminal *rbcL* matrix (discussed further below), analyses had to be stopped prematurely after running for a period of time, often several months (e.g. Chase *et al.* 1993; Rice *et al.* 1997).

Of course, adherence to the mantra of swapping to completion does not in any way guarantee the discovery of shortest trees, due to problems of "tree islands" (D. Maddison 1991), so the benefits of this approach are at best illusory. Because of the widespread adoption of these methods, many cladograms published with PAUP over the years, whether or not the trees were swapped to completion, are not actually among the most-parsimonious trees that could have been found easily with concurrent versions of Hennig86 or Nona (e.g. Chase

et al. 1993; Rice *et al.* 1997), or represent only a subset of the complete set of most-parsimonious trees, resulting in spurious resolution in the reported consensus (e.g. Donoghue and Doyle 1989).

The program Nona (Goloboff 1993b) was developed by the Argentine arachnologist Pablo Goloboff, while a graduate student at Cornell University, as a companion to his program *Pee-Wee* (Goloboff 1993c), which was designed to conduct implied-weighting tree searches. *Pee-Wee* was available as a beta version from 1991, and Nona from 1993. Nona has many similarities to Hennig86, but allows more precise control over search strategies, and by using the defaults it is very easy to implement customized two-stage searches as described above (e.g. *mult* + max**). This has resulted in more experimentation with different search strategies, including those described in this chapter.

The belief that data sets with 100 or more taxa are virtually intractable to analyze was common through the 1990s and persists even today, particularly among users of PAUP and PAUP*. This belief was voiced strongly by Rice *et al.* (1997), who expressed the goal of exploring "methodological and theoretical issues raised by very large data sets" (p. 554), yet conducted a one-stage reanalysis of the 500-terminal *rbcL* seed plant matrix of Chase *et al.* (1993).

Much of the attitude about intractability of large data sets rests on the misguided belief that an analysis must swap to completion in order to be valid. However, as illustrated in the present chapter, the methods and thus conclusions of the Rice *et al.* paper were flawed. By performing eight one-stage analyses while holding large numbers of trees in RAM, Rice *et al.* merely repeated the ineffective analytical strategies that were employed as the default settings of PAUP (and PAUP*). The only difference was the total amount of time devoted to the analyses, which consumed 11.6 months of computer time on three separate Sun workstations (Rice *et al.* 1997). Other supposed new search strategies proposed by Rice *et al.* were either not implemented or not shown to be effective, and thus the overall result of the paper was to reinforce the common myth that data sets of this magnitude were intractable. The current chapter, using the same data set as analyzed by Rice *et al.*, and using software available at the time, and computers of comparable speed, shows how ineffective the Rice *et al.* approach actually was at the time it was published.

7.3 Newer methods for parsimony analysis

The first breakthrough in analyzing what might now be considered large (>150 terminal) data sets came with the introduction of the parsimony jackknife by Farris *et al.* (1996), which remains the fastest method by which to undertake a parsimony analysis. Using the parsimony jackknife, Källersjö *et al.* (1998) analyzed a data set of 2538 terminals, and their results include, as far as we are aware, the largest cladogram ever published. Rice *et al.* (1997, p. 559), referring to the parsimony jackknife and to analogous approaches for rapid maximum likelihood analysis, incorrectly characterized the use of such methods as "abandoning the notion that maximum parsimony or maximum likelihood is the criterion that we are optimizing."

More recently, tree-search strategies and new algorithms that are more effective than simple two-stage methods have been developed, utilizing randomization of character weights during successive searches (the parsimony ratchet; Nixon 1999), temporary changes in optimality criteria (tree drifting; Goloboff 1999), as well as tree fusion (Goloboff 1999) and swapping on only a portion of a tree (sectorial searches; Goloboff 1999). Detailed descriptions of these methods are beyond the scope of this chapter, and the reader is referred to the original publications for more information on the algorithms and how they have been implemented.

The parsimony ratchet, originally available only in WinClada (running Nona as a daughter process), has now been implemented as the program PAUPRat (Sikes and Lewis 2001), which produces batch files that can be analyzed with PAUP*. The parsimony ratchet, tree fusion, tree drifting, and sectorial searches are available in the latest version of the computer program TNT (Goloboff *et al.* 2004). Besides raw speed (the TBR swapper is estimated to be at least 10 times faster than that of the latest version of Nona), TNT allows the user to combine methods (conventional, ratchet, tree drifting, and sectorial searches) to produce optimal or near-optimal trees that can then be subjected to tree fusion (driven searches). Since the success of tree fusion is dependent on the input trees collectively having all of the clades found in the shortest trees, the selection of methods to produce these input trees is very important.

With TNT it is possible to analyze large data sets very quickly. For example, an analysis of a matrix of 1553 small-subunit ribosomal DNA sequences, with a combination of the ratchet (using Nona as the search engine, because the ratchet was not yet available in TNT) and tree fusion (using TNT), took approximately 2 months of computer time on a 1.3 GHz Xeon processor, and resulted in a minimum of five independent discoveries of presumed shortest trees (Tehler *et al.* 2003). Attempts to use conventional searches, tree drifting, and sectorial searches to produce input trees for tree fusion did not yield trees as short as those obtained when the ratchet was used to produce the input trees, but it is not known if this would be the case with other matrices of comparable size.

The availability of new programs that implement advanced tree-search strategies (e.g. TNT) does not eliminate the need to fully understand the idiosyncrasies and factors influencing the efficiency of conventional tree-search strategies. With the exception of tree fusion, these new strategies (the parsimony ratchet, tree drift, and particularly sectorial searches) use standard branch-swapping techniques, applied in an iterative fashion. Although tree fusion does not utilize branch swapping *per se*, the trees which form the population to fuse must be found by conventional searches or more often with the ratchet or sectorial searches (e.g. Tehler *et al.* 2003). Thus, the detailed information provided here on conventional searches, in combination with further study of the nature of tree islands, and further exploration of the effectiveness of different combinations of tree-search methods, has utility in a general theoretical sense, and may facilitate the development of new tree search strategies.

7.4 Challenges of large data sets

With the rise of molecular systematics, which involves the scoring of thousands of characters for a taxon in the course of a single set of operations (i.e. the sequencing of a genomic region), it has become possible over the past two decades to assemble numerous characters for a single taxon in a short order of time, and thus to assemble cladistic data matrices with increasingly large numbers of taxa. As noted above, a milestone in this progression was reached more than 10 years ago, when Chase *et al.* (1993) generated and analyzed the *zilla* matrix, which consists of 500 *rbcL* sequences, though they did not discover most-parsimonious trees for the matrix. This point was demonstrated by the discovery of shorter trees by Rice *et al.* (1997), whose re-analysis of the data set consumed a total of nearly a year of computer time on three Sun workstations (i.e. an average of about one-third of a year on each of the three computers), and yielded trees five steps shorter than those that had been described by Chase *et al.* However, it soon became evident that Rice *et al.* also had not discovered the shortest trees for this matrix.

Nixon (1999) and Goloboff (1999) analyzed the same matrix, the former author using the parsimony ratchet and the latter using tree fusion, tree drifting, and sectorial searches (alone and in combination), and both authors found trees two steps shorter than the shortest trees that had been discovered by Rice *et al.* Trees of this length have since been discovered numerous times, in re-analyses by the authors of the present chapter, using the methods of Nixon and Goloboff. Although one cannot be certain that these are most-parsimonious trees for this matrix, it seems likely that they are, since trees of this length are discovered rapidly with these methods, in the course of many hundreds of searches. Even if shorter trees still remain to be discovered, trees of the length discovered by Nixon and Goloboff still represent a benchmark against which other analytical methods can be compared.

Here we present results of a set of conventional analyses (i.e. using only standard branch-swapping techniques) of the 500-terminal data set and, as detailed below, we have discovered trees of the same length as those previously found by Nixon and Goloboff using other techniques. These trees, with uninformative characters removed from the matrix, are 16 218 steps in length, and sets of trees of this length have been discovered more

than 200 times during the course of the present study. During the same period, approximately 10 times as many sets of trees of length less than or equal to 16 220 (the shortest length discovered by Rice *et al.* 1997) were discovered.

The analyses that constitute this overall study were conducted on several different personal computers of various processor speeds, over the course of a period of about 6 years. With a PC that is fast by current standards (3 GHz), and using Nona with the optimum software settings, as determined by the current analysis, one set of trees of length 16 218 can be discovered in about 1 day (details below). This does not imply that a single day's analysis is sufficient to declare the analysis of this or any other matrix of this magnitude complete, because there may be multiple "islands" (D. Maddison 1991) of equally parsimonious trees that cannot be discovered by conventional branch swapping from one tree of this length. Thus, the true consensus tree for a matrix may be less resolved than the one that is discovered following the initial discovery of a set of most-parsimonious trees.

The estimate of 1 day of data analysis also can be misleading because it is based on the use of optimum settings for this matrix, and does not include the preliminary analyses that are required to determine these settings. However, when this matter is taken into consideration it is reasonable to state that a thorough conventional analysis of the 500-terminal matrix, including the preliminary analyses, and resulting in a consensus tree that includes no erroneously resolved nodes, can be completed over the course of a few weeks. Thus, if it is assumed that the shortest trees currently known for the 500-terminal *rbcL* matrix actually are the shortest possible trees for this matrix, it is possible at present to analyze this matrix in a reasonable amount of time on one personal computer using conventional search methods. Desktop computers with processors approximately one-tenth the speed of those currently available have been available since 1997, as was the software that was used in the present study (Nona; Goloboff 1993b and 1993), so it was possible at that time that Rice *et al.* were conducting their analysis of this matrix to discover at least one set of shortest trees for the 500-terminal matrix over the course of a few weeks, and possibly, with some luck, as early as 1993, when the matrix first was assembled.

Although the 500-terminal *rbcL* matrix still is a relatively large data set, matrices with more than 1000 terminals have been generated and analyzed in recent years (e.g. Källersjö *et al.* 1998; Tehler *et al.* 2003), and the present availability of nucleotide sequences of growing numbers of genes from hundreds or even thousands of accessions suggests that additional matrices of this magnitude and greater will be generated during the next few years.

One might ask, then, whether we have reached the limits of conventional cladistic analysis. Faster processors become available in personal computers at regular intervals, but a doubling in rates occurs, at best, over a period of 1 year to a few years, and the required processing power for cladistic matrices is growing much faster than that. Although substantially more efficient approaches to cladistic analysis are now available (i.e. the methods of Nixon and Goloboff, as discussed above), those methods resemble conventional analytical techniques in employing heuristic tools such as branch swapping. Thus, empirical study of the limits of conventional analysis, as described in the present chapter, should provide insights that are applicable to the development of analytical strategies that may help to maximize the effectiveness of these and other methods that may be developed. Also, by establishing benchmarks and limits, the current analysis will help to establish a basis for comparisons among current and future methods.

We refer to the general methods of heuristic search techniques that have been used for cladistic analysis over the past several years, as described above, as conventional search methods. Similar methods have been used for maximum likelihood searches, and those too can be called conventional search methods. These searches vary in certain details, but they follow a basic multistep pattern, and it is useful to review this pattern. Typically, an initial or 'starting' tree is generated (usually a Wagner tree), and this tree is then subjected to a methodical regimen of branch rearrangements (i.e. branch swapping).

When swapping results in the discovery of a tree that is more optimal than the tree that is being subjected to swapping, the tree that is being swapped is abandoned, as are all other trees of equal length that are held in memory, and swapping is then initiated on the new tree, which can be regarded as a new parent tree. In this manner, conventional analyses proceed from a starting tree through sets of successively shorter trees, often spending considerable periods of time at one length or another, accumulating and swapping through a large number of trees, before shorter trees are discovered. Eventually, every search of this sort results in the discovery of a set of one or more trees of some length (which may or may not be most-parsimonious for the matrix), and the search culminates in a complete round of branch swapping through this set of trees.

With matrices for which the discovery of most-parsimonious trees is difficult, the starting tree for each search often is far from optimal, and there is an initial period of swapping during which there is a relatively rapid approach towards optimality, with a substantial portion of the trees that are subjected to swapping being discarded before being swapped completely. Later in the process the approach to optimality slows, with large numbers of trees accumulated and subjected to swapping, while little or no progress is made towards the discovery of more optimal trees.

In light of this pattern, a tradeoff in potential search strategies is apparent. If the investigator chooses to retain relatively few trees in memory, each search ends relatively quickly (i.e. stalls at some length after failing to find shorter trees within the imposed limits), but a large number of searches can be conducted in a given period of time. With few trees retained, the thoroughness of each search is minimal, and each search is relatively unlikely to result in the discovery of most-parsimonious trees.

Alternatively, if greater numbers of trees are retained in memory, each tree search is more thorough, and thus more likely to result in the discovery of most-parsimonious or nearly most-parsimonious trees, but each search also consumes a great deal of time, and few searches can be conducted. The latter situation is illustrated by the analysis of the 500-terminal *rbcL* matrix that was conducted by Rice *et al.* (1997), who allowed large numbers of trees to be held in memory, but conducted only eight individual searches during the course of nearly a year of computer time, with none of those searches yielding most-parsimonious trees for the data matrix. All eight of the searches were aborted prior to completion (i.e. an *ad hoc* limit on the number of trees to be swapped in each search was interposed during the course of the analysis) because more trees had been accumulated than could be subjected to swapping in a reasonable amount of time.

In light of the tradeoff between rapidity and thoroughness of individual searches, one would need to evaluate various combinations of settings between two extremes to determine the optimal point(s) of balance between the conflicting goals of conducting numerous searches and of having these searches be thorough. One of the goals of the present analysis was to examine this tradeoff, and we have determined that there can be multiple peaks of search efficiency for a matrix, and that one of the peaks of search efficiency can occur, even with a homoplasious matrix of 500 terminals, with as few as 50 trees held in memory during each search.

Apart from the number of trees held in memory and subjected to branch swapping during individual searches, several additional factors can affect the overall efficiency of conventional tree searches. We refer to the general approach described in the previous paragraphs (one or more independent search initiations, each followed by branch swapping from a starting tree, and terminating after a set of trees of some length has been accumulated and subjected to swapping) as a one-stage search. In fact, many practitioners now conduct two-stage searches (as with the *mh*∗ and *bb*∗ commands of Hennig86), with the first stage corresponding to a one-stage search in which a relatively large number of searches are conducted, each with a relatively small number of trees held in memory. Following the completion of this stage, the second stage proceeds as the most optimal sets of trees obtained during the first stage are subjected to additional swapping with greater numbers of trees held in memory. The two-stage approach has the advantage of confining the most intensive and time-consuming branch swapping (i.e. with numerous trees held in memory and subjected to swapping) to those sets of trees that are relatively optimal to begin with and, as detailed below, there are two key advantages to this approach.

First, trees that are nearly optimal for the matrix are more likely to yield most-parsimonious trees under intensive branch swapping than are those that are less optimal. Thus, use of the two-stage search strategy focuses the most-intensive swapping on sets of trees that are among the most likely to yield most-parsimonious trees. Second, the two-stage search strategy allows a greater number of sets of trees to be subjected to intensive swapping during a given period of time than does the one-stage search strategy. This is because the trees that are subjected to intensive swapping during the second stage of a search are relatively optimal to begin with, and therefore have a relatively shorter path to follow to their point of completion, whether or not they succeed in discovering most-parsimonious trees.

For example, swapping with 2 000 trees held in memory, when initiated with trees two steps longer than the most-parsimonious, involves branch swapping through a maximum of 6 000 trees (up to 2 000 trees each at the initial length and at one and two steps shorter), and a maximum of only 4 000 trees in searches that fail to yield most-parsimonious trees. However, if intensive branch swapping is initiated with trees that are 10 steps longer than the most-parsimonious, as many as 20 000 trees may be subjected to branch swapping in searches that do not yield most-parsimonious trees.

Hence, a key advantage of two-stage searching lies in its ability to minimize the time that is spent in intensive swapping during each search, which thereby allows a greater number of searches to be conducted in a given period of time. In other words, the limitations inherent in one-stage searches, as are evident in the tradeoff between conducting a small number of intensive searches, and a larger number of individually less-intensive searches, are ameliorated by the very structure of two-stage searches. As described below, with reference to the 500-terminal matrix, almost any two-stage search that is conducted with reasonable software settings is superior to even the most efficient one-stage search.

If two-stage searches are more efficient than one-stage searches, as a general rule (i.e. with appropriate software settings in each case), it still remains to determine how many trees should be held in memory during each stage, and what percentage of tree sets obtained during the first stage should be subjected to additional swapping during the second stage. In addition to these points, there are other significant factors. There are different branch swapping procedures, ranging from relatively cursory (e.g. nearest-neighbor interchange) to much more thorough (e.g. BB, or TBR). Also, trees with polytomies may or may not be recognized as acceptable trees for evaluation during a branch swapping procedure, and when polytomies are allowed, there are different rules that can be applied to determine whether the various possible dichotomous resolutions of a polytomy are recognized as being supported by a given matrix. All searches in the present analysis involved BB/TBR swapping, but we examined various combinations of the numbers of trees held, and of the search settings relating to polytomies.

If the optimum search conditions for one data matrix differ from those for another, as seems reasonable to assume, it is important to develop methods for estimating the appropriate settings to be used with any particular data set. This goal may be difficult to attain for large matrices, because the analysis of any data set that is recognized as a large one is, almost by definition, computationally demanding. The problem, then, is that if the most efficient search conditions for a given data set can be determined only by conducting a series of preliminary searches under a variety of conditions, and if most-parsimonious trees are discovered only infrequently under even the best of conditions, and furthermore, if the results of preliminary searches are evaluated in terms of success in the discovery of most-parsimonious trees, the computational demands of the required preliminary searches will be nearly as burdensome as a thorough analysis itself. However, a preliminary phase of searching, preceding an actual search, would be feasible if the effectiveness of the various settings that are explored during the preliminary phase could be evaluated in terms of a result that is correlated with efficiency in the discovery of most-parsimonious trees, and that could be determined on the basis of a relatively small number of preliminary searches.

We consider various correlates of search efficiency with the 500-terminal matrix, and demonstrate that one useful indicator is the ratio between the number of tree rearrangements required during a particular phase of a search and the minimum that would be required by an unsuccessful search, i.e. an excess branch-swapping ratio (see below).

We have conducted preliminary searches and applied these calculations to a second and larger data set, the three-gene matrix of 567 terminals of Soltis *et al.* (2000). As indicated by those authors, the shortest trees discovered by conventional searches with PAUP*, conducted over a period of several weeks, were six steps longer than those that were discovered in the course of a few hours with the parsimony ratchet, and the ratchet eventually yielded trees that were one further step shorter. Since that time, numerous additional ratchet searches have been conducted with the three-gene matrix and, as with the 500-terminal matrix, we are confident but not certain that the shortest trees obtained with the ratchet, and reported by Soltis *et al.*, are most-parsimonious trees for this matrix.

However, as noted above with reference to the 500-terminal matrix, even if shortest trees for the three-gene matrix have not yet been discovered, the tree length obtained with the ratchet still provides a basis for comparison among search strategies. It might be argued that an optimal method for the discovery of trees slightly longer than most-parsimonious is not the best method for obtaining most-parsimonious trees themselves, but if this were the case we would expect any differences in methods for the efficient discovery of most-parsimonious and nearly most-parsimonious trees to be relatively minor.

We conducted preliminary analyses with the three-gene matrix, using a range of settings, calculated excess branch-swapping ratios, used these ratios to select software settings for more thorough searches of this matrix, and then conducted additional searches using those settings.

7.5 Methodological matters

7.5.1 The *zilla* data set

Analyses were conducted using the *rbcL* data set that was assembled and first analyzed by Chase *et al.* (1993). Chase *et al.* analyzed two different versions of the same general data set, using different analytical procedures. In their search II they employed a data matrix comprising 500 terminals, including two sequences from *Canella*, so the matrix can be described as comprising 499 taxa or 500 terminals. Their search II was conducted with all transformations between nucleotides weighted equally, and therefore it represents the earliest analysis of a matrix of this magnitude using analytical techniques that are generally accepted today. Rice *et al.* obtained a copy of the search II matrix and nicknamed it *Treezilla* (attributing the name to A. Yoder), but this name often is replaced by the shorter name *zilla*, which we use for the balance of this chapter. Rice *et al.* (1997) re-analyzed the *zilla* matrix, using equal character weights, and the present analysis is also based on that version of the data.

The data matrix used here was downloaded in January 1998, as posted by Rice *et al.* (at www. herbaria.harvard.edu/~rice/treezilla/), converted from NEXUS format (Maddison *et al.* 1997) to Nona format, and stripped of cladistically uninformative characters, leaving 759 informative characters. The aforementioned Harvard website is unavailable at the time of writing, but the NEXUS-format version of the *zilla* matrix is available currently at a different website (www.cis.upenn.edu/~krice/treezilla/). In addition to the *zilla* matrix, the downloaded file includes a tree of length 16 533, which is the shortest tree length reported by Rice *et al.* (1997). We undertook extensive evaluation of this data set to establish that it was identical to the published 500-taxon data set of Chase *et al.* and Rice *et al.* Details of these tests are available upon request from the senior author of the present chapter (J. I. D.).

7.5.2 Comparability of reported results with *zilla* across software platforms

The present analyses of *zilla*, and prior analyses by other authors, have been conducted with a variety of programs, and in order to compare results of the various studies it is important to determine the comparability of the results reported by the various packages. Several different versions of PAUP, MacClade, Nona, and WinClada have been used in these various analyses, and two principal items of interest reported by these programs are tree lengths and numbers of trees examined during

branch swapping. Reported tree lengths vary, in part, because uninformative characters have been included in some calculations and excluded from others. Another source of variation lies in the use of different criteria of character informativeness by the various programs. This problem was noted by Rice *et al.*, who used PAUP and MacClade, and who observed that the programs reported different tree lengths when commands were invoked to exclude uninformative characters from consideration.

The following software packages have been employed in this and previous studies of the *zilla* data set: Chase *et al.* (1993) used PAUP version 3.0s (Swofford 1990) for their search II analysis; Rice *et al.* (1997) used PAUP versions 3.0s and 3.1.1 (Swofford 1993), plus MacClade version 3.04 (Maddison and Maddison 1992); and we used Nona version 1.5.1 (compiled September 4, 1996; Goloboff 1993b), the multi-thread version of Nona version 1.6 (Goloboff 1993, i.e. Paranona, compiled February 26, 1998), WinClada version 1.00.08 (Nixon 2002), PAUP version 3.1.1, PAUP* version 4.0b10 (Swofford 2002), and MacClade version 4.03 (Maddison and Maddison 2001).

Because of different reported lengths for the trees found by Chase *et al.* and Rice *et al.*, due to errors in excluding uninformative characters in older versions of PAUP, we use here the tree lengths as determined by MacClade version 3.04 (Maddison and Maddison 1992), Nona, and WinClada, which are all consistent (details of the comparisons that were made are available from J.I.D.). These programs determine the shortest trees found by Rice *et al.* to be of length 16 220 with uninformative characters excluded (80 steps fewer than the number reported by older versions of PAUP, and 313 steps fewer than the number obtained with uninformative characters included). Rice *et al.* included one of these trees in their posted matrix, and we confirm, using PAUP* version 4.0b10, that this tree is of length 16 533 with the data matrix in that file, when all characters are included. This version of PAUP* also detects 759 cladistically informative characters in the matrix, which is the same number detected by Nona and WinClada in the matrices that we derived from this file (see above). Also, this version of PAUP* reports a length of 16 220 for the Rice *et al.* shortest tree when uninformative characters are excluded from consideration. Finally, we have determined that PAUP* version 4.0b10 and MacClade version 4.03 both agree with Nona and WinClada in determining that the shortest trees discovered by the present analysis are of length 16 218 (uninformative characters excluded) or 16 531 (uninformative characters included). In both cases these trees are recognized as two steps shorter than those discovered by Rice *et al.*, and seven steps shorter than those discovered by Chase *et al.*

The preceding comparisons substantiate an overall consistency in data structure among the various *zilla* files that exist in NEXUS and Nona formats, and among Nona, WinClada, PAUP* version. 4.0b10, and MacClade in the determination of character informativeness and in the calculation of tree lengths with and without uninformative characters included. In contrast, it appears that older versions of PAUP employ a definition of character informativeness that differs from the one discussed by Farris (1989a), and employed by all current versions of the programs mentioned above. On the basis of these results we recommend that investigators using older versions of PAUP, or consulting publications that used those versions, verify all results that are based on distinctions between informative and uninformative characters, including items such as consistency indices, which are inflated when uninformative characters are interpreted as informative.

7.5.3 Three-gene matrix

In addition to the *zilla* data set, we conducted analyses with another large data matrix, the three-gene data set of Soltis *et al.* (2000). This data set comprises 567 taxa scored for *rbcL, atp*B, and 18 S rDNA. As noted by Soltis *et al.*, they found shortest trees for this matrix only with the parsimony ratchet (Nixon 1999), as implemented in WinClada, using Nona as a daughter process for tree searches (the Nona-format version of the matrix employed in those searches is available at www.cladistics. com/). Analyses of this data set for the present study were conducted using copies of the matrix downloaded from that website. This matrix includes 567 terminals scored for 2 153 informative characters, and most-parsimonious trees are of

length 44 163, or, as reported by Soltis *et al.*, 45 100 steps when uninformative characters are included.

7.5.4 One-stage tree searches

Except where specified otherwise, cladistic analyses were conducted with the multi-thread version of Nona version 1.6 cited above. All analyses were conducted as one-thread tree searches (using the default setting *thread 1*). Searches were performed with the *mult** command, which conducts a set of replicate tree searches (e.g. 10 replicates with *mult*10*), and most searches were conducted in sets of 10 or 20, with trees saved after each set of searches. The *mult** command initiates each replicate search by generating a Wagner tree, assembled with a taxon entry sequence determined by a seed number that is input with the command *rseed*, then conducts a round of subtree pruning-regrafting (SPR) swapping on the Wagner tree, with one tree held in memory, followed by BB/TBR branch swapping with a user-determined number of trees held. We used the command *rseed 0*, which uses the computer's clock to generate a random seed number, which is saved with the results of each search, thereby facilitating repeated searches using identical taxon entry sequences.

In each replicate search, a predetermined number of trees (x, as set with the *hold/x* command) is retained in memory and branch swapping is conducted successively on each of these trees. If a shorter tree is discovered, all trees except the new one are discarded, swapping is initiated on the new tree, and a new set of daughter trees is accumulated. For reasons discussed below, we also note that if a shorter tree is not discovered, the search terminates after all x trees in memory have been subjected to swapping, and that the final phase of every replicate search therefore consists of a complete round of swapping through x trees of the shortest length discovered in that search, unless this set of trees constitutes an island (D. Maddison 1991) of fewer than x trees. In the latter case, the final phase of the search would consist of swapping through all trees in the island. However, every search of the *zilla* matrix conducted during the course of the present analysis did lead to the accumulation of the maximum number of trees set by the *hold/* command, so each of these searches did end with a complete round of swapping through that number of trees.

The output generated by each set of replicate searches includes a record of the seed number that was used to initiate each search, the minimum tree length obtained by the search, and the number of trees of that length that were accumulated. Also, use of the *tcount* command causes the total number of tree rearrangements examined in each set of replicate searches to be reported (e.g. for *mult**10, the total for the 10 constituent replicates).

To determine the effectiveness of tree searches under a variety of conditions, analyses were conducted using a variety of settings for other commands (see below). The output from all searches conducted under each set of conditions was combined into a common file, and duplicate searches (i.e. those based on the same randomly drawn seed number, which invariably yielded identical results) were removed from consideration.

One setting that was varied among sets of one-stage searches was the number of trees retained and subjected to branch swapping, as specified with the *hold/* command. Analyses were conducted with the following numbers of trees retained in each search: 1, 2, 5, 10, 20, 50, 100, 200, 500, 1 000, 2 000, and 5 000. Two other settings that were varied among analyses are those that determine whether polytomies are allowed in trees (using the *poly* command) and, if so, the conditions for recognition of potential resolutions as being supported by the data (using the *ambiguous* or *amb* command).

A complete set of analyses (i.e. using each of the *hold/* settings listed above) was conducted with the default settings in Nona for these commands, *poly =* (polytomies allowed), and *ambiguous-* (when polytomies are allowed, branch collapsed to form a polytomy if its length is zero under at least one possible character optimization, i.e. a branch is resolved only if it has unambiguous support). Additional analyses also were conducted with the alternative settings for these commands, either *poly-* (polytomies disallowed), or *ambiguous =* (when polytomies are allowed, branch collapsed only if its length is zero under all possible optimizations, i.e. a branch is resolved if it has ambiguous or unambiguous support). When the *poly-* command

is invoked, the setting for the *ambiguous* command is irrelevant, because polytomies are disallowed in any case; therefore, apart from the default settings of these commands, preliminary analyses were conducted using only two of the three remaining combinations of the settings determined by these commands (*poly=*, *ambiguous=*, and *poly- ambiguous-*). On the basis of results from the preliminary analyses, intensive analyses were conducted with these polytomy and ambiguity settings with 50, 100, and 2 000 trees held in memory (i.e. using the commands *ho/50, ho/100,* and *ho/2 000*).

In addition to tree length, Nona and PAUP* both report the number of tree rearrangements conducted during the course of an analysis. In order to compare these programs, including one of the two versions of PAUP used by Rice *et al.* (1997), with respect to the numbers of tree rearrangements that they report when conducting comparable actions, we selected one putative most-parsimonious tree (i.e. length 16 218) from each of 10 sets that were generated during the course of this study. Each of the 10 trees was subjected to branch swapping, with 100 trees retained (i.e. 99 new trees propagated from the first, and all 100 from each set subjected to branch swapping), using Nona version 1.6, PAUP version 3.1.1, and PAUP* version 4.0b10.

Because this swapping never yielded shorter trees, this procedure resulted in a complete round of branch swapping through exactly 1 000 trees with each program, and the average number of rearrangements required to swap through one tree (signified as r_{avg}^1) was calculated from the results of these analyses. With Nona, the default polytomy and ambiguity settings were used (*poly=* and *amb-*), and BB (i.e. TBR) branch swapping was conducted, using the command *max**. With PAUP, TBR swapping was conducted with only minimal-length trees retained, and 'zero-length branches collapsed,' and with PAUP*, TBR swapping was conducted with branches collapsed if 'maximum length is zero.' It should be noted that most-parsimonious trees are not actually required for this procedure to be conducted, because a starting tree of any length can be used, with any number of trees retained and subjected to branch swapping, under any particular swapping regime, if it has been determined by a prior round of swapping (or after the fact, from the results of such a procedure) that this particular tree does not yield shorter trees when swapped under the specified conditions.

Results from the various one-stage analyses were compared in terms of two principal metrics, the frequency of success and the search efficiency, both of which are expressed with reference to a particular tree length. *Frequency of success* is the percentage of searches under a given set of conditions that result in the discovery of trees as short as the specified length. In most cases we specify success with reference to most-parsimonious trees, but when applied to a tree length greater than that, such as two steps longer than most-parsimonious, we apply this measure in a comprehensive manner, so that it includes all searches that yielded trees of the specified length or shorter.

Tree-search efficiency describes the results of a search in terms of the number of tree rearrangements required to discover a set of trees of a specified length. Like frequency of success, it can be specified with respect to most-parsimonious trees or longer trees, and in the latter case it includes searches that result in the discovery of trees of length equal to or shorter than the specified length. In order for greater efficiency to be specified by higher numbers, search efficiency is defined as the ratio of number of searches that result in trees of a given length or shorter to the number of rearrangements required to discover those sets of trees. Thus, if an average of 1 billion branch swaps is required to conduct each replicate search under a particular set of conditions, and if an average of one of every four such searches yields trees of a specified length or shorter, the frequency of success for this tree length is 25%, and the search efficiency for this length is one set of trees for every 4 billion tree rearrangements.

In this chapter we sometimes discuss efficiency informally, in terms of billions of rearrangements required per successful search, and it should be noted that with this parlance a higher number refers to a lower efficiency. It should be noted as well that frequency of success and tree-search efficiency are expressed in terms of the number of searches that obtain trees of a given length, not in terms of the number of trees that are obtained. Thus, if two searches are conducted, one with 50 trees retained, and the other with 500 trees retained, each search

that yields one set of trees of a specified length is counted as a single successful search, even though each of the latter searches accumulates 10 times as many trees as each of the former.

Search efficiency, as expressed in terms of tree rearrangements, can be converted to efficiency in terms of computer time elapsed, with the latter expression reflecting processor speed and other factors. Calculation of these indices in terms of the number of rearrangements facilitates the pooling of results generated here from several different computers, and also facilitates comparisons between these results and those obtained with other computer platforms.

7.5.5 Phases of one-stage searches

For the purpose of interpretation of preliminary results it is useful to distinguish two phases of branch swapping that occur during conventional searches. As noted above, conventional one-stage searches (as conducted in Nona, PAUP*, and other programs) begin with the generation of a Wagner tree, after which branch swapping is conducted, starting with this tree. Also noted above is the observation that for any number x of trees held in memory and subjected to swapping in searches of this sort (as determined by the Nona command *hold/x*), searches typically end immediately after a phase of unsuccessful swapping in which a set of x trees of some length is generated and subjected to branch swapping without yielding shorter trees. When several searches are conducted, this is the point at which each search ends, prior to the initiation of the next one. The number of trees accumulated and subjected to swapping during this phase of the search can be less than x if there are fewer than x trees in an island, but with large data sets, and with x set at a number that is likely to be efficient for tree searches, this may occur only rarely, and during the present study the accumulation of fewer than the set maximum number of trees never occurred with the *zilla* matrix, and occurred only once with the three-gene matrix.

The following discussion assumes that every search with *hold/x* involves a final phase of unsuccessful swapping through x trees. Although this is not necessarily the case, appropriate adjustments could be made for other situations.

Consider a search that is conducted with one tree held in memory (*hold/1*). In this case, swapping is initiated on a Wagner tree, and it proceeds as a succession of trees are subjected to swapping, each yielding a shorter tree before being subjected to a complete round of swapping. The search ends after it reaches the first tree in the progression that is subjected to a complete round of swapping without yielding a shorter tree.

Retrospectively, two phases of branch swapping in this search can be recognized. The first (phase 1) is the productive phase of the search, during which shorter trees are discovered, and the second (phase 2) is the nonproductive phase, during which the single shortest tree discovered by the search is swapped completely without yielding any shorter trees. If another search is initiated with the same seed number, and with two trees retained (i.e. *hold/2*), it proceeds through an identical sequence, except that one additional tree is retained in memory, and is subjected to swapping following the completion of swapping on the solitary tree that did not yield a shorter tree during the *hold/1* search.

If phase 2 is again recognized as being initiated when swapping begins on the first tree that is swapped unsuccessfully, and if the second tree, like the first, is swapped completely without yielding a shorter tree, the search ends with exactly one additional tree having been subjected to branch swapping. In this case, the search with *hold/2* is no more successful than the search with *hold/1*, and phase 1 of the two searches is identical, while phase 2 for the second search is precisely twice as long as phase 2 for the first search.

However, if swapping on the second tree yields a shorter tree, the total number of rearrangements during phase 2 of the two-tree search (which now includes some successful swapping) is greater than if it had not yielded a shorter tree. Thus, phase 2 of any search with *hold/1* is always unsuccessful, by definition (being the phase during which the first and only tree in the search is swapped unsuccessfully), but phase 2 of a search with more than one tree held in memory may or may not be successful. The final portion of phase 2 of any search ends when a complete round of swapping is conducted through a set of trees that is equal to the number held in memory, but when more than one tree is

held in memory this portion of phase 2 may be preceded by a period of successful swapping.

Thus, for any search that is conducted with a given seed number, with x trees held in memory, the number of rearrangements required during phase 1 is identical for all numbers x, and the minimum possible number of rearrangements during phase 2 is equal to the number that is required to swap through x trees. Similarly, the average number of rearrangements required during phase 1, for sufficiently large numbers of searches conducted with randomly selected seed numbers, should be identical during phase 1 with any number x trees held in memory, and the degree to which the average number of rearrangements required during phase 2 exceeds the average number required to swap through x trees is a reflection of the overall degree of success of these searches in discovering shorter trees.

To formalize these expressions, let $r_{\mathrm{avg}}^{\mathrm{hold}/x}$ be the average number of rearrangements required to conduct a complete search with any number x trees retained, and let r_{avg}^{x}, which is the average number of rearrangements required to swap through x trees without discovering shorter trees, be the minimum possible number of rearrangements required during phase 2 of any search (i.e. when phase 2 is unsuccessful). Because phase 2 of any search commences when branch swapping is initiated on the first tree that does not yield shorter trees, the average number of rearrangements during phase 1 is identical for any x, and it equals $r_{\mathrm{avg}}^{\mathrm{hold}/1} - r_{\mathrm{avg}}^{1}$. The average number of rearrangements actually conducted during phase 2, for any x, is the average total for a search with x trees held, minus the average number of rearrangements required during phase 1, or $r_{\mathrm{avg}}^{\mathrm{hold}/x} - (r_{\mathrm{avg}}^{\mathrm{hold}/1} - r_{\mathrm{avg}}^{1})$.

We define the *excess branch-swapping ratio* as the ratio of this number (i.e. the average number of re arrangements during phase 2) to the minimum possible for phase 2, or

$$\frac{r_{\mathrm{avg}}^{\mathrm{hold}/x} - (r_{\mathrm{avg}}^{\mathrm{hold}/1} - r_{\mathrm{avg}}^{1})}{r_{\mathrm{avg}}^{x}} \tag{1}$$

This ratio is 1 when $x = 1$, by definition, because phase 2 of such a search is the phase during which unsuccessful swapping is conducted on a single tree, and the number of rearrangements conducted during this phase is precisely the minimum number possible. When $x > 1$, the minimum value for the ratio in a given search is 1, and this ratio occurs when all swapping during phase 2 is unsuccessful, i.e. when a search with $x > 1$ finds trees no shorter than are found when $x = 1$. However, this ratio exceeds 1 to the extent that the additional swapping with $x > 1$ results in the discovery of shorter trees.

With the *zilla* data set (see results, below), we have determined that the second term in the numerator of the excess branch-swapping ratio is trivial in magnitude relative to the first when x is greater than about 20, so (1), above, is approximately equal to $r_{\text{avg}}^{\text{hold}/x}/r_{\text{avg}}^{x}$. However, when x lies between 1 and 20 the second term of the numerator has a substantial effect on the ratio that is obtained. The numbers required to compute this ratio, for any number x, are simply the average number of rearrangements required for searches with x trees held in memory, and the average number of rearrangements to swap unsuccessfully through x trees. Both can be obtained easily from Nona or PAUP*.

7.5.6 Two-stage searches

Two-stage analyses were conducted by subjecting sets of trees obtained from various one-stage searches to additional branch swapping, with greater numbers of trees retained for swapping than in the original searches. Of the available sets of trees from the one-stage searches, the sets subjected to additional swapping were those derived from the most successful one-stage searches, i.e. the shortest trees obtained from those searches, plus sets of successively longer trees, as is common in conventional two-stage analyses, where second-stage swapping is conducted on the best available trees obtained from first-stage swapping. This creates certain complications in evaluating the results. First, trees as short as 16 218 steps (i.e. trees believed to be most-parsimonious for the *zilla* matrix) were obtained from some of the one-stage searches conducted with 50 trees held in memory. These sets of trees, and sets up to four steps longer, were subjected to additional swapping with various larger numbers of trees held in memory, to determine the frequency with which continued swapping on trees of various lengths yielded trees of length 16 218.

Although the trees of length 16 218 were not expected to yield shorter trees (and did not), and although subjecting them to additional swapping diminished the efficiencies that were ultimately computed for the two-stage searches, because this constituted additional but fruitless swapping, they were nonetheless included in the second round of swapping so that this procedure would accurately reflect the additional swapping that normally would be conducted in the analysis of a matrix for which the length of shortest trees is not known. It should also be noted that each episode of second-stage swapping began with a set of 50 trees that had already been subjected to swapping during the first stage, and that all 50 of these trees were swapped again, as part of the larger task of swapping through a larger set of trees. This action also diminishes the efficiency of the two-stage searches, but it too reflects actions that normally are taken in a two-stage conventional search.

The results of these searches were used to calculate the efficiency of two-stage searches by summing the number of rearrangements required for the one-stage search that generated the sets that were subjected to additional swapping, along with the additional searches conducted in the second stage, and dividing the total number of sets of trees obtained of length 16 218 by this sum. In many cases it was not feasible to conduct second-stage swapping on all available sets of trees of a given length obtained from the initial one-stage analysis, and in these cases the sets actually subjected to second-stage swapping were selected randomly from the available sets, and success and efficiency rates were calculated by pro-rating the observed results. In this manner, the efficiencies of two-stage searches were calculated for several combinations of settings (number of trees held during primary and secondary searches, various percentages of trees from the primary searches subjected to secondary swapping, and various combinations of polytomy and ambiguity settings).

7.5.7 Preliminary analyses to estimate optimum search conditions

After most of the one- and two-stage analyses of *zilla* had been conducted, and the most efficient settings had been determined, we computed excess branch-swapping ratios for the various combinations of settings to determine if this ratio is a useful predictor of search efficiency. Having determined that it is, we conducted preliminary analyses with the three-gene matrix, in most cases involving only 10 replicate searches for each number of trees retained, and used these results to determine the settings for more extensive one- and two-stage analyses.

7.6 Comparability of branch swapping with Nona and PAUP

BB (or TBR) swapping through 10 sets of 100 most-parsimonious trees for the *zilla* data set, with the multi-thread version of Nona version 1.6, using one thread, and with the default settings *poly=* and *amb-*, involved an average of ca. 9.75×10^6 rearrangements to swap through each tree. TBR swapping through 10 sets of 100 most-parsimonious trees propagated from the same 10 starting trees required an average of ca. 9.95×10^6 rearrangements per tree with PAUP version 3.1.1, and 9.74×10^6 with PAUP* version 4.0b10. On the basis of these comparisons it appears that Nona, PAUP, and PAUP* count tree rearrangements in approximately the same manner when conducting comparable branch-swapping processes, or at least those processes which constitute the bulk of the branch swapping reported here. Hence, it appears that the numbers reported by Nona for this overall study can be compared directly with those reported by PAUP and PAUP*.

Comparisons of other sorts also can be made with the analysis conducted by Rice *et al.* (1997). Those authors reported that they conducted a total of 27.9 billion tree rearrangements in the course of eight one-stage searches with large numbers of trees held in memory. Their searches apparently were aborted at various stages, but the overall average for the eight searches was ca. 3.5 billion tree rearrangements conducted per search. For the present analysis, an average of ca. 3.8 billion rearrangements was required for the completion of each one-stage search with 200 trees retained in memory, under the default polytomy and ambiguity settings in Nona (Table 7.1). With these set-

tings, and with 200 trees held in memory, about 4% of all searches yielded trees of length 16 220 or shorter. Thus, there is a rough equivalence between our results and those reported by Rice *et al.*, who discovered trees of length 16 220 in at least one of their eight one-stage searches (they reported the discovery of trees of this length, but not the number of times that this occurred among their eight searches).

The present analysis involved over 2.17×10^{14} (i.e. 217 trillion) tree rearrangements during one-stage searches (Table 7.1), or about 7 778 times the number of tree rearrangements reported by Rice *et al.* We also conducted 4.80×10^{13} (i.e. 48 trillion) tree rearrangements in the course of our two-stage analyses and the various preliminary analyses of the *zilla* matrix (details available on request), for a total of more than 2.65×10^{14} (i.e. 265 trillion) tree rearrangements, or about 9 500 times the number reported by Rice *et al.* In addition to the analysis of the *zilla* matrix, we conducted 1.85×10^{13} (i.e. 18.5 trillion) tree rearrangements with the three-gene matrix of Soltis *et al.* (2000), for a total of ca. 284 trillion tree rearrangements, or more than 10 000 times the number reported by Rice *et al.*

7.7 One-stage searches

A series of one-stage searches of the *zilla* matrix conducted in Nona using the default polytomy and ambiguity settings, with various numbers of trees held in memory, indicates that while the number of rearrangements required per search increases with the number of trees held in memory (Table 7.1, Fig. 7.1a), this increase is uneven in rate. On a log-log graph of the relationship between these variables, the increase in tree rearrangements required per search is greatest between 10 and 50 trees held, and between 500 and 2 000 trees held. Similarly, the average length of trees obtained by these searches drops most steeply in the same two regions (Fig. 7.1b), and the greatest excess branch-swapping ratio occurs in both of these regions, with one peak corresponding to 50 trees held in memory, and a secondary peak corresponding to 2 000 trees held (Fig. 7.1c). Preliminary searches conducted with alternative polytomy and

Table 7.1 Results of one-stage searches with various numbers of trees held in memory under various combinations of polytomy and ambiguity-of-support settings. Bold italics identifies the two search strategies in each column that are most successful as measured by average tree length obtained, percentage of successful searches, or search efficiency

Polytomy and ambiguity settings, and number of trees retained	Number of searches	Total number of trees examined in all searches (billions)	Average number of trees examined per search (billions)	Average tree length	Discovery of trees of length 16 218		Discovery of trees of length ≤ 16 220	
					Number (and frequency, in percent) of successful searches	Number of trees examined per successful search (billions)	Number (and frequency, in percent) of successful searches	Number of trees examined per successful search (billions)
poly=, amb-								
1	3 664	216.8	0.059	16 237.881	0 (—)	—	0 (—)	—
2	11 640	838.8	0.072	16 237.380	0 (—)	—	0 (—)	—
5	3 614	414.9	0.115	16 236.707	0 (—)	—	0 (—)	—
10	3 606	673.4	0.187	16 236.101	0 (—)	—	0 (—)	—
20	3 221	1 673.7	0.520	16 233.084	0 (—)	—	5 (0.155)	334.7
50	3 225	5 979.6	1.854	16 227.383	4 (0.124)	*1 494.9*	99 (3.070)	*60.4*
100	3 209	8 670.1	2.702	16 226.249	4 (0.125)	2 167.5	134 (4.176)	*64.7*
200	3 200	12 244.0	3.826	16 226.134	3 (0.094)	4 081.3	124 (3.875)	98.7
500	2 708	19 845.6	7.329	16 225.790	4 (0.148)	4 961.4	130 (4.801)	152.7
1 000	850	14 119.7	16.611	16 224.866	5 (0.588)	2 823.9	83 (9.765)	170.1
2 000	765	31 952.1	41.767	*16 222.868*	20 (*2.614*)	*1 597.6*	189 (*24.706*)	169.1
5 000	765	55 919.6	73.098	*16 222.841*	14 (1.830)	3 994.3	190 (*24.837*)	294.3
poly=, amb=								
50	3 224	5 218.3	1.619	16 228.978	0 (—)	—	35 (1.086)	149.1
100	1 628	4 926.7	3.026	16 226.187	2 (0.123)	2 463.4	55 (3.378)	89.6
2 000	765	29 881.8	39.061	16 223.529	18 (*2.353*)	1 660.1	145 (18.954)	206.1
poly-, amb-								
50	3 226	4 953.5	1.535	16 229.513	1 (0.031)	4 953.5	17 (0.527)	291.4
100	1 629	5 315.6	3.263	16 226.012	2 (0.123)	2 657.8	73 (4.481)	72.8
2 000	388	14 651.9	37.763	16 224.054	6 (1.546)	2 442.0	70 (18.041)	209.3
Totals	51 327	217 496	—	—	83 (n/a[d])	—	1 349 (n/a)	—

[d] n/a, not applicable.

ambiguity settings reveal the same general relationship between number of trees held in memory and average tree length obtained (Fig. 7.1b), as well as between the number of trees held and the greatest excess in rearrangements required per search (Fig. 7.1c).

The average tree length obtained should decrease with every increase in the number of trees held, if all other factors are constant, but this was not always observed to be the case. For example, the average tree length obtained with the settings *poly=* and *amb=* was substantially greater with five trees held than with two trees held (Fig. 7.1b). We

attribute anomalies such as these to our use of random taxon entry sequences, in combination with limited sample sizes. Note that the most complete set of analyses was conducted with the settings *poly=* and *amb-*, and that the only anomaly observed for this set of analyses was in the slightly higher average tree length obtained with 5 000 trees held than with 2 000 trees held (Table 7.1, Fig. 7.1b); in both of these cases only 765 analyses were conducted. Fewer replicate searches were conducted with the alternative settings, in some cases amounting to only a small fraction of the number of searches conducted with *poly=* and

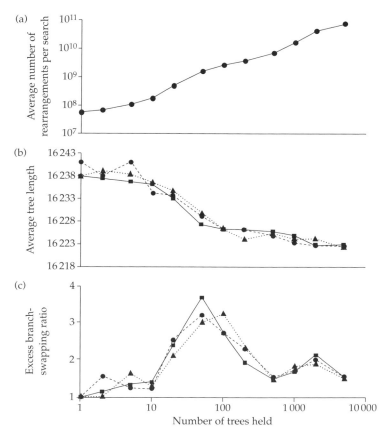

Figure 7.1 Rearrangements required, lengths of shortest trees discovered, and excess branch-swapping ratios (see text) for one-stage searches of the 500-terminal *rbcL* data set (*zilla*), as conducted with Nona, with various numbers of trees held in memory and with various combinations of polytomy and ambiguity-of-support settings. Most of the results depicted represent preliminary sets of 10–20 searches with each combination of settings, but in some cases (particularly analyses with *poly=* and *amb-*) the results of more-extensive searches conducted during the course of the overall study are depicted. (a) Average number of tree rearrangements required per search with *poly=* and *amb-*. (b) Average tree length discovered in each search, with all three combinations of polytomy and ambiguity settings that were examined (■, *poly=* and *amb-*; ●, *poly=* and *amb=*; ▲, *poly-* and *amb-*). (c) Excess branch-swapping ratio (key as in b).

amb-, and it is in these cases that the anomalies are most prevalent.

Although the peak excess branch-swapping ratio with the settings *poly=* and *amb-* was observed with 50 trees held in memory, as it was with the settings *poly=* and *amb=*, the peak with the settings *poly-* and *amb-* occurred with 100 trees held in memory (Fig. 7.1c). This difference initially was observed among preliminary sets of searches, but it continued to be observed after more thorough sampling was conducted with both 50 and 100 trees held in memory for all three

combinations of polytomy and ambiguity settings (Table 7.1). Additional sampling with the two non-default combinations of polytomy and ambiguity settings was also conducted with 2 000 trees held in memory (Table 7.1), and comparable series of one-stage searches therefore exist for all three of these combinations of polytomy and ambiguity settings with 50, 100, and 2 000 trees held during searches. With 50 and 2 000 trees held, the success rates and efficiencies of searches conducted with the settings *poly=* and *amb-* exceeded those with the other two combinations of settings in the discovery of

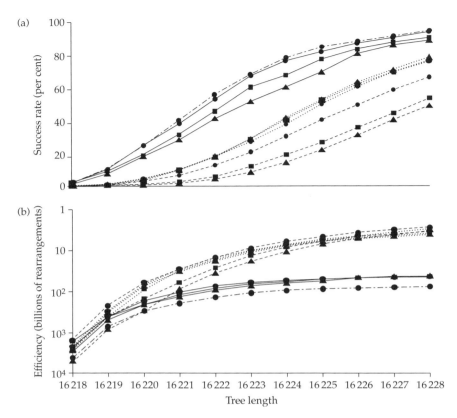

Figure 7.2 Success rates and efficiencies of one-stage searches of the 500-terminal *rbcL* data set (*zilla*), as conducted with Nona with various numbers of trees held in memory and with various combinations of polytomy and ambiguity-of-support settings. Results are depicted for searches with 50, 100, and 2 000 trees held in memory under all three combinations of polytomy and ambiguity settings that were examined, and with 5 000 trees held in memory with the settings *poly=* and *amb-* (see text); results of these searches are also provided in Table 7.1. Dashed lines, 50 trees held in memory; dotted lines, 100 trees held; solid lines, 2 000 trees held; alternately dashed/dotted lines, 5 000 trees held; ●, *poly=* and *amb-*; ■, *poly=* and *amb=*; ▲, *poly-* and *amb-*. (a) Success rates for all tree lengths from 16 218 to 16 228, with success defined as the frequency of discovery of trees of a given length or shorter among the searches conducted. (b) Search efficiencies for all tree lengths from 16 218 to 16 228, with efficiency defined as the number of tree rearrangements required per successful search, with success for each tree length including sets of that length and shorter.

most-parsimonious trees (Table 7.1) and for tree lengths up to six steps longer than most-parsimonious (Fig. 7.2), though in some cases slightly greater efficiencies were observed with other polytomy and ambiguity settings for the discovery of longer trees (e.g. length 16 228 with 2 000 trees held in memory, Fig. 7.2b). With 100 trees held in memory, success rates in the discovery of most-parsimonious trees were nearly identical under all three combinations of polytomy and ambiguity settings (ca. 0.125% of searches yielded trees of this length; Table 7.1), and similar results also were

obtained for sets of trees of greater length (Fig. 7.2). However, the efficiency of searches conducted with 100 trees held in memory, in the discovery of most-parsimonious trees, was greatest with the settings *poly=* and *amb-* (Table 7.1). Thus, for all three numbers of trees held in memory that were examined intensively, the default polytomy and ambiguity settings in Nona yielded the greatest efficiency in the discovery of most-parsimonious trees (though in some cases by narrow margins), and consequently additional analyses were conducted only with these settings.

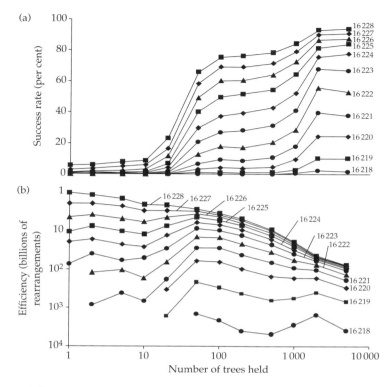

Figure 7.3 Success rates and efficiencies of one-stage searches of the 500-terminal *rbcL* data set (*zilla*), as conducted with Nona with various numbers of trees held in memory, from 1 to 5 000, with the settings *poly=* and *amb-*; results from the same searches are also provided in Table 7.1. (a) Success rates, as in Fig. 7.2. (b) Search efficiencies, as in Fig. 7.2.

Comparisons of success rates and efficiency rates of one-stage searches, in the discovery of most-parsimonious trees and trees up to 10 steps longer, with the settings *poly=* and *amb-*, and for a range of numbers of trees held in memory, are depicted in Fig. 7.3. The greatest efficiency of searches occurs with 50 trees held, and the efficiency of searches with 2 000 trees held is slightly less (Table 7.1; ca. 1.5 vs. 1.6 trillion rearrangements required per set of most-parsimonious trees). Success rates of searches with 2 000 and 5 000 trees held in memory are nearly equal for all tree lengths examined (i.e. there is a plateau between these points; Figs 7.2a and 7.3a), but greater efficiencies consistently are observed with 2 000 trees held than with 5000 (Fig. 7.3).

In light of the plateau in success rates between 2 000 and 5 000, and the limited number of searches conducted with these settings, it is not surprising that anomalies are evident in the slightly greater success rates obtained for length 16 218 with 2 000 trees held in memory than with 5 000 trees held (Table 7.1, Fig. 7.3a). This anomaly is also evident in the relative success rates for some tree lengths greater than the most-parsimonious (e.g. lengths 16 221 and 16 222, Fig. 7.3a). With fewer than 800 searches conducted with these particular settings (though more than 110 trillion rearrangements were conducted in these two sets of searches), we attribute this anomaly to the effects of random sampling of taxon addition sequences.

As is evident from the shapes of the success and efficiency curves in Fig. 7.3, success rates in the discovery of trees longer than the most-parsimonious rise unevenly with an increase in the number of trees held, with the regions of steepest increase between 10 and 100 trees held, and between 500 and 2 000 trees held (Fig. 7.3a), and with nearly flat plateaus observed outside these two regions.

As already noted, there are two peaks in efficiency for searches yielding most-parsimonious trees, corresponding to 50 and 2 000 trees held (Table 7.1, Fig. 7.3b). Two peaks in efficiency are also observed for tree lengths of 16 219 (i.e. one step longer than the most-parsimonious; Fig. 7.3b), with the peak for 50 trees held (ca. 214 billion rearrangements required per set of trees) substantially higher than that for 2 000 trees held (ca. 404 billion rearrangements). For trees of length 16 220, the peak with 2 000 trees held is nearly absent and, for greater tree lengths it is absent. Meanwhile, a second peak in efficiency appears with fewer than 50 trees held in memory (e.g. for length 16 221, there is a peak with five trees held). This peak shifts to lower numbers of trees held (i.e. to two and then to one) for successively longer trees and, eventually, for trees around nine to ten steps longer than the most-parsimonious, it supplants the peak at 50 trees held, with the single peak efficiency in the discovery of trees of length 16 228 corresponding to one tree held during searches.

Thus, the greatest search efficiencies for trees longer than the most-parsimonious occur with relatively few trees held, and for most-parsimonious trees the greatest search efficiency is observed with 2 000 trees held in memory. Beyond this point (i.e. with 5 000 trees held) individual searches are only slightly more likely to yield most-parsimonious trees, and because these searches involve many more tree rearrangements they are substantially less efficient. Many more searches would need to be conducted to determine with confidence whether the most efficient searches for most-parsimonious trees occur with 50 or 2 000 trees held in memory (or possibly with somewhat different numbers of trees held). The similarity in overall efficiency between these two settings, in contrast with the substantially greater frequency of successful searches with 2 000 trees held than with 50 trees held (ca. 1.8 vs. 0.12%; Table 7.1) is indicative of the tradeoff between the number of searches conducted and the intensiveness of each search.

The excess branch-swapping ratio appears to be a reliable indicator of tree-search efficiency. With the settings $poly=$ and $amb-$, the peak search efficiencies of one-stage searches are observed with 50 and 2 000 trees held in memory, and these are the settings at which the greatest excess branch-swapping ratios are observed (Fig. 7.1c). Fewer combinations of settings were examined in depth with alternative polytomy and support-ambiguity settings, and the results of those analyses should be interpreted with caution, but the excess branch-swapping ratios are generally highest for those settings that yielded the most efficient searches (cf. Table 7.1, Fig. 7.1c). Thus, the amount of excess swapping required during phase two of one-stage searches appears to be a useful indicator of tree-search efficiency, and it can be determined with fewer searches conducted than are required to discover most-parsimonious trees. This having been said, it should be noted that the location of the peak in the excess branch swapping ratio is of greater importance than its height. With the settings $poly=$ and $amb-$ the excess branch-swapping ratio is substantially greater for 50 trees held in memory than for 2 000 trees held, but the actual search efficiencies of these settings are quite similar in magnitude. Also, the excess branch-swapping ratio is substantially greater with 100 trees held in memory for the settings $poly-$ and $amb-$ than for $poly=$ and $amb-$, but the latter combination exhibits a greater search efficiency.

7.8 Two-stage searches

With two-stage searches, the principal settings include the number of trees held in memory during each of the two stages, and the percentage of tree sets obtained during the first stage that are subjected to additional branch swapping during the second stage. In light of the many possible combinations of these settings, plus the various polytomy and support-ambiguity settings, and the many individual searches that must be conducted to examine the actual search efficiency with each of the possible combinations of these settings, we were not able to examine all possible combinations. We concentrated our efforts on those combinations that appeared to be the most promising on the basis of results of the one-stage searches, and we also conducted less-intensive searches with a range of alternative settings to provide preliminary indications of variation patterns in success and efficiency rates.

Nine combinations of numbers of trees held during the first and second stages of two-stage searches eventually were examined in detail with the settings *poly=* and *amb-* (Table 7.2, Fig. 7.4), and one combination (50 trees held followed by 2 000 trees held) was examined with the settings *poly=* and *amb=*. In most cases, second-stage swapping was conducted on tree sets of various lengths that had been obtained during the one-stage searches. However, the numbers of sets of trees from the one-stage searches that had been conducted with two trees held in memory (i.e. those in the first six rows of combinations presented in Table 7.2) were insufficient to conduct thorough two-stage analyses, so the available tree sets obtained from the one-stage searches were supplemented by the results of an additional series of one-stage searches with two trees held.

Also, when only a few sets of trees of a given length were available for second-stage branch swapping (e.g. just seven sets of length 16 222 from the one-stage searches with two trees held in memory, and the settings *poly=* and *amb-*), they were combined with sets of successively greater length into a more inclusive set for examining search success and efficiency. Hence, the shortest tree sets subjected to second-stage swapping after first-stage swapping with two trees held in memory are those of length less than or equal to 16 224 (i.e. from the most-parsimonious to six steps longer, a category that included only 0.43% of available sets of trees; Table 7.2), and the shortest tree sets subjected to second-stage swapping after first-stage swapping with 50 trees held in memory are those of length less than or equal to 16 219 (Table 7.2).

As second-stage swapping proceeded on trees of successively greater length (e.g. up to length 16 229 for tree sets derived from first-stage swapping with two trees held in memory; Table 7.2), the number of available tree sets for second-stage swapping was substantially greater than the number of tree sets of shorter length. For these longer sets of trees, swapping was conducted only with randomly selected tree sets from among those that were available, and success rates and effi-

ciencies for these categories were calculated by extrapolating from the results of these searches.

Calculations of search efficiencies for two-stage searches took into account the rearrangements required during the first stage of each search, the percentage of trees available in each tree-length category, the percentage of trees within each of these categories that actually were subjected to second stage swapping, and the number of rearrangements required during the second stage. The latter factor varies substantially among tree-length categories.

For example, for two-stage searches with 50 trees held during the first stage and 2 000 during the second, with the settings *poly=* and *amb-*, the average number of rearrangements required per set of trees during the second stage was ca. 22.3 billion for those that started at length 16 219, and ca. 33.2 billion for those that started at length 16 222. Although these numbers are substantially different, both of them compare favorably with the ca. 41.8 billion rearrangements required for one-stage searches with 2 000 trees held and the same ambiguity and polytomy settings.

Thus, when second-stage branch swapping in a two-stage search is concentrated on a small sample of relatively optimal trees derived from the first stage, the number of rearrangements required per tree set during the second stage can be small (additional details available on request). However, the total number of rearrangements required per set of trees subjected to second-stage swapping also includes the swapping that must be conducted during the first stage to create the pool of sets of trees for the second stage. When only a small percentage of the shortest tree sets obtained during the first stage are subjected to second-stage swapping, the swapping required during the first stage is apportioned across a small number of sets of trees, and therefore the total number of rearrangements required per set (including first- and second-stage swapping) is relatively large (Fig. 7.4a).

As the percentage of available sets of trees that are subjected to second-stage branch swapping increases, the average number of second- stage rearrangements required per set increases, because trees of successively greater lengths are now being subjected to intensive swapping. However, the total amount of branch swapping that was conducted

Table 7.2 Results of two-stage searches with various combinations of polytomy and ambiguity-of-support settings, various numbers of trees held in memory during each stage, and various percentages of available sets of trees from the first stage of each search subjected to branch swapping during the second stage. Bold italics in each row identifies the combination that yielded the greatest percentage of successful searches and the combination that yielded greatest overall tree-search efficiency. Success rate is given as the percentage of second-stage searches yielding trees of length 16 218, and efficiency as billions of trees examined per set of trees discovered of length 16 218. Numbers in parentheses give the percentage of total sets available

Number of trees held, stage 1/stage 2	Number of sets swapped in stage 2	Success rate and efficiency											
		Success	Efficiency	Success	Efficiency	Success	Efficiency	Success	Efficiency	Success	Efficiency	Success	Efficiency
poly=, amb-													
		≤16 224 (0.43)		≤16 225 (0.92)		≤16 226 (1.85)		≤16 227 (3.55)		≤16 228 (5.93)		≤16 229 (9.07)	
2/50	1 339	*2.0*	*902.8*	1.0	962.7	0.5	1 081.2	0.5	*725.9*	0.3	926.0	0.2	1 221.9
2/100	1 239	*2.0*	*934.4*	1.3	708.4	0.9	615.6	0.7	*562.1*	0.4	751.6	0.4	662.3
2/500	167	4.0	*567.6*	1.9	744.3	0.9	1 088.0						
2/1 000	442	*6.9*	*416.2*	3.3	630.8	3.0	591.3	2.2	733.2	1.3	1 216.6	1.8	866.8
2/2 000	167	*20.0*	*233.9*	14.0	275.1	8.7	413.9						
2/5 000	107	*20.0*	*394.0*	15.0	464.7								
		≤16 219 (0.87)		≤16 220 (3.07)		≤16 221 (6.51)		≤16 222 (12.81)					
50/1 000	214	*39.3*	571.8	13.4	*532.0*	7.2	556.6	4.4	607.5				
50/2 000	214	*53.6*	439.4	27.9	*302.2*	18.3	306.1	11.7	383.8				
50/5 000	128	*53.6*	496.9	32.4	*356.4*	20.5	408.8						
poly=, amb=													
		≤16 219 (0.40)		≤16 220 (1.09)		≤16 221 (2.67)		≤16 222 (6.22)					
50/2 000	41	*46.2*	923.5	28.6	*612.0*	14.6	612.1	7.9	739.0				

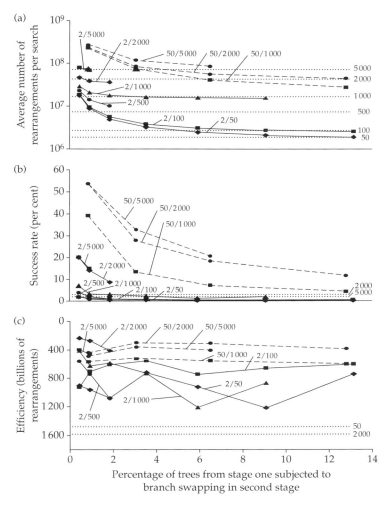

Figure 7.4 Rearrangements required, success rates, and efficiencies of two-stage searches of the 500-terminal *rbcL* data set (*zilla*), as conducted with Nona using the settings *poly=* and *amb-*, with various numbers of trees held in memory during each stage, and with various percentages of tree sets from the first stage subjected to swapping during the second stage. Baseline data from comparable one-stage searches (cf. Figs 7.1–7.3, Table 7.1) also are presented, as horizontal dotted lines through each panel, with numbers on the right indicating number of trees held; all one-stage searches that were conducted but are not depicted in Fig.7.4b and 7.4c had lower success rates and efficiencies, respectively, than those that are depicted. Portions of the data on two-stage searches are also presented in Table 7.2. Dashed lines, two-stage searches conducted with 50 trees held in memory during the first stage; solid lines, two-stage searches with two trees held in memory during the first stage. (a) Average number of tree rearrangements required per set of trees subjected to two-stage search, with the number of trees held during each stage indicated in the figure body for each set of results, and with number of trees held for baseline one-stage searches indicated on the right. (b) Success rates for searches resulting in the discovery of trees of length 16 218, with labels as in (a). (c) Efficiencies of searches resulting in the discovery of trees of length 16 218, with labels as in (a).

during the first stage is now apportioned among a greater number of sets of trees that are subjected to second-stage swapping, so the average number of total rearrangements per tree set subjected to second-stage swapping actually diminishes.

With increasing percentages of the available tree sets subjected to second-stage branch swapping, the average number of total rearrangements required for each set approaches the average number required for one-stage searches conducted with the same number of trees held as during the second stage of two-stage searches, and in at least some cases it overshoots that number and fewer branch rearrangements are actually required (e.g. Fig. 7.4a, in which the curve representing a two-stage search with two trees held in memory in the first stage and 1 000 in the second crosses the line corresponding to a one-stage search with 1 000 trees held). The limit of this trend would occur if second-stage branch swapping were conducted on all tree sets derived from the first stage of a search.

For example, if a two-stage search were conducted with 50 trees held during the first stage and 2 000 held during the second, and all tree sets derived from the first stage were subjected to second-stage swapping, the overall amount of swapping (and the average per set swapped with 2 000 trees held) would be slightly greater than that which is conducted during a one-stage search with 2 000 trees held, with the excess corresponding to the 50 trees derived from each first-stage search that are subjected to swapping a second time at the beginning of each second-stage round of branch swapping.

A similar relationship exists between the percentage of available tree sets that are subjected to second-stage swapping and the frequency of success in the discovery of most-parsimonious trees. When only the shortest trees obtained by first-stage swapping are subjected to second-stage swapping, the percentage of tree sets that yield most-parsimonious trees can exceed the percentage obtained with one-stage searches by an order of magnitude or more (Fig. 7.4b). The greatest frequency of success obtained in one-stage searches, specifically in those in which 2 000 or 5 000 trees were retained in memory, is ca. 2% (Table 7.1, Fig. 7.3a), while frequencies as great as 7%, 20%, and more were obtained in two-stage searches when as many as 12% of available tree sets were subjected to second-stage searches, with the greatest success rates observed when only the shortest trees obtained during the first stage were subjected to second-stage swapping (Table 7.2, Fig. 7.4b).

This increase in the percentage of searches yielding most-parsimonious trees, relative to the numbers obtained in single-stage searches, is also evident in searches in which the greatest number of trees held for branch swapping is as few as 50. In two-stage searches using the default polytomy and ambiguity settings in Nona, with two trees held during the first stage, 50 trees held during the second stage, and only the shortest 0.43% of available sets of trees from the first stage subjected to second-stage swapping (i.e. only tree sets of length 16 224 and less), 2% of the tree sets subjected to second-stage swapping yielded most-parsimonious trees (i.e. length 16 218; Table 7.2), while only 0.12% of single-stage searches with 50 trees held and with the same polytomy and ambiguity settings yielded trees of this length. In this case, the use of a two-stage search strategy increases the success rate by a factor greater than 16, and in other cases (e.g. with 50 trees held in the first stage, and 2 000 trees in the second, as compared to a one-stage search with 2 000 trees held) the increase in success rate is by a factor greater than 20.

It is evident, then, that the overall efficiency of two-stage searches reflects a complex set of tradeoffs between the various factors of the search, with a critical role played by the percentage of tree sets obtained during the first stage that are subjected to second-stage branch swapping. When only a small percentage of the available sets is subjected to second-stage swapping, the total amount of swapping that is required for each of these sets is large, because a complete accounting of the swapping required with this approach includes the large amount of first-stage swapping that is required to generate each tree set that is subjected to second-stage swapping. However, the success rate for second-stage swapping that is conducted on only a small percentage of the available tree sets (i.e. the shortest trees available from first-stage swapping) is also quite high.

These factors are integrated in the calculation of overall search efficiencies (Table 7.2, Fig. 7.4c). Before considering the results of particular search strategies, a general feature of this figure should be noted, which is that all of the two-stage searches that were conducted were more efficient than any of the one-stage searches. The combinations of settings that were tested in two-stage searches were not chosen at random, as the choice was based on the results of one-stage searches, but on the basis of those results we are confident that no combination of settings for single-stage searches would yield results that were substantially superior to the most efficient ones described above. Hence, it appears that many combinations of settings for two-stage searches, if chosen according to reasonable criteria, will yield results that are superior to even the most efficient one-stage searches.

Two-stage searches conducted with relatively few trees held during the first stage (i.e. two trees), and with relatively large numbers of trees (e.g. 2 000 or 5 000) held during the second stage exhibit peak efficiencies when a relatively small percentage of the tree sets obtained during the first stage are subjected to second-stage swapping (Table 7.2, Fig. 7.4c). For example, the peak efficiency for a search with two trees held in the first stage and 500 in the second (i.e. a '2/500' search) occurs when only the most optimal 0.43% of tree sets derived from the first stage are subjected to second-stage searching. The peak efficiencies of 2/1 000, 2/2 000, and 2/5 000 searches also occur when the most optimal 0.43% of tree sets from first-stage searching are subjected to second-stage searching. This overall pattern is not surprising, because the strategy implicit in searches of this sort is to conduct thorough searches on each of a very small number of tree sets that have been examined in only a cursory way during the first stage. Of the various two-stage searches conducted, the most efficient one is in this category (the 2/2 000 search) with an efficiency of ca. 234 billion tree rearrangements required per set of most-parsimonious trees obtained.

An alternative and similarly efficient strategy for two-stage searches is to conduct relatively thorough first-stage searches (i.e. with 50 trees held in memory), and to conduct second-stage searching on a relatively large percentage of the tree sets that are obtained in the first stage. Among the combinations that were examined, the 50/2 000 search yielded the highest efficiency (ca. 302 billion rearrangements per set of most-parsimonious trees obtained), with second-stage swapping conducted on the most optimal 3.1% of tree sets obtained during the first stage, and with second-stage swapping on the most optimal 6.5% of tree sets yielding nearly equivalent results. In searches of

this sort (including 50/1 000 and 50/5 000 searches), the first stage of branch swapping yields a relatively large percentage of sets of most-parsimonious and nearly most-parsimonious trees, though at the cost of a considerable amount of branch swapping. In light of these facts, it is not surprising that the best results are obtained when a relatively large percentage of these tree sets is subjected to second-stage swapping.

It is notable that branch swapping during the second stage of the two most efficient two-stage searches (2/2 000 and 50/2 000) was conducted with the same number of trees held in memory, and that this number corresponds to one of the two points of peak efficiency for one-stage searches (Figs 7.1 and 7.3b). With 50 trees held during the first stage of the 50/2 000 two-stage analysis, both settings (i.e. 50 and 2 000) correspond to the two points of peak efficiency. For the 2/2 000 search, the greatest efficiency occurred when tree sets of length 16 224 and shorter were subjected to second-stage swapping (Table 7.2, Fig. 7.4c), and a secondary point of peak efficiency for the discovery of trees of this length and shorter occurs when two trees are held in memory (Fig. 7.3b).

Thus, the two most efficient two-stage searches utilize settings that are predicted by observed points of peak efficiency for one-stage searches, and two of these settings (50 and 2 000) correspond to peaks in the excess branch-swapping ratio (Fig. 7.1c). We note that one of the settings for the most efficient two-stage search (i.e. two trees held) was not predicted by the excess branch-swapping ratio. It may be the case, however, that the most efficient two-stage searches, even for large data sets, often involve only cursory branch swapping during the first stage. If so, the utility of large numbers of such searches in identifying a few promising starting points for second-stage branch swapping may have more to do with breadth of coverage (i.e. among possible taxon-entry sequences) than with search efficiency *per se*.

7.9 Three-gene matrix

Preliminary analyses of the three-gene matrix, using the same range of number of trees held in memory as were examined with the *zilla* matrix (i.e. a range of settings from 1 through 5 000), and the default Nona settings *poly=* and *amb-*, identified peaks in the excess branch-swapping ratio corresponding to 20 and 200 trees held in memory, with no evidence of a peak between 200 and 5 000. When BB swapping is conducted, an average of ca. 1.34×10^7 tree rearrangements are required to swap through a single tree, or about 1.4 times the number required with the *zilla* matrix. On the basis of the observed peaks in the excess branch swapping ratio, two-stage searches with a variety of settings were conducted, and most-parsimonious trees were discovered eight times.

We regard the number of successful searches sufficient to provide only a general estimate of the optimal conditions for conducting a conventional analysis of this matrix, and the maximum efficiency obtainable by these methods, but the precision of these estimates is limited. The most efficient two-stage searches require approximately 1.5 trillion tree rearrangements for the discovery of each set of most-parsimonious trees, or about five to eight times as many as are required for the *zilla* matrix. This efficiency is obtained by two-stage searches conducted with 200 trees held in memory during the first stage, followed by second-stage swapping, with 2 000 trees held in memory, on the most optimal 3% of tree sets obtained during the first stage. The preliminary analyses had suggested a search-efficiency peak with 200 trees held in memory, but not the apparent peak at 2 000. It is possible that there is another peak that corresponds to 5 000 trees held in memory, but we have not conducted analyses with settings in that range.

7.10 Real time

The analyses of the *zilla* and three-gene matrices, as described above, were conducted with Nona on several different computers with a range of processor speeds. One of these computers, a Pentium 4 with a 1.7 GHz chip speed, can conduct and evaluate about 1.65×10^6 tree rearrangements per second. With a minimum of 2.34×10^{11} rearrangements required for the discovery of one set of most-parsimonious trees (i.e. for a two-stage 2/2 000 search), this computer can discover a set of most-parsimonious trees about once every

1.6 days. On the fastest standard desktop PCs currently available, which have processor speeds at least twice that of this computer, it should be possible to discover approximately one set of most-parsimonious trees for the *zilla* matrix per day. Another computer that was used in the present study, with a 75 MHz Pentium I chip, was manufactured in 1994. On this computer, Nona conducts about 8.10×10^4 rearrangements per second (about 5% of the number conducted with the 1.7 GHz Pentium 4), and a set of most-parsimonious trees for the *zilla* matrix can be discovered with this computer, using conventional search methods, about once per month. Thus, by 1997, when computers with processors about three times as fast as the 75 MHz chip were available, as was Nona, it was possible to discover a set of most-parsimonious trees approximately once every 2 weeks with the proper settings. Under those circumstances, a year of processor time on a conventional desktop computer would have been sufficient to conduct a fairly thorough analysis of the *zilla* matrix, including time for preliminary analyses to determine the most efficient settings, and afterward, for the discovery of a dozen or more sets of most-parsimonious trees.

Why, then, did Rice *et al.* (1997), using a total of about a year of processor time on three Sun workstations, fail to discover shortest trees? We will leave aside the matter of the search efficiency of Nona vs. that of PAUP, and examine search strategies. First, it is likely that Rice *et al.* used the default ambiguity setting of PAUP, which closely resembles the *amb=* setting in Nona. On the basis of the results presented here, we would urge investigators to use settings corresponding to the *amb-* setting of Nona, in association with those corresponding to *poly=*.

Second, Rice *et al.* conducted one-stage searches and, as demonstrated here, almost any two-stage search with reasonable settings should outperform a one-stage search. With the *zilla* matrix, the most efficient two-stage searches are approximately five times as efficient as the most efficient one-stage searches (ca. 200–300 billion vs. ca. 1.5–1.6 trillion rearrangements required for the discovery of one set of most-parsimonious trees). In fact, the exclusive reliance on one-stage searches is perhaps the greatest problem with the analysis of Rice *et al.* As discussed below, we believe that investigators should never rely on conventional one-stage searches when analyzing matrices for which multiple sets of presumed shortest trees cannot be found in a reasonable amount of time. Using currently available hardware and software this corresponds to matrices of ca. 150–250 taxa.

Third, Rice *et al.* conducted their one-stage searches with large numbers of trees held in memory, which allowed very few individual searches to be conducted during the course of their study. As demonstrated by our results, the most efficient searches of the *zilla* matrix are those with as few as 50 trees or even two trees held during the first stage of a two-stage search, and the most efficient one- and two-stage analyses never involved swapping with more than 2 000 trees held in memory.

7.11 Conclusions and recommendations

The present analysis of the 500-terminal *rbcL* matrix, conducted with a variety of software settings, demonstrates (unsurprisingly) that the amount of branch swapping that is required to complete a one-stage analysis increases with the number of trees held in memory. However, the rate of increase in the required amount of branch swapping is uneven, and we have demonstrated that the intervals in which the branch swapping requirements ascend most steeply are those in which the average tree length obtained descends most steeply. The regions of greatest change in these relationships correspond in turn to the points at which the greatest excess branch-swapping ratios are obtained, and these ratios are themselves predictive of peaks in search efficiency. This overall set of relationships demonstrates that the optimal settings for tree searches are those that require substantially more branch swapping than when fewer trees are held for swapping, and hence those that consume substantially more processor time.

Perhaps it should not be surprising that the settings that require the greatest amount of branch swapping are also those that are most efficient, because the reason that these settings require so much branch swapping is that they are successfully

discovering shorter trees, and subjecting them to additional swapping. What might not have been predicted, however, is that the increase in required processor time—and tree-search efficiency—rises so unevenly with an increase in the number of trees retained for branch swapping. The general significance of this phenomenon is that search efficiencies under some software settings may differ substantially, and in unpredictable ways, from those obtained with what may appear to be fairly similar settings (e.g. one-stage searches conducted with 50 or 2 000 trees held in memory are about three times as efficient in the discovery of most-parsimonious trees as those conducted with either 200 or 500 trees held; Table 7.1, Fig. 7.3b).

This study also demonstrates that optimal settings for one-stage searches are predictive to some degree of optimal settings for two-stage searches. This should not be surprising, since both stages of a two-stage search are themselves one-stage searches. However, this relationship is important because it allows a preliminary series of one-stage searches to be predictive of optimal settings for two-stage searches, which appear to be more efficient than one-stage searches, in general, for the analysis of large data sets. When conducting preliminary one-stage analyses, searches with very few trees held in memory (e.g. two) should be included, because settings in this range may be useful during the first stage of a two-stage search.

Using these various relationships, we have discovered that the *zilla* matrix is quite amenable to analysis using conventional methods. With Nona running on a standard PC, and with appropriately chosen software settings, a set of most-parsimonious trees can be discovered in a day or so of computer time. However, with matrices substantially larger than *zilla,* conventional searches on a single PC become impractical. The 567-terminal three-gene matrix appears to lie near the current limits of conventional cladistic analysis with a single PC. Fortunately, alternative methods such as tree fusion, tree drifting, sectorial searches, and the parsimony ratchet are available.

On the basis of the present analysis, and our experiences with other data sets, we believe that conventional single-stage analyses are sufficient only with relatively small data sets. With a matrix of up to 100 or perhaps 150 taxa, a good starting point for exploratory analyses would be the equivalent of approximately 100 one-stage analyses with ca. 20 trees held in memory (i.e. *hold/20; mult*100* in Nona), with polytomies allowed, and with unambiguous support required for each dichotomous resolution (i.e. *poly=* and *amb-* in Nona; if a user wishes to examine the diversity of trees obtained under alternative polytomy or ambiguity settings, additional branch swapping might be conducted under those conditions after a thorough search has been conducted using the recommended settings). There can be no guarantee that shortest trees have been discovered, even with matrices this small, but if the shortest trees found by the overall analysis are discovered by 10% or more of the 100 searches in each set of analyses, and if this occurs (with the same shortest tree length discovered in each set of 100 analyses) on repeated runs of *mult*100,* there is a good likelihood that these are most-parsimonious trees for the matrix. It may be advisable at that point to run a series of additional one-stage analyses, with greater numbers of trees held in memory (e.g. *hold/50* or *hold/100*), and if shorter trees are not discovered, those that have been discovered can be accepted provisionally as shortest trees for the matrix, but more-extended swapping with a few of the optimal sets obtained (i.e. limited two-stage searching) still would be advisable. Note that this set of recommendations is contingent upon the discovery of trees of a given length by multiple individual searches within each set of 100 one-stage searches.

If the shortest trees obtained during the initial sets of one-stage analyses are found in only a small percentage of the individual searches, or if the number of taxa exceeds 150 or so and is less than 300 or so, it is advisable to consider running two-stage searches from the outset. Because the initial stage of a two-stage search is a series of one-stage searches, the analysis still begins, of course, with one-stage searches, but will not end with them. Preliminary analyses can be conducted with the intention of calculating excess branch-swapping ratios or, alternatively, a minimum of several hundred one-stage searches should be conducted, as described in the previous paragraph, with

perhaps 2–20 trees held in memory, and with the best 5–10% of the tree sets obtained by these searches subjected to second-stage searching, with five to ten times as many trees held in memory during the second stage as were held during the first. Results from all searches that yielded shortest trees eventually should be combined into a single tree file, from which duplicate trees are eliminated (e.g. using the *unique* or *best* commands of Nona), and an attempt to swap through all of these trees should be made, with the goal of discovering all most-parsimonious trees for the matrix.

With matrices of more than 200 or so taxa, any results obtained by conveniential searches should be verified by conducting parallel searches using the fastest available methods, such as those described by Nixon (1999) and Goloboff (1999); this approach was taken by Davis *et al*. (2004), who found identical sets of trees for a series of matrices with up to 218 terminals using conventional searches as well as the parsimony ratchet.

With matrices larger than 200–250 taxa, the use of these alternative methods should be regarded as essential, but conventional searches also should be conducted whenever practicable, for there appear to be some matrices in this size range that are more amenable to analysis by conventional methods than by the use of the more recently developed methods. Indeed, it should be noted that a parsimony ratchet analysis, as implemented in WinClada (Nixon 2002), is initiated with a short one-stage conventional search.

Thus, optimal approaches to the analysis of large data sets may involve successive stages in which various methods are employed (e.g. Tehler *et al*. 2003), including conventional searches during early stages (e.g. numerous one-stage searches) to generate sets of relatively short trees that are then subjected to additional searching using other methods. Goloboff (1999), for example, combined multiple search methods, and of the strategies he tested, the most efficient searches were those that used tree drifting and sectorial searches to produce suboptimal trees that were then subjected to tree fusion to produce optimal trees. A similar strategy was used by Tehler *et al*. (2003), but with their matrix the parsimony ratchet, rather than conventional searches, tree drifting, or sectorial searches, was required to produce suboptimal trees with appropriate qualities (likely a sample of trees from multiple islands) amenable to the discovery of shortest trees with tree fusion.

When only one or a few personal computers are available, it is currently possible to discover shortest trees for matrices with up to 500–700 terminals using conventional methods alone, over the course of several days or a few weeks. With larger data sets, however, the success of such an endeavor would be questionable. Thus, conventional analytical methods are useful with smaller data sets, and with larger ones they are likely to continue to play an important role during the preliminary stages of searches that also invoke the more recently developed methods.

CHAPTER 8

Parsimony and Bayesian phylogenetics

Pablo A. Goloboff and Diego Pol

8.1 Introduction

Methods of phylogeny reconstruction are often divided into statistical methods (which require an explicit model of evolution) and non-statistical methods. Among methods with an explicit statistical justification, the most widely used are the methods of maximum likelihood, resulting from Felsenstein's (1973, 1981c) work, and more recently, Bayesian phylogenetic methods based on Monte Carlo Markov chains, following Li (1996), Mau and Newton (1997), and Larget and Simon (1999).

The aim of a statistically based method is to estimate tree topologies and values of possibly relevant parameters, as well as the uncertainty inherent in those estimations. A method that could do that with reasonable accuracy would be attractive indeed. It is often claimed that it is advantageous for a method to be based on a specific evolutionary model, because that allows incorporating into the analysis the 'knowledge' of the real world embodied in the model. Bayesian methods have become very prominent among model-based methods, in part because of computational advantages, and in part because they estimate the probability that a hypothesis is true, given the observations and model assumptions. Early work on phylogenetics suggested the desirability of probabilifying the falsehood or truth of hypotheses. This includes early papers by Farris (1973, 1977, 1978), who later reconsidered the question of whether phylogeny estimation is to be viewed as a statistical problem or not, and moved to the position that phylogenetic inference is best viewed in non-statistical terms (Farris 1983). When he first approached phylogeny as a statistical

problem, Farris (1973, p. 250) pointed out that the tree to be selected "should be the most probable tree on the basis of available data," and that (for tree T and data X) this probability (normally called *posterior probability*) can be calculated with Bayes' Theorem:

$$\Pr(T|X) = \frac{\Pr(X|T)\Pr(T)}{\Pr(X)}$$

where $\Pr(T)$ is the prior probability of the tree being analyzed (i.e. the probability, *a priori* of any observation, of the tree being the true one), the factor $\Pr(X|T)$ is the likelihood of the topology (i.e. the probability of the data, given the tree), and the denominator $\Pr(X)$ is the prior probability of the observed data (calculated as $\sum \Pr(X|T)\Pr(T)$ for all possible topologies). Farris (1973) noted that because the prior probability of each tree topology can be assumed to be the same (equal prior probabilities are usually called a *flat prior*), and because $\Pr(X)$ is fixed for the given observations, the choice, equivalent to parsimony, depends only on the likelihood of the tree. Farris (1973) developed a very general model, with minimal assumptions; under that model, the most likely tree is equivalent to the most-parsimonious tree. In the very same issue of *Systematic Zoology*, Felsenstein (1973) laid the basis for his subsequent developments of a very different model, with much more specific assumptions (including assumptions of Markovian evolution and Poisson substitution), conceived mostly as applicable to the evolution of DNA sequences. In the approach of Felsenstein (1981c) the values of parameters as well as branch lengths are jointly estimated in order to maximize the likelihood function of a tree.

Bayesian approaches to phylogenetics have taken Felsenstein's methods a step further, and instead of producing point estimations of all parameters to maximize the likelihood, they have suggested integrating the likelihood across the different possible parameter values (i.e. branch lengths and substitution model parameters):

$$\Pr(X|T) = \int_{B_T} \int_{\Phi} \Pr(X|T, \beta_T, \varphi) f(\beta_T, \varphi)\, d\varphi d\beta_T$$

where B_T is the set of possible branch lengths (β_T) of topology T, Φ is the set of all possible substitution parameter values (φ) of the model, and $f(\beta_T, \varphi)$ is the prior distribution of these parameters. Both Farris (1973) and Felsenstein (1973) had considered such a type of integration desirable, but noted that a major problem with this approach is that it involves the calculus of a multidimensional integral for every possible topology, which is exceedingly complex and computationally demanding.

In order to overcome this problem, some researchers (e.g. Farris 1973; Hasegawa and Kishino 1989; Smouse and Li 1989) have attempted to compute the Bayesian posterior probability of a topology using the parameter values that maximize its likelihood factor (e.g. the maximum likelihood estimate of branch lengths). However, this approximation (as noted by Goloboff 2003, for the case of maximum likelihood) ignores an infinite number of additional hypotheses that result from alternative sets of branch lengths (or other parameter values) for that topology.

Others, instead, have suggested calculating the exact probabilities, integrating the likelihood of a topology across all possible sets of branch lengths (e.g. Rannala and Yang 1996) or other parameters (e.g. Sinsheimer et al. 1996). The complexity of this procedure, however, precludes its applicability to data sets of more than a few sequences, and therefore these methods were hardly ever used.

8.2 Markov chain Monte Carlo

Recently, three independent groups originally applied Markov chain Monte Carlo methods (MCMC) to approximate the posterior probabilities of trees (Li 1996; Mau 1996; Mau and Newton 1997; Yang and Rannala 1997; Larget and Simon 1999; Mau et al. 1999; Newton et al. 1999; Li et al. 2000).

The idea in a MCMC is to make computationally feasible the integration of the posterior probabilities across the parameters of interest (e.g. topology, branch lengths, substitution parameters). The chain uses a proposal mechanism, which consists of gradual modifications from a starting point (ideally, randomly chosen), and it alternatively changes some parameter values (e.g. topology, branch lengths, substitution parameters), stochastically and aperiodically. These proposals or transitions are accepted with a probability given by the Metropolis–Hastings algorithm (Metropolis et al. 1953; Hastings 1970; see Larget and Simon 1999 or Huelsenbeck et al. 2002 for details) and the Markov chain proceeds until it reaches a stationary state.

If the Markov chain is irreducible (i.e. it is possible for the chain to visit every possible set of parameters and tree topologies) its stationary state converges to the joint posterior probability distribution of the parameters being modified (Tierney 1994). Thus, the frequency with which a given topology is visited in the Markov chain approximates its marginal posterior probability (Mau et al. 1999). Thus, the results of MCMC are directly interpreted in probabilistic terms; they can estimate the probability that a particular tree is the true tree for these sequences, conditional on the stochastic model of substitution (Li et al. 2000). Additionally, since the posterior probability distribution is simultaneously estimated, measures of uncertainty can be derived from the Markov chain.

The outcome of the stationary state of the MCMC is a set of phylogenetic trees (with their associated parameters). In phylogenetic applications of MCMC, the relative frequency of each topology (irrespective of branch length and substitution parameter values) is interpreted as its posterior probability (given the stochastic model and data). Therefore, it seems straightforward to take the topology with the highest posterior probability as the point estimate of the true topology. This was clearly recognized by several authors (Li 1996; Rannala and Yang 1996;

Yang and Rannala 1997; Larget and Simon 1999; Mau *et al.* 1999). The estimated tree was referred to as the maximum posterior probability (MAP) tree by Rannala and Yang (1996).

However, the availability of the approximation of the posterior distribution of trees also allows the evaluation of the variability of the estimates (e.g. topology or any other parameter integrated by the MCMC). As noted by Mau *et al.* (1999), summarizing the distribution of MCMC trees indeed presents a challenge and several methods have been proposed for such purpose.

For instance, Mau *et al.* (1999) and Rannala and Yang (1996) noted that a Bayesian "credible set" can be obtained as the collection of topologies having the sum of their posterior probabilities constrained to be no less than a specified value (e.g. 0.95). Li (1996) also considered alternative ways to estimate the posterior probability of the true phylogeny from the MCMC results, such as the use of the tree that has the minimum topological distance to the majority (e.g. 90%) of the MCMC trees, or the use of a majority-rule consensus of the set of topologies generated by MCMC.

In the latter option, the frequency of the clades has been interpreted by most Bayesian phylogeneticists (e.g. Huelsenbeck *et al.* 2002) as the posterior probability that the clade is true, following ideas of Newton *et al.* (1999) and Larget and Simon (1999). These authors propose summing the posterior probabilities of the trees in which each clade of the MAP is present as a way to summarize uncertainty in the tree topology estimate. This approach, which sums the frequency with which a particular clade appears in the Markov chain in order to estimate its posterior probability, is certainly the most commonly used way to summarize MCMC results. This approach is implemented in available software packages (e.g. MrBayes of Huelsenbeck and Ronquist 2001; BAMBE, of Simon and Larget 1998), and is frequently reported in empirical analyses using Bayesian methods. Here we will focus on some undesirable properties found on this frequently used option to summarize MCMC results. Other alternative ways to summarize the results are less frequently used, and they differ from this one in depending much more

on whether the chain has succeeded in finding the actual MAP(s). As the chain is not conceived as a search mechanism, but instead as a sampling mechanism, it is extremely unlikely that it will find the individual trees of maximum *a posteriori* probability, except in very small data sets.

8.3 Problems with estimations of monophyly by MCMC

In this section, the discussion will be within the realm of the rules and goals postulated by defenders of model-based methods. We also have other general concerns about model-based methods; these reflect a viewpoint not shared by Bayesians, and are therefore discussed in the following section. While the MCMC can be used to estimate any parameter of the evolutionary process, we are concerned here with the estimates that are relevant for phylogenetic studies: estimations of monophyly of groups. Other parameters, such as transition:transversion (ts:tv) ratios, while possibly the primary interest for other evolutionary studies, are only of secondary interest to the phylogeneticist. Part of the attraction of MCMC Bayesian methods is that the values estimated for those other parameters, such as ts:tv ratios, do not rely on estimation of a tree topology, an advantage for such studies which we do not dispute. However, the fact that our examples show that there are problems when the estimations of monophyly are carried out in a certain way suggests that establishing proper estimations from MCMC is far from automatic, and raises concerns about the validity of the inferences of those other parameters as well.

The most common approach to estimating probability of monophyly of a group *X* is by summing the posterior probabilities of all the trees where group *X* is monophyletic. This can be done for the groups present in the individual tree of highest posterior probability (as proposed in Larget and Simon 1999), or for each of the groups found in the analysis; these options make no difference for our argument.

Huelsenbeck *et al.* (2002, p. 674) claimed that, since "Bayesian inference is based on the likelihood function, it should inherit many of the nice statistical properties of the maximum-likelihood method." The "nice statistical property" for which likelihood has been held superior to parsimony is, quintessentially, statistical consistency. Statistical consistency has been proven for maximum likelihood, but only as a byproduct of the consistent estimation of the branch lengths between taxa (see Rogers 1997; Chang 1996; with discussion in Goloboff 2003). If the tree topologies are estimated without estimating branch lengths—integrating branch lengths for a given tree topology, as done in the Bayesian methods—then statistical consistency might be lost (as discussed in Goloboff 2003). And even if Bayesian analysis used optimal branch lengths (which would slow it down considerably), the fact that posterior probabilities of individual clades are estimated from sums of posterior probabilities of the trees having the clade still creates problems. So, the idea that Bayesian analysis should automatically "inherit the nice statistical properties of maximum likelihood" is no more than wishful thinking; Bayesian analysis with MCMC involves substantial modifications to maximum-likelihood.

Estimating the posterior probability for monophyly of a given group as the sum of posterior probabilities of the trees with that group may create serious problems, and it is easy to see why. Imagine that there is a single tree of highest likelihood[1], where group X is not present. Imagine that there are many trees of a likelihood only slightly inferior, where group X is monophyletic. The sum of the likelihoods of the trees with the group may exceed the likelihood of the one tree without the group, and then the method would conclude that the group has a relatively large probability of being monophyletic. While there are many situations under which such an asymmetry could occur, some of them are surprisingly simple. Consider the case of Fig. 8.1, a 25-taxon data set, with taxon A having only missing entries. The data determine a perfectly pectinate tree, except for the placement of A. The strict consensus for these data (analyzed with either parsimony or likelihood) is

---

[1] Whether the likelihood is calculated as the likelihood for optimal branch lengths, or the sum of the likelihoods for all the branch lengths for the given topology, makes no difference to our argument.

an unresolved bush, which does express the fact that the monophyly of *no* group is actually supported by the data. Note that A can float in the skeleton tree of the remaining taxa; each of the 45 trees with alternative placements of A has exactly the same likelihood (and thus, under a flat prior on tree topologies, the same posterior probability). However, of all those trees, only two (A sister to B, or A sister to C) make the group BC non-monophyletic; the proportion of trees with the group BC monophyletic is thus $43/45 = 0.955$. That is almost exactly the posterior probability for monophyly of BC estimated by MrBayes (see Fig. 8.1; values on the branches are values reported by MrBayes, numbers above the branches are the frequencies of the groups in the 45 most-parsimonious trees). The group BCD, instead, is made non-monophyletic by two additional locations of taxon A, so it is monophyletic in a proportion of $41/45 = 0.911$. As one moves towards the middle of the tree, the proportions of locations which make the group non-monophyletic decreases: $6/45$ for group $BCDE$, $8/45$ for group $BCDEF$, etc. Past the middle of the tree, the proportions start increasing again. This is reflected almost exactly in the posterior probabilities reported by MrBayes. Since this is perfectly expected, the proposal mechanism used by MrBayes seems—at least for data sets as simple as this one—to provide a sample of the tree space adequate to estimate the sums of posterior probabilities for different groups; our criticism has nothing to do with sampling problems, but simply with the quantity that is being estimated. The (estimated) sum of posterior probabilites of the trees with and without a group provides a measure with no apparent utility. Using such a measure leads to the unfounded conclusion that, even when what we know about A is nothing, we can still estimate with some precision its placement in the tree! That location of A is determined rather by the priors on trees, but that means that the priors on groups are highly unequal. That an equal prior on trees may mean an unequal prior on groups has been discussed by Pickett and Randle (2005); Pickett and Randle also note that using flat priors on some aspects of a simulation may impose non-flat priors on other aspects. While the non-flat priors on groups (which undoubtedly exist) influence the posterior

```
ROOT  AAAAAAAAAAAAAAAAAAAAAAAAAAAAAAAAAAAAAAAAAAAAAAAAAAAAAAAAAAAAAAAAAAAAAAAAAAAAAAAAAAAAAAAAAAAAAAAAAA
X     AAAAAAAAAAAAAAAAAAAAAAAAAAAAAAAAAAAAAAAAAAAAAAAAAAAAAAAAAAAAAAAAAAAAAAAAAAAAAAAAAAAAAAAAAAAAAAAAAA
W     GGGGAAAAAAAAAAAAAAAAAAAAAAAAAAAAAAAAAAAAAAAAAAAAAAAAAAAAAAAAAAAAAAAAAAAAAAAAAAAAAAAAAAAAAAAAAAAAAA
V     GGGGGGGGAAAAAAAAAAAAAAAAAAAAAAAAAAAAAAAAAAAAAAAAAAAAAAAAAAAAAAAAAAAAAAAAAAAAAAAAAAAAAAAAAAAAAAAAAA
U     GGGGGGGGGGGGAAAAAAAAAAAAAAAAAAAAAAAAAAAAAAAAAAAAAAAAAAAAAAAAAAAAAAAAAAAAAAAAAAAAAAAAAAAAAAAAAAAAAA
T     GGGGGGGGGGGGGGGGAAAAAAAAAAAAAAAAAAAAAAAAAAAAAAAAAAAAAAAAAAAAAAAAAAAAAAAAAAAAAAAAAAAAAAAAAAAAAAAAAA
S     GGGGGGGGGGGGGGGGGGGGAAAAAAAAAAAAAAAAAAAAAAAAAAAAAAAAAAAAAAAAAAAAAAAAAAAAAAAAAAAAAAAAAAAAAAAAAAAAAA
R     GGGGGGGGGGGGGGGGGGGGGGGGAAAAAAAAAAAAAAAAAAAAAAAAAAAAAAAAAAAAAAAAAAAAAAAAAAAAAAAAAAAAAAAAAAAAAAAAAA
Q     GGGGGGGGGGGGGGGGGGGGGGGGGGGGAAAAAAAAAAAAAAAAAAAAAAAAAAAAAAAAAAAAAAAAAAAAAAAAAAAAAAAAAAAAAAAAAAAAAA
P     GGGGGGGGGGGGGGGGGGGGGGGGGGGGGGGGAAAAAAAAAAAAAAAAAAAAAAAAAAAAAAAAAAAAAAAAAAAAAAAAAAAAAAAAAAAAAAAAAA
O     GGGGGGGGGGGGGGGGGGGGGGGGGGGGGGGGGGGGAAAAAAAAAAAAAAAAAAAAAAAAAAAAAAAAAAAAAAAAAAAAAAAAAAAAAAAAAAAAAA
N     GGGGGGGGGGGGGGGGGGGGGGGGGGGGGGGGGGGGGGGGAAAAAAAAAAAAAAAAAAAAAAAAAAAAAAAAAAAAAAAAAAAAAAAAAAAAAAAAAA
M     GGGGGGGGGGGGGGGGGGGGGGGGGGGGGGGGGGGGGGGGGGGGAAAAAAAAAAAAAAAAAAAAAAAAAAAAAAAAAAAAAAAAAAAAAAAAAAAAAA
L     GGGGGGGGGGGGGGGGGGGGGGGGGGGGGGGGGGGGGGGGGGGGGGGGAAAAAAAAAAAAAAAAAAAAAAAAAAAAAAAAAAAAAAAAAAAAAAAAAA
K     GGGGGGGGGGGGGGGGGGGGGGGGGGGGGGGGGGGGGGGGGGGGGGGGGGGGAAAAAAAAAAAAAAAAAAAAAAAAAAAAAAAAAAAAAAAAAAAAAA
J     GGGGGGGGGGGGGGGGGGGGGGGGGGGGGGGGGGGGGGGGGGGGGGGGGGGGGGGGAAAAAAAAAAAAAAAAAAAAAAAAAAAAAAAAAAAAAAAAAA
I     GGGGGGGGGGGGGGGGGGGGGGGGGGGGGGGGGGGGGGGGGGGGGGGGGGGGGGGGGGGGAAAAAAAAAAAAAAAAAAAAAAAAAAAAAAAAAAAAAA
H     GGGGGGGGGGGGGGGGGGGGGGGGGGGGGGGGGGGGGGGGGGGGGGGGGGGGGGGGGGGGGGGGAAAAAAAAAAAAAAAAAAAAAAAAAAAAAAAAAA
G     GGGGGGGGGGGGGGGGGGGGGGGGGGGGGGGGGGGGGGGGGGGGGGGGGGGGGGGGGGGGGGGGGGGGAAAAAAAAAAAAAAAAAAAAAAAAAAAAAA
F     GGGGGGGGGGGGGGGGGGGGGGGGGGGGGGGGGGGGGGGGGGGGGGGGGGGGGGGGGGGGGGGGGGGGGGGGAAAAAAAAAAAAAAAAAAAAAAAAAA
E     GGGGGGGGGGGGGGGGGGGGGGGGGGGGGGGGGGGGGGGGGGGGGGGGGGGGGGGGGGGGGGGGGGGGGGGGGGGGAAAAAAAAAAAAAAAAAAAAAA
D     GGGGGGGGGGGGGGGGGGGGGGGGGGGGGGGGGGGGGGGGGGGGGGGGGGGGGGGGGGGGGGGGGGGGGGGGGGGGGGGGAAAAAAAAAAAAAAAAAA
C     GGGGGGGGGGGGGGGGGGGGGGGGGGGGGGGGGGGGGGGGGGGGGGGGGGGGGGGGGGGGGGGGGGGGGGGGGGGGGGGGGGGGGGGGGGGGGGGGGGG
B     GGGGGGGGGGGGGGGGGGGGGGGGGGGGGGGGGGGGGGGGGGGGGGGGGGGGGGGGGGGGGGGGGGGGGGGGGGGGGGGGGGGGGGGGGGGGGGGGGGG
A     ??????????????????????????????????????????????????????????????????????????????????????????????????
```

```
        ROOT
        X
          W
    95      V
   93  91     U
   96  88  86   T
         88  82  82   S
              88  78  77   R
                   82  74  73   Q
                        78  69  68   P
                             74  64  64   O
                                  68  59  60  A
                                       64  55  46   N
                                            59  52  51   M
                                                 56  58  55   L
                                                      50  60  K
                                                      54  64  64   J
                                                           58  68  68   I
                                                                64  72  73   H
                                                                     69  77  77   G
                                                                          73  81  82   F
                                                                               78  85  86   E
                                                                                    82  90  91   D
                                                                                         87  93  95   C
                                                                                              92  96  B
                                                                                                   96
```

Figure 8.1 A data set with a taxon (A) scored only with missing entries. No group has any actual support, since the monophyly of any group can be violated at no cost. The numbers on the branches are the posterior probabilities of monophyly, estimated by MrBayes with 100 000 generations, using four chains, with a sampling frequency of 100, and a 'burn-in' of 250 (i.e. discarding the first 25 000 generations). The numbers above the branches indicate group frequency in the most-parsimonious (dichotomous) trees. The numbers below the branches show the bootstrap frequencies, as calculated by PAUP* (with 100 replications, analyzing each resampled data set with a branch-and-bound solution). Tree topology corresponds to the analysis with MrBayes.

probabilities reported by MrBayes, that is only part of the picture; the other aspect is the shape of the likelihood landscape, which is what our examples show.

Admittedly, the example of Fig. 8.1 is contrived in that no worker will attempt to analyze a matrix where a taxon is represented only by missing entries. But the same effects may come in much more subtle flavors; for example, a sub-clade of a larger clade that can connect with different rootings (all with about the same likelihood) to the rest of a tree will produce, inside the clade with an undetermined root, the same effect observed for Fig. 8.1. This can also happen even for groups of a relatively large size (which non-flat priors on groups of different sizes do not easily explain), as in the example of Fig. 8.2, analyzed with MrBayes under the No Common Mechanism model (=parsimony). Under such a model, group N–Z is well supported by the data, and group T–Z is not: each of the characters that might support the monophyly of T–Z becomes an ambiguous synapomorphy when taxon M is the sister group of N–Z (so that there are trees of best fit that do not have T–Z as monophyletic). However, this happens only when M is the sister group of N–Z; for each of the other (numerous) possible locations of M in the rest of the tree, the group T–Z is required to provide the best fit to the data. The group T–Z is present in about 91% of the most-parsimonious

trees for the data set[2]. Not surprisingly, MrBayes reports unsupported group T–Z as strongly supported, with a posterior probability of 0.93.

The examples of Figs 8.1 and 8.2 were not derived from any model, and for this reason may perhaps be dismissed by Bayesians as being irrelevant. But the same effect can appear even in simulated data, where there are no violations of the model. The easiest way to produce the effect is to mimic the conditions of Fig. 8.1. For this, we used as the model tree a perfectly pectinate tree, as in Fig. 8.3, with taxa A and B forming a monophyletic group at the tip of the tree, and successive terminals appearing as successive sister groups. All the branches in the tree had a length of

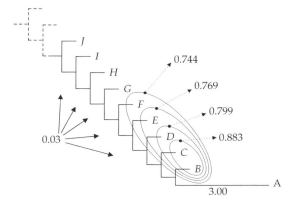

Figure 8.3 Tree shape used in the simulations (results reported in Fig. 8.4). Data were generated for trees with different numbers of taxa, using a Jukes–Cantor model. All branches were of length 0.03, except the branch leading to A, with a length of 3.0. The simulations generated 1 000 characters each. MrBayes analyses used 50 000 generations, with three chains, sampling every 50 generations, and a burn-in of 250 (i.e. discarding the first 12 500 generations). The posterior probabilities are shown for four incorrect groups (for 50 taxa, average posterior probability for 20 replications); note that the posterior probabilities decrease towards the middle of the tree, just as in Fig. 8.1.

ROOT	AAAAAAAAAA	AAAAA
A	AAAAAAAAAA	AAAAA
B	AAAAAAAAAA	AAAAA
C	AAAAAAAAAA	AAAAA
D	AAAAAAAAAA	AAAAA
E	AAAAAAAAAA	AAAAA
F	AAAAAAAAAA	AAAAA
G	AAAAAAAAAA	AAAAA
H	AAAAAAAAAA	AAAAA
I	AAAAAAAAAA	AAAAA
J	AAAAAAAAAA	AAAAA
K	AAAAAAAAAA	AAAAA
L	AAAAAAAAAA	AAAAA
M	AAAAAAAAAA	GGGGG
N	GGGGGGGGGG	AAAAA
O	GGGGGGGGGG	AAAAA
P	GGGGGGGGGG	AAAAA
Q	GGGGGGGGGG	AAAAA
R	GGGGGGGGGG	AAAAA
S	GGGGGGGGGG	AAAAA
T	GGGGGGGGGG	GGGGG
U	GGGGGGGGGG	GGGGG
V	GGGGGGGGGG	GGGGG
W	GGGGGGGGGG	GGGGG
X	GGGGGGGGGG	GGGGG
Y	GGGGGGGGGG	GGGGG
Z	GGGGGGGGGG	GGGGG

Figure 8.2 A data set with a group (T–Z) unsupported but found in many optimal trees, and thus with a high estimated posterior probability. See text for details.

[2] We calculated the frequency of group T–Z in most-parsimonious trees by taking a pseudo-random sample of 1 000 most-parsimonious trees. We generated each by a Wagner tree where both the insertion and addition sequences were randomized (as implemented in TNT; Goloboff *et al.* 2004), followed by tree bisection and reconnection (TBR) branch swapping. Randomizing the insertion sequence means that, for each taxon to be added to form the Wagner tree, the pre-existing locations to insert the new taxon are tried in a random order; this eliminates bias in tree shapes in the resulting Wagner trees for poorly informative data.

0.03 (thus, a probability of no change along the branch of 0.978; we used a Jukes–Cantor model), except for the branch leading to taxon *A*, which was very long (with a length of 3; that is, a probability of no change of 0.287). The model tree was used to generate simulated data sets, with 1000 characters, for different numbers of taxa. Since *A* has a very long branch, it connects to the rest of the tree with about the same likelihood at every possible location. The effect is therefore the same as that of Fig. 8.1. In most of the simulations, MrBayes reports a high posterior probablity that the group *BC* is monophyletic, which is in fact false. The estimated probability of monophyly of the wrong group *BC* actually increases with the number of taxa, since then the alternative locations of *A* that make a significant contribution to the sum of posterior probabilities for group *BC* also increases. Because there is significant variability in different simulated data sets, we used 20 replications for each of 5, 10, 20, 30, 40 and 50 taxa. The results are shown in Fig. 8.4. While there is of course some sampling error in our measurements, the trends evident in Fig. 8.4 make it clear that the high posterior probability attributed to the wrong

groups (often over 0.90) is not the effect of sampling error or lack of convergence in the chains. As the number of taxa increases, so does the apparent confidence on the false groups (the more so for the smaller groups), while the confidence on the true groups decreases (the more so for the smaller groups). Whereas the reference to smaller and larger groups makes sense in these examples, with pectinate trees, this does not mean that MCMC analysis will in general favor groups of a given size; the problem arises because of the relative differences in likelihood (= posterior probabilities, since we used a flat prior on trees) of those trees with and without each group, and this effect could potentially happen for groups of any size. These results could possibly derive as well from violations of the model, or from examining data for several genes where some of the taxa have not been sequenced for all the genes.

The examples are not intended to be realistic, but they show unequivocally that the estimations of posterior probabilities of individual groups may lead to grossly mistaken conclusions, and in real cases such an effect can easily be confounded by other factors. For the simulated examples, a taxon

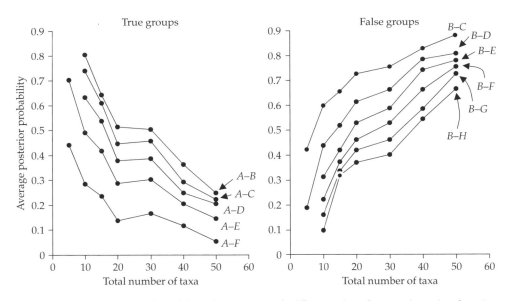

Figure 8.4 Results for the simulations, using the model tree shown in Fig. 8.3, for different numbers of taxa. As the number of taxa increases, so does the estimated posterior probability of the false groups (*BC, BCD*, etc.), the more so the smaller the group. All the averages reported correspond to 20 replications for each number of taxa.

with a branch as long as the branch leading to *A* cannot be confidently placed anywhere in the tree; every location will have roughly the same fit. The most serious problem faced by Bayesian analysis is not that it places *A* in some definite location (i.e. in the middle of the tree), but rather that it leads one to conclude that there is a very high probability that *A* is not placed as sister to *B*, which is the one true placement. A proper method should recognize, in cases like Fig. 8.3, that no conclusion is possible. Note that our criticism of Bayesian analysis, in this case, is not equivalent to Siddall's (1998) criticism of likelihood; Siddall (1998) criticized likelihood because in his simulations it separated long branches that were in fact sisters; Swofford *et al.* (2001) showed in their reply that, while likelihood indeed separates long sister branches (for small numbers of characters), the likelihoods of the alternative trees that place those long branches together is only slightly inferior, so that the maximum likelihood analysis actually implies that no decision is possible. That is not the case for the Bayesian results; they attribute a high probability to false groups that should at least be recognized as ambiguous.

8.4 Potential problems of the statistical approach

Statistically justified, model-based approaches to phylogeny have come to dominate the field in the last decade, but many authors still feel that those model-based justifications miss the mark. The controversy, not surprisingly, has often involved criticism and even misrepresentation from both sides. Among the topics on which the debate has centered are the questions of statistical consistency, the complexity of the evolutionary models used, the possible empirical basis of the evolutionary models on which the inferences are to be based, and whether epistemological considerations support the use of specific models of evolution.

The issue of consistency has been discussed mostly in relation to the likelihood vs. parsimony controversy (Steel *et al.* 1993; Siddall 1998; Farris 1999; Swofford *et al.* 2001). We consider that consistency, since it is relevant, at best, only under

unrealistic conditions (i.e. infinite, or at least massive amounts of data evolving under the same model[3], with a perfect fit to the model), is not a very relevant property at the time of deciding among possible methods of phylogenetic inference. Even some statistically inclined phylogeneticists hold this point of view (e.g. Kim 1996; Sanderson and Kim 2000). The focus of this chapter is on Bayesian methods: Bayesians, with their claim that Bayesian analysis, being based on likelihood methods, should inherit the "nice statistical properties" of likelihood (see above), have adhered (implicitly, at least) to the notion that consistency is desirable. However, as we show later, the only feasible implementation of Bayesian phylogenetic analyses is likely to suffer from inconsistency.

The complexity of the inferential models used has also appeared in the likelihood vs. parsimony controversy. Although several likelihoodists (Goldman 1990; Steel and Penny 2000; Lewis 2001) had suggested that parsimony requires estimation of many more parameters than maximum likelihood, Goloboff (2003) reconsidered the problem and concluded that, if anything, parsimony requires estimation of fewer parameters than traditional maximum likelihood methods (a similar conclusion had been reached by Farris 1986, p. 22). Bayesian phylogenetic methods could in theory integrate uncertainty in some parameters during the MCMC (thus not requiring 'estimation' of those parameters). The problem is that the parameter space to be explored then becomes more complex, so that the chain would have to be run for much longer to insure convergence and an adequate sampling.

Finally, the epistemological questions about using evolutionary models and their empirical basis are perhaps the problems that have been less openly discussed in the literature. Several authors have presented the controversy in terms of which of the two approaches can be justified under Karl

---

[3] Note that we say here "under the same model." While it is true that current-day techniques allow sampling of very long DNA sequences, the chances of all the sites still obeying to the same model decrease as the sequences become longer and include different genes or gene regions (as pointed out by Pol and Siddall 2001). The amount of data available for a given model will always be in the order of a few kilobases.

Popper's philosophy (Popper 1968). Most notable among recent philosophical defenses of model-based approaches is perhaps the paper by de Queiroz and Poe (2003). They argue that parsimony can be justified as a Popperian approach only by reference to specific models of evolution. de Queiroz and Poe (2003) say that it is not true that characters provide falsifiers of phylogenies (which Farris 1983 had used to characterize phylogenetic hypotheses as falsifiable), because a phylogeny cannot *per se* make impossible any particular character distribution. de Queiroz and Poe (2003) argue that, having disposed of other possible justifications, the only way to falsify a phylogeny is to show that it is less probable than its rival, and that parsimony can only be justified as Popperian if coupled with specific evolutionary models that specify those probabilities. But de Queiroz and Poe (2003) have not actually disposed of other possible justifications: they present only part of Farris' arguments. Farris (1983) had made it clear that no character could provide absolute falsification, and that the relationship between falsifier and hypothesis is purely logical. That is, if a given apparent character-state homology between two taxa is truly due to common ancestry, then it follows that a phylogeny that places them apart is truly false. Contra de Queiroz and Poe (2003), a strictly logical justification of parsimony, made without reference to a specific evolutionary model, is possible. Probabilistic models are necessary only to interpret the results of a parsimony analysis probabilistically; they are unnecessary otherwise.

The question of whether some apparent homologies are more probably truly homologous than others only enters the picture when we accept statistical justification and specific models. Contrary to some defenders of parsimony (e.g. Siddall and Kluge, 1997; Kluge 1997, 2001; Kluge, Chapter 2 of this volume), we do not argue that such a type of justification is philosophically and intrinsically flawed. Our concerns have to do with common sense, more than with philosophy. If one knew with certainty that sequence evolution is driven exclusively by a reduced set of parameters, and that those parameters remain very stable over time, then model-based methods of phylo-

geny reconstruction would be perfectly justified. Using those model-based methods would have the advantage that they make it possible to provide measures of uncertainty with a direct interpretation.

The alternative is considering that sequence evolution is driven by too many parameters, which may change too much over time, and that the samples (of sequences) we may expect to ever obtain are far below what could reasonably allow accurate inferences. Of course no one expects inferences that are 100% error-free; but the problem is how much is too much. Philosophical positions aside, many people who use parsimony place themselves at the 'too much' side of the scale, and tend to think that the probabilities estimated by using specific models are likely to be so far off that there is no point in trying to consider the results in terms of real probabilities. All we can expect is to simply provide the best explanation of the data, and it is best to remain silent about the probability of the resulting hypothesis being true. When de Queiroz and Poe (2003) claim that parsimony can be justified only by reference to some specific model, they mean "parsimony *as a statistical method* can be justified only by reference to some specific model," which is true in itself, but then most proponents of parsimony do not view parsimony as attempting to provide the tree with the highest posterior probability: any attempt to provide a figure representing an actual probability, in the case of a process as complex as phylogeny, is no more reliable than a guess[4]. In this sense, the number of things that model-based methods try to estimate (statistically speaking) is much greater, and then it is natural that researchers with no previous experience in the field are attracted by estimation methods which are apparently omnipotent.

To some extent, the two aims—providing the best possible rationalization of the data by means of a phylogeny, or providing the best statistical

---

[4] Measures of support such as the Bremer support (Bremer 1994; Goloboff and Farris 2001) or resampling (Farris *et al.* 1996; Goloboff *et al.* 2003b) are often interpreted as somehow measuring the truth content of the hypothesis, but this is not correct: all they measure is how much evidence supports the hypothesis.

estimation of the phylogeny—are both defensible in their own right. The difference is not only regarding whether a probabilistic model forms the basis for inferences or not; the difference is also about how the results are to be interpreted. Which aim a particular worker prefers and pursues may depend on many factors, actual personal interests, or even peer pressure, among others. But, fashions in science aside, the decision of whether using phylogenetic methods based either on models or pure logic depends also to a good extent on the dose of skepticism the researcher holds. Several defenders of model-based methods (e.g. Swofford *et al.* 2001) have suggested that, in different fields of science, the first approach is based on intuitive methods and that, as the field becomes mature, explicit statistical justifications replace the original intuitive ones. This claim is not strictly true (or testable, at least), and it is presented as if somehow the current use of statistically justified methods was evidence of maturity—when the alternative interpretation, namely that the use of statistical methods in phylogenetics is still premature and unjustified, may be much more reasonable to some workers. But how is one to decide whether the field is mature enough, or whether our knowledge of the mechanisms of evolution is detailed enough, to justify using those models? The answer to this need not be an all-or-none answer; there is instead a gray area between those who believe that our ignorance of evolutionary mechanisms is almost total (a view which some supporters of parsimony seem to hold), and those who believe that our knowledge is so complete as to guarantee even the most detailed inferences (as some likelihoodists and Bayesians seem to believe). At what particular point of this scale a particular worker finds himself/herself will depend on how he/she resolves a large number of subtle issues; such a decision requires reason and logic, but it cannot be accomplished with a statistical test. This, of course, is not to say that anything goes; for example, a model-based method that may produce incorrect estimations—even without violations of the model from which it is derived—is clearly to be avoided. Such is the case with estimations of posterior probabilities of monophyly by MCMC, as we have already seen.

Furthermore, whether the data can be modeled reasonably will depend on the nature of the data. While a Poisson model of substitution seems reasonable for some types of DNA sequence data, it seems unfounded to apply such a model to morphological data (although it has been attempted, by Lewis 2001). It is unfounded because there is very little ground to think that all characters of the given organisms have about the same probability of changing along a given branch of a tree, and that alternative states in morphological characters are like units turned on and off. Some vertebrates have mammary glands, and some arthropods have chelicerae, but within a given group of tetrapods, who could claim that the chances of gaining mamary glands are the same as—or even comparable to—the chances of gaining chelicerae? For genomic data, which include so many different types of transformations (insertions, deletions, translocations, inversions, etc.), postulating reasonable models is also very difficult or impossible. Therefore, for these types of data, only the parsimony aproach has been used so far (starting from Sankoff and Blanchette 1998), for no other approach seems reasonable. Even the simplest case, insertions/deletions, presents a serious challenge to modeling; although some programs (like POY; Wheeler *et al.* 2003) have implemented "maximum likelihood models" for insertions/deletions; these Poisson "models" are based simply on attributing some probability to an insertion as a function of other parameters (like branch "length"). Wheeler *et al.* (personal communication[5]) present the likelihood methods in POY as "interpretive tools, without any necessary relationships to the actual process of change in nature." They point out themselves that this has a meaning quite different from the use of Poisson models in DNA sequences: in those models, a base that is to replace another one exists outside the sequence and therefore, given a certain chance of replication error, there is a certain chance for each possible base to be inserted. Thus, Poisson

[5] Wheeler, W., Aagesen, L., Arango, C., Faivovich, J., Grant, T., D'Haese, C., Janies, D., Smith, W., Varon, A. and Giribet, G. (unpublished manuscript). *Dynamic Homology and Phylogenetic Systematics: A Unified Approach using POY.*

substitution models are based on more than just attributing arbitrary probabilities of change to events; the probabilities postulated by those models are based on some knowledge of the mechanisms that govern the process of DNA substitution, at least in the absence of selection and constraints (whether the model is factually correct, of course, is a different matter, but it is plausible and coherent in itself). Gaps, on the other hand, are not units to be incorporated into a string of DNA being synthesized. A Poisson model for gaps, while it seems natural given the widespread use of Poisson models for DNA substitutions, may be totally inadequate. Other likelihood 'models' of insertions/deletions (e.g. Thorne et al. 1992; Miklós et al. 2004; see De Laet, Chapter 6 in this volume, for comments on these) do not use Poisson models, but still are based on arbitrarily assigning probabilities to the possible events. In those models, the final probability is no more real than are the figures obtained by parsimony analysis. On the other hand, analyzing sequences of unequal length by prealigning them and then discarding positions with gaps (a practice common among likelihoodists, and Bayesians) is probably even more inadequate, so that we are again in a situation where probabilities cannot really be assigned meaningfully. Much the same can be said of other types of chromosomal rearrangement.

8.5 Discussion

Strictly speaking, our simulations do not demonstrate that the estimations of posterior probabilities of individual groups produced by MrBayes are inconsistent. That would require either running data sets with infinite numbers of characters, or an analytical treatment of the multidimensional integral across all possible trees. Neither of those is possible. Admittedly, in cases like our simulations, as the number of characters increases, the difference in likelihood between the correct and alternative placements of the long branch increases. Eventually this difference might be so great as to make the likelihood of the individual best placement of the long branch (the correct one) higher than the sum of the likelihoods of the alternative (wrong) placements. However, as the number of taxa increases,

this situation becomes less and less likely, for the sum of likelihoods of the alternative placements increases as well. So, it is hard to predict what would happen for infinite numbers of characters in cases with very large numbers of taxa. However, even if there is the potential for Bayesian estimations of monophyly to provide correct topological estimations for infinite numbers of characters, there is still the problem that Bayesian analysis claims to do much more than simply producing consistent estimations: it also claims to measure the degree of support of the conclusions, in a statistical sense. Our examples show that it does not.

Several recent papers (Suzuki et al. 2002; Alfaro et al. 2003; Cummings et al. 2003) have compared bootstrap and clade credibility values. In terms of the problem discussed above, some of those comparisons could never have been very informative, despite large amounts of computational effort. For example, the study of Cummings et al. (2003) used over 15 years of CPU time, but examined only data sets with four taxa. The problems pointed out here with Bayesian analyses can arise only with larger numbers of taxa, so Cummings et al.'s effort could never have led to discovery of those problems. Moreover, even for larger numbers of taxa, the problem pointed out here could not have been discovered by comparing posterior probabilities with the bootstrap values produced by PAUP* (the program used in essentially all published comparisons between bootstrap and Bayesian credibilities; Swofford 2002). When bootstrapping or jacknifing, in the case of multiple trees for a resampled matrix, PAUP* weights each group found according to its frequency. This produces exactly the same results as summing posterior probabilities of monophyly: groups that are very frequent in optimal or quasi-optimal trees always appear as highly supported, regardless of their actual support. Fig. 8.1 shows, below the branches, the bootstrap values estimated by PAUP*; they are almost exactly the same as Bayesian estimates. The relative (although not universal) agreement between bootstrap and Bayesian estimations has been taken as mutual confirmation, but in fact MrBayes and PAUP*'s implementation of bootstrapping and jacknifing share similar biases. Alternative implementations of resampling methods (such as the one in TNT; see Goloboff *et al.* 2004) avoid this problem by producing the strict consensus for each resampled matrix.

What happens with other ways to summarize the results of a MCMC? As noted above, they depend on whether the chain succeeded in finding the individual trees of highest posterior probability. For larger numbers of taxa, it is extremely unlikely that the chain will ever pass through the optimal tree(s), let alone pass through the optimal tree(s) enough times to estimate their posterior probability with any accuracy. Although tree bisection reconnection (TBR) forms the basis for both tree search and MCMC algorithms, rearrangements leading to worse trees are often accepted under MCMC, while they are normally rejected in a tree search. For equivalent numbers of rearrangements, then, a tree search (specially one combining different algorithms, like the methods in TNT; see Goloboff 1999, for details) is much more likely than a MCMC to find an optimal tree. Even if MCMC and a tree search had the same chances of finding an optimal tree for the same number of rearrangements, the numbers of rearrangements required to find optimal trees during a search cannot ever be achieved in Bayesian analyses. For example, in the case of a relatively small matrix of 84 taxa (from Goloboff 1995), TNT requires at least 5–10 million rearrangements to produce the first hit to minimum length. For Chase *et al.*'s (1993) data set (*zilla*, 500 taxa), it takes TNT an average of about 500 million rearrangements to find an optimal tree for the first time[6]; for the 854 taxa used in Goloboff (1999), it takes about 5 000 million rearrangements (about 18 min in an 800 MHz machine). Running 5 000 000 000 generations of a MCMC is impossible (in practical terms).

Suppose anyway that the chain succeeds in finding the trees of maximum *a posteriori* probability a certain number of times. The posterior probability of each individual tree will thus be negligible (the more so the more taxa are included in the analysis). In our view, such a low posterior probability is perfectly reasonable, and illustrates the fact that the statistical significance of phylogenetic conclusions cannot be meaningfully assessed in real cases. But statistically minded phylogeneticists will likely show continued interest in making probabilities more robust, i.e. in producing more 'acceptable' values. The alternative is to identify a credible set of trees. A strict consensus of the credible set of trees may contain exclusively well-supported groups, but only to the extent that the chain was run for long enough to find some trees that are relatively close to optimal trees. For the simulations carried out here (small numbers of taxa, very clean data without violations of the model, chains quickly converging), credibility sets of 90% still display, in many cases, false groups. Only running very large numbers of generations would avoid that problem, but in the case of larger data sets this will be impossible.

8.6 Acknowledgments

It is our pleasure to contribute to a book in the honor of James S. Farris, whose work in the field of phylogenetic systematics we greatly admire and respect. We also thank comments from Victor Albert, Jan De Laet, Mark Simmons, and Ward Wheeler. Financial support from FONCyT and CONICET (PAG), and The American Museum of Natural History (DP), is gratefully acknowledged.

---

[6] Note that the figure of 500 million trees corresponds to analyses using sectorial searches, tree drifting, and tree fusing. Davis *et al.* (Chapter 7) report the numbers of rearrangements required to find optimal trees for *zilla* using only TBR; those numbers are much larger.

IV

Mathematical attributes of parsimony

CHAPTER 9

Maximum parsimony and the phylogenetic information in multistate characters

Mike Steel and David Penny

9.1 Introduction

In this chapter we investigate some of the statistical issues that surround the maximum parsimony (MP) method. Such issues have long been of interest, since the pioneering work of Farris (1973) and Felsenstein (1978). The latter was particularly interested in the question of statistical consistency: would MP select a correct tree under a simple finite-state Markov model, as the number of characters became large? Although much more is now known about the (necessary and sufficient) conditions for this to occur there is still a lot that isn't. More recently, there has also been interest in other types of statistical questions. For example, when will MP and maximum likelihood (ML) select the same tree on any given data, and under what sort of model(s) is MP an ML method?

This chapter considers this last question, and describes some new sufficient conditions for such an equivalence. We are particularly interested here in settings that involve a large state space. Traditionally most of the biological studies involving MP have involved a state space that is small (typically 2 or 4 or 20) and fixed (independent of the number of taxa). Indeed much standard software for parsimony (including PAUP*) appears to have problems dealing with a state space that has size of more than (say) 64. However increasingly there is interest in genomic characters such as gene order where the underlying state space may be very large (Rokas and Holland 2000; Moret *et al.* 2001, 2002; Gallut and Barriel 2002). For example, the order of k genes in a signed circular genome

can take any of $2^k(k-1)!$ values. In these models whenever there is a change of state—for example a re-shuffling of genes by a random inversion (of a consecutive subsequence of genes)—it is likely that the resulting state (gene arrangement) is a unique evolutionary event, arising for the first time in the evolution of the genes under study. At this point the reader may object that the observed number of states in such a situation can never exceed the number n of extant species and so this is the only bound that matters. However when we come to investigate the stochastic properties of MP under simple models of state transition, it is the potential rather than the observed number of states that is important. Having a large state space allows for a low level of predicted homoplasy, leading to one of the links we report below between MP and ML.

A related central question we consider in this chapter is how many characters are needed to unambiguously recover a phylogenetic tree? We consider this both for random models of state transition, and in the deterministic setting. We also consider the question of when, on a fixed tree, we can expect the most-parsimonious reconstruction of a character to correspond exactly with its actual evolution.

This chapter is organized so the first three sections are largely 'model-free' (beyond the assumption of evolution on a tree), and the remaining three sections are based on simple Markov models of character evolution. We begin by recalling some background and definitions that are required to state our results, and by reviewing some basic combinatorial properties of MP.

9.2 Preliminaries

Throughout this chapter, X will denote a set of n extant species or individuals. A *character (on X, over a set R of character states)* is any function χ from X into some finite set R. Throughout this chapter, we let r denote the size of R. Suppose we have a tree $T = (V, E)$. We say that T is a *tree on X* if X is a subset of V, and all vertices of T of degree 1 or 2 are contained in X. If, in addition, X is precisely the set of leaves of T we say that T is a *phylogenetic X-tree*, and if, furthermore, every vertex of T has degree 3 we say that T is *fully resolved*. Two phylogenetic X-trees are regarded as equivalent if the identity mapping from X to X induces a graph isomorphism between the two trees. Further background and mathematical details concerning phylogenetic trees can be found in Semple and Steel (2003).

The MP method for reconstructing a tree on X from a collection of characters on X can be described as follows.

Suppose we have a tree $T = (V, E)$ on X, and a character $\chi : X \to R$. A function $\bar{\chi} : V \to R$ is said to be an *extension* of χ since it describes an assignment of states to *all* the vertices of T that agrees with the states that χ stipulates at the leaves.

Let $\operatorname{ch}(\bar{\chi}, T) := |\{e = \{u, v\} \in E : \bar{\chi}(u) \neq \bar{\chi}(v)\}|$. Given a character $\chi : X \to R$, the *parsimony* score of χ on T, is defined by

$$l(\chi, T) := \min_{\bar{\chi}:V \to R,\, \bar{\chi}|X=\chi} \{\operatorname{ch}(\bar{\chi}, T)\}$$

where $\bar{\chi}|X$ denotes the restriction of $\bar{\chi}$ to X. A map $\bar{\chi}$ that extends χ and which minimizes $\operatorname{ch}(\bar{\chi}, T)$ is called a *minimal extension* (or most-parsimonious extension) of χ on T. Let

$$h(\chi, T) = l(\chi, T) - |\chi(X)| + 1$$

be the *homoplasy* of χ on T. By necessity, $h(\chi, T) \geq 0$ and when $h(\chi, T) = 0$ we say that χ is *homoplasy-free* on T. This condition is exactly equivalent to a statement that, informally, says the following: regardless of where T is rooted, one can evolve

states down the tree (from the root to the leaves) in such a way that (1) the leaf states are specified by χ and (2) there is no convergent or reverse evolution (for a more formal rendition of this equivalence, see Semple and Steel 2002).

Suppose we are given a sequence $\mathcal{C} = (\chi_1, \dots, \chi_k)$ of characters on X. The *parsimony* score of $\mathcal{C}$ on T, denoted $l(\mathcal{C}, T)$, is defined by

$$l(\mathcal{C}, T) := \sum_{i=1}^{k} l(\chi_i, T)$$

Any tree T on X that minimizes $l(\mathcal{C}, T)$ is said to be a *maximum parsimony* (MP) tree for $\mathcal{C}$, and the corresponding l-value is the *parsimony* or MP score of $\mathcal{C}$, denoted $l(\mathcal{C})$. Similarly, we may define

$$h(\mathcal{C}, T) := \sum_{i=1}^{k} h(\chi_i, T)$$

the total homoplasy of $\mathcal{C}$ on T, and the tree(s) on X that minimize h are precisely the MP trees (since $h(\mathcal{C}, T) = l(\mathcal{C}, T) + \text{constant}$, where the constant depends on $\mathcal{C}$ and not T). This minimal value of h we write as $h(\mathcal{C})$.

As is well known, the problem of finding an MP tree for $\mathcal{C}$ is computationally intractable (NP-hard), as shown by Foulds and Graham (1982). One might therefore ask for a more reasonable goal. For example, is it possible to determine splits that are shared by all (or some) MP trees? One sufficient condition that allows for the identification of such splits was described recently by David Bryant, and can be stated as follows (from Bryant 2003, Lemma B6). Recall that two binary characters are *compatible* if there exists a tree T on which they are both homoplasy-free (this is equivalent to the condition that at most three (of the four possible) pairs of states are assigned by these two characters).

Proposition 9.2.1. Suppose $\mathcal{C} = (\chi_1, \dots, \chi_k)$ is any sequence of binary characters on X. Let χ be any nontrivial binary character that is compatible with all the characters in $\mathcal{C}$. Then there exists an MP tree T for $\mathcal{C}$ that contains the X-split defined by χ. Furthermore, if χ is one of the characters in $\mathcal{C}$ then *every* MP tree for $\mathcal{C}$ contains the X-split defined by χ.

9.2.1 Bounds on the MP score of data

For a single character $\chi : X \to R$ it is easily shown that

$$\min_{T}\{l(\chi, T)\} = |\chi(X)| - 1 \tag{1}$$

and that

$$\max_{T}\{l(\chi, T)\} = |(X)| - m \tag{2}$$

where m is the largest number of species in X that are assigned the same state (formally $m = \max\{|\chi^{-1}(\alpha)| : \alpha \in R\}$). In (1) and (2) T ranges over all phylogenetic X-trees (or, equivalently, over all fully resolved phylogenetic X-trees).

For a collection $\mathcal{C}$ of characters it is also useful to determine lower bounds on $l(\mathcal{C})$. We first recall an easily computed lower bound. Form a graph by taking X as the set of vertices, and placing an edge between each pair of vertices (this produces the 'complete graph on X'). Weight each edge $\{x,y\}$ by the number of characters f in $\mathcal{C}$ for which $f(x) \neq f(y)$, then construct a minimum-length-spanning tree for this graph. This last task can be accomplished using one of the well-known polynomial-time techniques, such as Kruskal's algorithm or Prim's algorithm. Let $L(\mathcal{C})$ denote the sum of the weights of the edges in this tree. Then, $l(\mathcal{C}) \geq \frac{1}{2}L(\mathcal{C})$. Furthermore, the factor of $\frac{1}{2}$ is (asymptotically) optimal for a lower bound based on this approach due to Foulds (1984); however by adopting a more complex polynomial-time approach a slightly better approximation to $l(\mathcal{C})$ is possible (see Prömel and Steger 2000). Here we describe a quite different type of lower bound, which has the advantage of coinciding with $l(\mathcal{C})$ when the homoplasy $h(\mathcal{C})$ is zero (in contrast to the minimum-length-spanning tree bound, which does not have this property in general).

Let $\mathcal{F}$ be a family of subsets of $\{1, \ldots, k\}$ with the property that each number $1, 2, \ldots, k$ appears in the same number of sets from $\mathcal{F}$. In this case we say that $\mathcal{F}$ is *uniformly covering*. Let $v(\mathcal{F})$, or more briefly v, denote this number of sets from $\mathcal{F}$ that each number appears in (formally, $v(\mathcal{F}) = |\{S \in \mathcal{F} : j \in S\}|$ for each $j \in \{1, \ldots, k\}$). One natural example of such a family is the collection $\mathcal{F}^p$ of all subsets of $\{1, \ldots, k\}$ of fixed size

p (i.e. $\mathcal{F}^p := \{S \subseteq \{1, \ldots, k\} : |S| = p\}$), for which $v(\mathcal{F}^p) = \binom{k-1}{p-1}$. A second class of examples is where $\mathcal{F}$ is a partition of $\{1, \ldots, k\}$ into nonoverlapping subsets in which case $v(\mathcal{F}) = 1$.

Given a sequence $\mathcal{C} = (\chi_1, \ldots, \chi_k)$ of characters on X, and a set $S \subseteq \{1, \cdots, k\}$, let $\mathcal{C}_S = (\chi_j : j \in S)$ and let

$$h^{\mathcal{F}} := \sum_{S \in \mathcal{F}} h(\mathcal{C}_s)$$

The following result extends the 'partition theorem' of Hendy *et al.* (1980).

Proposition 9.2.2. Let $\mathcal{F}$ be a uniformly covering family $\mathcal{F}$ of subsets of $\{1, \ldots, k\}$, let $\mathcal{C}$ be a sequence of characters. Then,

$$h(C) \geq \frac{1}{v(\mathcal{F})} h^{\mathcal{F}}$$

Proof. Let T_0 denote an MP tree for $\mathcal{C}$, and let $h'(j) := h(\chi_j, T_0)$. For $S \subseteq \{1, \ldots, k\}$, let $h'(\mathcal{C}_s) := \sum_{j \in S} h'(j)$. Thus, $h'(\mathcal{C}_s) \geq h(\mathcal{C}_s)$ and so

$$h(\mathcal{C}) = \sum_{j=1}^{k} h'(j) = \frac{1}{v(\mathcal{F})}\sum_{S \in \mathcal{F}} h'(\mathcal{C}_s) \geq \frac{1}{v(\mathcal{F})}\sum_{S \in \mathcal{F}} h(\mathcal{C}s)$$
$$= \frac{1}{v(\mathcal{F})} h^{\mathcal{F}}$$

where the second equality is justified by the identity:

$$\sum_{S \in \mathcal{F}} h'(\mathcal{C}_s) = \sum_{S \in \mathcal{F}}\sum_{j \in s} h'(j) = \sum_{j=1}^{k}\sum_{S : j \in S} h'(j)$$
$$= \sum_{j=1}^{k} v(\mathcal{F}) h'(j)$$

For applications one would construct a family $\mathcal{F}$ of (small) subsets of X that cover each element of X the same number of times, and compute $h(\mathcal{C}_S)$ for each small subset. As a special case, if we take $\mathcal{F} = \mathcal{F}^{(2)}$ (so that $v = k - 1$) and note that $h(\mathcal{C}_S) \geq 1$ whenever $\mathcal{C}_S$ is incompatible, then we obtain the following bound for any collection $\mathcal{C} = (\chi_1, \ldots, \chi_k)$ of characters:

$$l(C) \geq \sum_{i=1}^{k}(|\chi_i(X)| - 1) + \frac{In(C)}{k-1}$$

where $In(\mathcal{C})$ is the number of pairs of characters in $\mathcal{C}$ that are incompatible. The 'partition theorem' from Hendy *et al.* (1980), which states that if $\mathcal{F}$ is a partition of $\{1, \ldots, k\}$ then $l(\mathcal{C}) \geq \sum_{S \in \mathcal{F}} l(\mathcal{C}_S)$, also follows directly from Proposition 9.2.2.

Note that the requirement of Proposition 9.2.2 that $\mathcal{F}$ covers each element of X the same number of times can be weakened by adopting a linear programming approach. That is, if we let h be the minimal value of $\sum_{i=1}^{k} x_i$ subject to the linear inequality constraints, $x_i \geq 0$ for all $i = 1, \ldots, k$, and $\sum_{j \in S} x_j \geq h(\mathcal{C}_S)$ for all $S \in \mathcal{F}$, then clearly $l(\mathcal{C}) \geq \sum_{i=1}^{k} (|\chi_i(X)| - 1) + h$.

9.3 How phylogenetically informative is a single *r*-state character?

In this section we consider the question of to how to quantify the phylogenetic information a single *r*-state character carries (*a priori*, without regard to other characters, or to the character's fit on an existing tree). Let $\chi : X \to R$ be a character. One measure of the phylogenetic information content of χ, based on compatibility, is the following:

$$I(\chi) = -\log(p(\chi)) \qquad (3)$$

where $p(\chi)$ is the proportion of fully resolved phylogenetic X-trees for which χ is homoplasy-free. For example, if χ assigns the same state to all species in X or, at the other extreme, a separate state to each species in X then $I(\chi) = 0$, as we should expect, since every such character is homoplasy-free on all trees.

A measure of phylogenetic content is only useful if it can be readily computed. For the measure I described in (3) it might seem tempting to approximate this quantity by simulation: simply generate fully resolved trees at random and count what proportion of them allow χ to be homoplasy-free. However this turns out to be generally impractical once X becomes large, for the obvious reason: even if you simulate a huge number of large trees at random, it is likely that few if any of them will provide a homoplasy-free fit for χ. Fortunately it turns out that I can be easily computed by a simple exact formula, and without

recourse to simulations. That such a formula exists is truly remarkable, and is due to a little-known but nontrivial result from Carter *et al.* (1980). Using straightforward algebra one can easily derive the following result from Theorem 2 of that paper.

Proposition 9.3.1. Suppose that a character χ partitions a set of n species into classes of size $a_1, a_2, \ldots, a_r.$ Then

$$I(\chi) = \sum_{j=3}^{n-r+1} (1 - b_j) \log(2j - 3)$$

where $b_j = |\{i : a_i \geq j\}|$.

For example, consider a character χ that partitions 20 species into classes of size 6, 4, 4, 3, 2, and 1. Then

$$I(\chi) = -3\log(3) - 2\log(5) + \log(11) + \log(13) + \cdots + \log(27)$$

In this example, $b_3 = 4$ (giving rise to the $-3 (= 1 - b_3)$ multiplier for $\log(3)$), $b_4 = 3$, $b_5 = b_6 = 1$, and $b_j = 0$ for $j > 6$.

Proposition 9.3.1 may be useful for deciding how to construct and select between possible character codings, for example for genomic data. Ideally we would like $I(\chi)$ to be as large as possible, and achieving this may assist in tuning certain coding procedures. Further aspects of this information measure have also been explored recently using simulations by Dezulian and Steel (2004). At this point we will simply note an interesting consequence of Proposition 9.3.1. Firstly, if we fix r, the number of classes that X is partitioned into, then $I(\chi)$ is largest when all of the classes have (approximately) the same size. Let $I_{\max}(n, r)$ be this largest value of $I(\chi)$ over all characters χ that partition a set of size n into r non-empty sets. We may ask how this quantity varies as a function of r. Clearly if $r = 1$ or $r = n$ then $I_{\max}(n, r) = 0$. Consequently, there is some intermediate value, between $r = 1$ and $r = n$, where $I_{\max}(n, r)$ is largest. A plot of $I_{\max}(120, r)$ is shown in Fig. 9.1. Under the I measure, the most informative character for

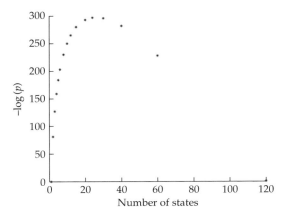

Figure 9.1 Distribution of −log of the number of fully resolved phylogenetic trees on 120 species for a homoplasy-free character that partitions the species into r equally sized sets.

$n = 120$ is one that partitions the taxa into 24 groups, each of size 5.

Other measures of the informativeness of a character are possible and have been proposed; for example, following Farris (1989), one can consider the difference

$$\delta(\chi) = \max_T\{l(\chi, T)\} - \min_T\{l(\chi, T)\}$$

where the terms on the right-hand side of this last equation are given by (1) and (2). Note that if, as above, we fix r, the number of classes that X is partitioned into by χ, then $\delta(\chi)$ is largest when all of the classes have (approximately) the same size. Let $\delta_{max}(n, r)$ be this largest value of $\delta(\chi)$ over all characters χ that partition a set of size n into r nonempty sets. We may ask how this quantity varies as a function of r. Clearly if $r = 1$ or $r = n$ then, as with the measure based on I, we have $\delta_{max}(n, r) = 0$. Note that if r divides n, then applying (1) and (2) gives

$$\delta_{max}(n, r) = n(1 - \frac{1}{r}) - r + 1$$

Maximizing this expression for r we find the maximal value of this expression as r varies (over the real numbers) occurs precisely when $r = \sqrt{n}$. For the example discussed above with $n = 120$ the character that maximizes δ partitions the taxa into fewer groups (namely 10 or 12) than the 24-fold partition that maximizes I.

9.3.1 Coding gene order as multistate character data

It is instructive to consider the types of genomic data for which we may expect, simultaneously, both low homoplasy due to a large state space, and yet phylogenetically informative characters.

For gene-order data, one approach (that has been called "maximum parsimony on multistate encodings") was proposed by Bryant (2000) and tested by Wang *et al.* (2002). Suppose one has n genomes. We will take these to be circular, and consider the genes as signed (oriented) and we will suppose that the genomes have been edited so that each of them contains the set of N genes, which we can label 1, 2, ..., N. A circular gene ordering then can be regarded as a signed circular permutation, for example (1, −4, −3, −2) (which is equivalent to (−4, −3, −2, 1) or to (4, −1, 2, 3), etc.). The coding procedure considered by Bryant (2000) and Wang *et al.* (2002) is based on the observation that each gene order induces a sequence of length $2N$ by considering the gene that immediately follows each given gene in either direction. Given a collection of genomes, this allows one to define a sequence of characters $\chi_1, \ldots, \chi_{2N}$ (on a state space of size $2N$) as follows. For each i between 1 and N, set $\chi_i(j) = \pm k$ if $\pm k$ immediately follows gene i in genome j; and for each i between $N+1$ and $2N$, $\chi_i(j) = \pm k$ if $\pm k$ immediately follows gene $-i$ in genome j. For example, for $j = (1, −4, −3, −2)$ the sequence $(\chi_i(j) : i = 1, 2, \ldots, 8)$ is $(−4, 3, 4, −1, 2, 1, −2, −3)$.

The method of Gallut and Barriel (2002) has a similar flavor. In their approach each gene is associated with the (unordered) pair of genes that appear on either side of it. Thus if there are n genomes, each consisting of N genes, then this coding method produces N characters that have a state space of size $\binom{N}{2}$.

Other methods of coding are also possible, and these are currently being investigated (Dezulian and Steel, unpublished work).

9.4 The smallest number of multistate characters required for tree reconstruction

In this section we consider two related questions: given a fully resolved phylogenetic tree T with

n leaves, what is the smallest possible number of characters for which (1) T is the unique MP phylogenetic tree for these characters, and (2) T is the unique phylogenetic tree for which the characters have no homoplasy? If we call these two numbers, respectively, $n_1(T)$ and $n_2(T)$ it is clear that $n_1(T) \leq n_2(T)$. It might be expected that both these quantities would grow with the size of the tree, yet it has recently been shown that this is not so, provided no bound is placed on the size of the state space. More precisely, we have the following result, from Huber *et al.* (2002).

Theorem 9.4.1. For any fully resolved phylogenetic tree T, on any number of species, the quantities $n_1(T)$ and $n_2(T)$ are at most 4.

When a bound is placed on the size of the state space, then an elementary counting argument shows that both $n_1(T)$ and $n_2(T)$ cannot be bounded by any fixed number that is independent of the number n of leaves of T. This begs the question: how fast must $n_1(T)$ and $n_2(T)$ grow with n? In the case of *binary* characters it is well known that

$$n_2(T) = n - 3$$

since every one of the $n-3$ interior edges of the fully resolved tree T must be distinguished by at least one of the binary characters. Furthermore, for r-state characters, it was shown by Semple and Steel (2002) that

$$n_2(T) \geq \frac{n-3}{r-1}$$

and it seems that this bound is fairly close to the true value. The behavior of $n_1(T)$ has received less investigation, and consequently little is known about how large $n_1(T)$ might be. However the following result shows that $n_1(T)$ must grow at least logarithmically with n (at least for some trees).

Proposition 9.4.2. For any given state space size r, there is a positive constant c such that for each n there exists a fully resolved phylogenetic tree T with n leaves, for which $n_1(T) \geq c \cdot \log(n)$.

Proof. Suppose that to each fully resolved phylogenetic X-tree T we can associate a sequences $\mathcal{C}_T$ of k characters on X for which T is the unique MP phylogenetic tree. Then the number $B(n)$ of fully

resolved phylogenetic trees on a set of size n must be less or equal to the number of sequences of k characters on a set of n species. This latter number is r^{nk} where $n = |X|$, which we may rewrite as $e^{nk \log(r)}$. Now $B(n) = \prod_{i=3}^{n} (2i - 5)$ and it can be shown (using Stirling's approximation for $n!$) that for a constant $\beta > 0$ we have $B(n) > e^{\beta n \log(n)}$. Thus $B(n) \leq r^{nk}$ implies that $k \geq c \log n$ where $c = \beta / \log(r)$. This completes the proof.

It seems plausible that this lower bound on $n_1(T)$ is not too far from the true value, even for binary characters, and so we offer the following.

Conjecture 9.4.3. There exists a constant $c > 0$ such that, for any fully resolved phylogenetic tree T, there exists a sequence of at most $\lfloor c \cdot \log(n) \rfloor$ binary characters on X for which T is the unique MP phylogenetic tree, where n denotes as usual the number of leaves of T.

Proposition 9.2.1 places interesting constraints on the sorts of sequences of characters that this last conjecture requires. Namely, any split that is not in T must be incompatible with at least one of the (at most) $c \cdot \log(n)$ characters in the collection promised by the conjecture. Can such a small set of binary characters be incompatible with virtually all other binary characters? We end this section by describing a result that shows that this is indeed possible. The proof is given in Appendix 9.1.

Proposition 9.4.4. There exists a set $\mathcal{C}$ of $\log_2(n)$ binary characters on a set X of size $n(= 2^k)$ with the following property: any binary character on X that is compatible with every character in $\mathcal{C}$ is a trivial character.

A further interesting feature of the type of data sets that would be required to verify Conjecture 9.4.3 is that many of the characters would need to have large homoplasy values on the tree T. The effectiveness of such data sets in recovering trees is in line with recent observations by Källersjö *et al.* (1999).

9.4.1 Reconstructing ancestral states

In the previous section we considered the question of defining a tree using parsimony. Now we will consider the analogous question for the 'small parsimony' (i.e. fixed-tree) problem. Given a phylogenetic tree X-tree, T, and a character $\chi : X \to R$ that has evolved on T, when are the states that were present at the ancestral vertices of the tree identical to the most-parsimonious reconstruction? We will present a sufficient condition (on the evolution of the character) that guarantees the historical accuracy of the ancestral-state reconstructions. Essentially this sufficient condition is that substitutions that occur are 'well-separated' in the tree (that is, they do not occur too close to each other in the tree). Apart from its intrinsic interest, this result will also be useful later in providing a limiting Poisson distribution for the parsimony score of a tree, under low substitution rates.

Theorem 9.4.5. Suppose that T is a phylogenetic X-tree, and consider the assignment of states $\bar{\chi} : V(T) \to R$ corresponding to the evolution of some character on T. Let $\chi = \bar{\chi}|X$ be the observed states on the extant set of species (leaves of T). Suppose furthermore that the evolution of the character is such that any two edges of T on which a net transition occurs are separated by at least three other edges of T. Then $\bar{\chi}$ is a minimal extension of χ on T; moreover it is the only minimal extension of χ on T.

Proof. Suppose that T is a phylogenetic X-tree, and $\bar{\chi} : V(T) \to R$. Suppose furthermore that for any two edges $\{u, v\}$ and $\{u', v'\}$ for which $\bar{\chi}(u) \neq \bar{\chi}(v)$ and $\bar{\chi}(u') \neq \bar{\chi}(v')$ there are at least three other edges separating $\{u, v\}$ and $\{u', v'\}$. Let $\chi = \bar{\chi}|X$. Then we claim that $\bar{\chi}$ is the unique minimal extension of χ on T. To establish this claim, let $\bar{\chi}'$ be a minimal extension of χ on T; we will show that for each vertex v of T we have $\bar{\chi}'(v) = \bar{\chi}(v)$.

Let us root tree T on vertex v and direct all the edges of T away from v. For any vertex u in this rooted tree, let $S(u)$ denote the set of states assigned to u by applying the first pass of the Fitch–Hartigan algorithm (Fitch 1971; Hartigan 1973) to the pair (T, χ). We will establish the following. Claim: suppose that u is an internal vertex of T and that $v_1, v_2, \ldots v_k$ are the vertices of T that are immediate descendents of u. Then

$$S(u) = \begin{cases} \{\bar{\chi}(v_1), \bar{\chi}(v_2)\}, & \text{if } k = 2 \text{ and } \bar{\chi}(v_1) \neq \bar{\chi}(v_2) \\ \{\bar{\chi}(u)\}, & \text{otherwise} \end{cases}$$

The proof of this claim is by induction on the height h of u (i.e. h is the number of edges separating u from a most distant descendant leaf). When $h = 1$ the claim holds, since the assumption on $\bar{\chi}$ implies that all but at most one (of the two or more) descendant leaves of u has the same state under χ. Suppose the claim holds for all internal vertices of height h and that u has height $h + 1$. By the assumption on $\bar{\chi}$ one of the following two cases applies: (i) $\bar{\chi}(v_i) = \bar{\chi}(u)$ for all $i \in \{1, \ldots, k\}$; (ii) $\bar{\chi}(v_i) = \bar{\chi}(u)$ for all but at most one i.

In case (i), we may apply the induction hypothesis to the vertices $v_1, \ldots v_k$ which each have height at most h. It follows that $\bar{\chi}(u) \in S(v_i)$ for all i. Furthermore there is at most one vertex v_i for which $S(v_i) \neq \{\bar{\chi}(u)\}$ since if there were two such vertices, then we would obtain two edges on which $\bar{\chi}$ changes state, yet which are separated by only two edges in T. Consequently, by the Fitch–Hartigan recursion we deduce that $S(u) = \{\bar{\chi}(u)\}$.

Consider now case (ii). We may suppose that $\bar{\chi}(v_1) \neq \bar{\chi}(u)$. Consider first the case where $k > 2$. Applying the induction hypothesis to $v_1, \ldots, v_k$ and invoking the assumption on $\bar{\chi}$ we have that $S(v_1) = \{\bar{\chi}(v_1)\}$, and for all $i > 1$ we have $S(v_i) = \{\bar{\chi}(u)\}$. It now follows by the Fitch–Hartigan recursion (remembering that $k > 2$) that $S(u) = \{\bar{\chi}(u)\}$. Thus we have established the second part of the claim. It remains to consider the other possibility for case (ii), namely $k = 2$. Again we apply the induction hypothesis on v_1, v_2 and invoke the assumption on $\bar{\chi}$ to deduce that $S(v_1) = \{\bar{\chi}(v_1)\}$ and $S(v_2) = \{\bar{\chi}(v_2)\}$; hence $S(u) = \{\bar{\chi}(v_1), \bar{\chi}(v_2)\}$, as required to justify the claim.

Now let us take $u = v$, the vertex we have selected as our putative root for T. Since T is a phylogenetic tree, v has degree at least three, so by the claim we have $S(v) = \{\bar{\chi}(v)\}$. However, since v is the root of the tree for the recursion, $S(v)$ is precisely the set of states that can occur at v across all possible minimal extensions of χ on T (Hartigan 1973). Thus we have shown that all such minimal extensions (in particular $\bar{\chi}'$) assign vertex v the state as that specified by $\bar{\chi}$. Since we can repeat this argument for any vertex v in T the theorem now follows.

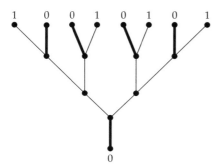

Figure 9.2 Example showing that two-edge separation does not suffice for Theorem 9.4.5.

Note that Theorem 9.4.5 is no longer true if we weaken the edge-separation requirement from three edges to two. For example, consider the tree and character χ shown in Fig. 9.2. Then the extension $\bar{\chi}$ of χ defined by making substitutions precisely on the five (bold) edges incident with leaves in state 0 as indicated in Fig. 9.2 is not a minimal extension of χ, even though each pair of bold edges is separated by at least two other edges. For this example, the minimal extension is provided by assigning state 0 to all the interior vertices of the tree. Note also that an ancestral-state reconstruction satisfying the requirements of Theorem 9.4.5 is not necessarily the 'true' reconstruction, it is merely the unique most-parsimonious reconstruction. Nevertheless, as we will see in the next section (Proposition 9.5.1), certain stochastic models of character evolution imply that this unique most-parsimonious reconstruction is also likely to be historically accurate, provided the substitution probabilities are uniformly small.

9.5 The Poisson model

In this section and the next we consider the simplest tree-based model for the evolution of characters with state space R, which we will refer to here simply as the *Poisson model on R* (with parameters (T, p)). In this model, we have a tree T on X, select any element $x_0 \in X$ as a reference vertex, and direct all edges of T away from x_0. We will regard the value from R assigned to vertex x_0 as being given (it would make little difference to the argu-

ments below if we allowed the state at x_0 to be random). The model assigns states from R recursively to the remaining vertices of the tree according to the following scheme: if $e = \{u, v\}$ is an edge of T directed from u to v and u has been assigned state α, then, with probability $1 - p(e)$ we assign v state α, otherwise, with probability $p(e)$ we select uniformly at random one of the other $r - 1$ states (different to α) and assign this state to v. The assignments are made independently across edges, and the value $p(e)$ is called the *substitution probability* associated with edge e. It is natural to constrain $p(e)$ to lie in the interval $\left[0, \frac{r-1}{r}\right]$; the reason for the upper bound is that, if we realise this model by a continuous-time Markov process, the probability of a net substitution over any period of time is always less than $\frac{r-1}{r}$. We will say that the mapping $e \to p(e)$ is *admissible* if the $p(e)$ values all lie within this allowed interval.

When $r = 4$, this model is essentially the same as what is often referred to as the Jukes–Cantor model. For general values of r, this model was investigated in 1970 by Jerzy Neyman (1971), and has more recently been studied by Paul Lewis (2001) as a starting framework for likelihood analysis for certain morphological characters. This model has been christened in the bioinformatics literature under a variety of titles, including the Neyman r-state model and the r-state Jukes–Cantor model.

Given the pair (T, p) where $T = (V, E)$ is a tree on X, and p is an admissible assignment of transition probabilities, and given a map $\bar{\chi} : V \to R$, let $\Pr(\bar{\chi}|T, p)$ denote the probability that the vertices in T take values specified by $\bar{\chi}$ under the Poisson model on R with parameters (T, p). More formally, $\Pr(\bar{\chi}|T, p) = \Pr(\cap_{v \in V - \{x_0\}} \{\eta(v) = \bar{\chi}(v)\})$, where $\eta(v)$ is the random variable state assigned to v under the model. By the assumptions of the model, we have

$$\Pr(\bar{\chi}|T, p) = \prod_{\{u,v\} \in E : \bar{\chi}(u) \neq \bar{\chi}(v)} \frac{p(e)}{r - 1} \prod_{\{u,v\} \in E : \bar{\chi}(u) = \bar{\chi}(v)} (1 - p(e))$$

(4)

For any character $\chi : X \to R$, let

$$\Pr(\chi|T, p) = \sum_{\bar{\chi} \in c(\chi)} \Pr(\bar{\chi}|T, p)$$

where $c(\chi) = \{\bar{\chi} : V \to R : \bar{\chi}|X = \chi\}$.

9.5.1 Distribution of the parsimony score

Theorem 9.4.5 has the following consequence for the (limiting) distribution of the parsimony score of a character under the Poisson model.

Proposition 9.5.1. Consider a process on a fully resolved phylogenetic tree T with n leaves, and let

$$h = \max\{p(e) : e \in E\}\sqrt{n}$$

and

$$\mu = \sum_{e \in E(T)} p(e)$$

Generate a character χ by this process on T and let $\bar{\chi}$ denote the states at all the vertices of T. Then, for small values of h, the most-parsimonious reconstruction of χ is likely to be both unique and historically accurate, and the parsimony score $\mathcal{L} = l(\chi, T)$ of a character χ generated by this process on T is closely approximated by a Poisson distribution with mean μ. More precisely, for any value of h we have (i) $\Pr[\bar{\chi}$ is the unique MP reconstruction of χ on $T] \geq 1 - 28h^2$, (ii) $|\Pr(\mathcal{L} = k) - e^{-\mu}\frac{\mu^k}{k!}| < 32h^2$, and (iii) $\Sigma_{k=0}^{\infty}|\Pr(\mathcal{L} = k) - e^{-\mu}\frac{\mu^k}{k!}| < 60h^2$.

To illustrate this result, suppose that a fully resolved phylogenetic X-tree has $n = 10\,000$ leaves, and the substitution probability $p(e)$ on each edge is (say) 2×10^{-4}. In this case we can take $h = 2 \times 10^{-2}$, and so we may approximate $\mathcal{L}$ closely by a Poisson distribution with mean 4.

Notice that, in Proposition 9.5.1, a small value of h does not necessarily imply a small value for μ if the number of leaves in the tree T is large.

Proof of Proposition 9.5.1

Let A be the event that substitutions occur on some pair of edges that are separated by two or fewer edges. The number of ordered pairs of edges that are separated by two or fewer edges is at most $(2n - 3) \cdot (4 + 8 + 16)$ since $(2n - 3)$ is the number of edges of T and since $(4 + 8 + 16)$ bounds the number of edges of T that are separated by 0, 1 or 2 other edges from any given edge of T. Thus the number of unordered pairs of edges that are separated by two or fewer edges is at

most $\frac{1}{2} \cdot (2n - 3) \cdot (4 + 8 + 16) < 28n$, and so, by the Bonferroni inequality,

$$\Pr(A) < 28n \cdot \left(\frac{h}{\sqrt{n}}\right)^2 \leq 28h^2 \tag{5}$$

which, together with Theorem 9.4.5 establishes part (i).

Let $\mathcal{L}^*$ denote the random number of edges of T on which there is a substitution. Thus $\mathcal{L} \leq \mathcal{L}^*$, and $\mathcal{L}^*$ has a limiting Poisson distribution since it is the sum of an increasing (with n) number of independent $0/1$ random variables, where the probability that each variable takes the value 1 converges to 0 (with n). Moreover, Le Cam's inequality (Le Cam 1960) gives

$$\sum_{k=0}^{\infty} |\Pr(\mathcal{L}^* = k) - e^{-\mu}\frac{\mu^k}{k!}| < 2\sum_{e} p(e)^2 \tag{6}$$

By the law of total probability,

$$\Pr(\mathcal{L} = k) = \Pr(\mathcal{L} = k|A^c)\Pr(A^c) \\ + \Pr(\mathcal{L} = k|A)\Pr(A) \tag{7}$$

and

$$\Pr(\mathcal{L}^* = k) = \Pr(\mathcal{L}^* = k|A^c)\Pr(A^c) \\ + \Pr(\mathcal{L}^* = k|A)\Pr(A) \tag{8}$$

where A^c is the complementary event of A.

Now, conditional on the event A^c, Theorem 9.4.5 guarantees that $\mathcal{L} = \mathcal{L}^*$ (with probability 1); that is, $\Pr(\mathcal{L} = k|A^c) = \Pr(\mathcal{L}^* = k|A^c)$. Applying this identity to (7) and (8) gives

$$|\Pr(\mathcal{L} = k) - \Pr(\mathcal{L}^* = k)| = |\Pr(\mathcal{L} = k|A) \\ - \Pr(\mathcal{L}^* = k|A)|\Pr(A) \leq \Pr(A) < 28h^2$$

where the last inequality is from (5). Furthermore, (6) implies that

$$|\Pr(\mathcal{L}^* = k) - e^{-\mu}\frac{\mu^k}{k!}| < 4h^2$$

Combining these last two inequalities gives

$$|\Pr(\mathcal{L} = k) - e^{-\mu}\frac{\mu^k}{k!}| < 32h^2$$

which establishes part (ii). Similarly,

$$\sum_{k=0}^{\infty} |\Pr(\mathcal{L} = k) - \Pr(\mathcal{L}^* = k)|$$

$$\leq \sum_{k=0}^{\infty} |\Pr(\mathcal{L} = k|A) - \Pr(\mathcal{L}^* = k|A)|\Pr(A)$$

$$\leq 2\Pr(A) < 56h^2$$

from which part (iii) now follows.

9.6 Links between MP and ML

Given a sequence $\mathcal{C} = (\chi_1, \ldots, \chi_k)$ of characters on X, we put

$$\Pr(\mathcal{C}|T, p) = \prod_{i=1}^{k} \Pr(\chi_i|T, p),$$

$$L(T|\mathcal{C}) = \sup_{p} (\Pr(\mathcal{C}|T, p)),$$

$$\Pr(\mathcal{C}|T, p)_{\mathrm{mp}} = \prod_{i=1}^{k} \max (\Pr(\overline{\chi}_i|T, p)|\overline{\chi}_i \in c(i))$$

$$L_{\mathrm{mp}}(T|\mathcal{C}) = \sup_{p} (\Pr(\mathcal{C}|T, p)_{\mathrm{mp}})$$

where the supremum is taken over all admissible choices of p and $c(i) = c(\chi_i)$ is the set of extensions of χ_i to V. Note that $\Pr(\mathcal{C}|T, p)$ is the probability of generating the k characters by independent and identical evolution under a Poisson model with parameters (T, p).

Similarly one has analogous definitions for the 'no common mechanism' Poisson model, in which each character evolves independently under a Poisson model on R but where p in the parameter pair (T, p) for this model takes admissible values that are permitted to vary freely between the characters. Specifically, let

$$\Pr(\mathcal{C}|T, (p_1, \ldots, p_k)) = \prod_{i=1}^{k} \Pr(\chi_i|T, p_i)$$

and

$$L_{\mathrm{ncm}}(T|\mathcal{C}) = \sup_{(p_1, \ldots, p_k)} (\Pr(\mathcal{C}|T, (p_1, \ldots, p_k)))$$

where the supremum is taken over all k-tuples $(p_1, \ldots, p_k)$ where each p_i is admissible.

Recall that $L(T|\mathcal{C})$ and L_{ncm} are referred to as the *maximum (average) likelihood* or ML score, and $L_{\mathrm{mp}}(T|\mathcal{C})$ as the *most-parsimonious likelihood* or MPL score, of T given $\mathcal{C}$ (cf. Barry and Hartigan 1987; Steel and Penny 2000).

The distinction between these two forms of likelihood is as follows: the ML score of T is the largest probability (over all admissible choices of substitition probabilities p) of generating the observed sequence of characters at the leaves of T but without specifying or conditioning on any particular assignment of sequences of characters at the interior vertices of the tree (these are effectively 'averaged over'). In contrast the MPL score of T is the largest probability (over all admissible choices of substitution probabilities p) of generating any particular assignment of sequence of characters to all the vertices of the tree, so that the sequences assigned to the tips are the observed sequences.

A tree T on X is said to be an ML tree or an MPL tree for $\mathcal{C}$ if $L(T|\mathcal{C}) \geq L(T'|\mathcal{C})$ or $L_{\mathrm{mp}}(T|\mathcal{C}) \geq L_{\mathrm{mp}}(T'|\mathcal{C})$, respectively, holds for all other trees T' on X. The problem of finding an MPL tree given only $\mathcal{C}$ was recently shown to be NP-hard by Addario-Berry *et al.* (2004) (where the method is referred to as "ancestral maximum likelihood",). Finding an MP tree from $\mathcal{C}$ is also NP-hard (Foulds and Graham 1982); most likely so too is the problem of finding an ML tree for $\mathcal{C}$.

We say that an MP, ML, or MPL tree for $\mathcal{C}$ is *irreducible* if we cannot collapse any edge of T to obtain another such tree for $\mathcal{C}$.

We now describe three links between two tree reconstruction methods, one of which (ML) is based explicitly on an underlying Markov model for the evolution of characters on a tree (the Poisson model), while the other method—MP—is based solely on a minimality principle.

9.6.1 Link 1: no common mechanism and an extension

MP is an ML estimator for phylogenetic trees under the 'no common mechanism' model described above. In particular, a tree T maximizes $L_{\mathrm{ncm}}(T|\mathcal{C})$ precisely if T is an MP tree for $\mathcal{C}$. This result, established in Tuffley and Steel (1997), extended the result for $r = 2$ that was described by Penny *et al.* (1994). Here we describe a further slight extension of this result where we allow the size of the state space of the Poisson model to vary from character to character. In this case it can be shown that a weighted form of MP is an ML estimator for a phylogenetic tree under the 'no common mechanism' model.

First recall that character-weighted parsimony is directly analogous to standard MP; given a sequence $(\chi_1, \ldots, \chi_k)$ of characters and a weighting function $w: \{1, \ldots, k\} \rightarrow \mathbb{R}^{\geq 0}$ we simply replace $l(\mathcal{C}, T)$ by its weighted version $l_w(\mathcal{C}, T) = \Sigma_{i=1}^k w(i) l(\chi_i, T)$. We then have the following result.

Theorem 9.6.1. Suppose $\mathcal{C} = (\chi_1, \ldots, \chi_k)$ are characters on X. Consider the model in which all characters evolve independently on a phylogenetic tree T and that each character χ_i evolves according to some Poisson model on a state space of size r_i according to admissible edge parameters that are free to vary from character to character. Then the (average) ML method ranks phylogenetic trees on X in exactly the same order as the weighted MP method provided that each character χ_i is assigned weight $\log(r_i)$.

Proof. The proof relies on a key result from Tuffley and Steel (1997): for any character $\chi : X \rightarrow R$, and any phylogenetic X-tree T' we have

$$\sup_{p'} \Pr(\chi | T', p') = r^{-l(\chi, T')} \tag{9}$$

where the supremum is over all admissible p'. Consequently,

$$\begin{aligned} L_{ncm}(T' | \mathcal{C}) &= \prod_{i=1}^k r_i^{-l(\chi_i, T')} \\ &= \exp\left(-\sum_{i=1}^k \log(r_i) l(\chi_i, T')\right) \\ &= \exp(-l_w(\mathcal{C}, T')) \end{aligned}$$

where w is the character weight function defined by $w(i) = \log(r_i)$. Consequently the tree(s) T' that maximize $L_{ncm}(T' | \mathcal{C})$ are precisely the tree(s) that minimize $l_w(\mathcal{C}, T)$, as claimed.

Note that if the size (r_i) of the state space for character χ_i is unknown for some or all values of i, then in an ML framework we might optimize these

variables (r_i) subject to the obvious constraint that $r_i \geq |\chi_i(X)|$. In that case Theorem 9.6.1 holds if we replace the character weight $\log(r_i)$ by $\log(|\chi_i(X)|)$.

9.6.2 Link 2: large state space

In this section, we describe a quite different link between MP and ML. In contrast to the aforementioned link we consider here the 'common mechanism' setting for which the two methods are in general quite different, since they may select different trees (Felsenstein 1973). However when the number of states is sufficiently large, then once again ML trees are always MP trees. As we will see this may be relevant to the use of certain genomic data (such as gene order) for inferring phylogenies, as in this case the underlying state space may be very large. The proof of the following result—which also relies on the identity (9)—can be found in Steel and Penny (2004).

Theorem 9.6.2. Suppose $\mathcal{C} = (\chi_1, \chi_2, \ldots, \chi_k)$ is a sequence of k characters on X over a state space R of size $r \geq 4^{nk}$. Under the model in which the characters evolve independently according to the same Poisson model on R, any ML tree for $\mathcal{C}$ is an MP tree for $\mathcal{C}$.

9.6.3 Link 3: dense sampling of sequences

Let $\mathcal{S} = \{S_1, S_2, \ldots, S_n\}$ be a collection of aligned sequences of length k on $r \geq 2$ states. Equivalently, we may view $\mathcal{S}$ as a sequence $\mathcal{C}_{\mathcal{S}} = (\chi_1, \ldots, \chi_k)$ where χ_i is an r-state character on X. If we write S_i as $S_i(1), \ldots S_i(k)$, then $S_i(l) = \chi_l(i)$ for all $i \in \{1, \ldots, n\}$ and $l \in \{1, \ldots, k\}$. Let d_H denote the Hamming metric on $\mathcal{S}$, defined by setting $d_H(S_i, S_j) = |\{l : S_i(l) \neq S_j(l)\}|$. We will suppose that the sequences in $\mathcal{S}$ are distinct: that is, $d_H(S_i, S_j) > 0$ for all $i \neq j$. Let $G_{\mathcal{S}}$ be the graph with vertex set $\mathcal{S}$ and with an edge connecting any two sequences that differ in exactly one coordinate. Equivalently, $G_{\mathcal{S}} = (\mathcal{S}, E)$ where

$$E = \{(S_i, S_j) : d_H(S_i, S_j) = 1\}$$

In the context of molecular genetics, $G_{\mathcal{S}}$ is the 'haplotype graph' described, for example, in

Excoffier and Smouse (1994). We say that $\mathcal{S}$ is *ample* if $G_{\mathcal{S}}$ is connected. It is easily shown that if $\mathcal{S}$ is an ample collection of sequences then the set of spanning trees of $G_{\mathcal{S}}$ (i.e. the trees in $G_{\mathcal{S}}$ on vertex set $\mathcal{S}$) is precisely the set of irreducible MP trees for $\mathcal{C}_{\mathcal{S}}$. Consequently, $\mathcal{C}_{\mathcal{S}}$ has MP score n − 1.

Theorem 9.6.3 below implies that when $\mathcal{S}$ is ample, then any spanning tree for $\mathcal{C}_{\mathcal{S}}$ is also an MPL tree for $\mathcal{C}_{\mathcal{S}}$ under this model. That is, we cannot improve the MPL score by introducing additional 'Steiner points' (hypothetical ancestral sequences). As an aside, this result provides another case where a particular instance of an NP-hard problem (namely that described by Addario-Berry *et al.* 2004) has a simple, polynomial-time solution. We note also that the Buneman complex (Buneman 1971) or, equivalently, the median network Bandelt *et al.* (1995) of a collection of *X*-splits provides natural examples of ample sets of sequences. The proof of the following result can be found in Steel and Penny (2004).

Theorem 9.6.3. Suppose that $\mathcal{S}$ is ample. Then, under the model in which the characters evolve independently under the same Poisson model on *R*, the MP trees and the MPL trees for $\mathcal{C}_{\mathcal{S}}$ coincide. Furthermore, the MPL value is given by

$$L_{\mathrm{mp}}(T|\mathcal{C}_{\mathcal{S}}) = \left[\frac{1}{k(r-1)} \left(1 - \frac{1}{k}\right)^{k-1} \right]^{n-1}$$

where *k* is the length of the sequences, and *r* is the size of the state space.

9.7 More general models; the probability of homoplasy-free evolution

In this section we investigate a more general class of Markov processes than the simple Poisson model. For these models we ask the question of how likely it is that a character has evolved without homoplasy. This question has been investigated for the two-state Poisson model (and pairs of taxa) by Chang and Kim (1996). Here we consider more general processes on a larger state space, and for many taxa. Consequently we obtain bounds rather than the exact expressions that are possible in the simpler setting of Chang and Kim (1996).

To introduce the more general class of Markov processes, we note that many processes involving simple reversible models of change can be modeled by a random walk on a regular graph. To explain this connection, suppose there are certain 'elementary moves' that can transform each state into some 'neighboring' states. In this way we can construct a graph from the state space, by placing an edge between state α and state β precisely if it is possible to go from either state to the other in one elementary move. The graph so obtained is said to be *regular*, or more specifically *d-regular* if each state is adjacent to the same number *d* of neighboring states.

For example, aligned sequences of length *N* under the *r*-state Poisson model can be regarded as a random walk on the set of all sequences of length *N* over *R*; here an elementary move involves changing the state at any one position to some other state (chosen uniformly at random from the remaining *r* − 1 states). Thus the associated graph has r^N vertices and it is $N(r-1)$-regular.

As another example, consider a simple model of (unsigned) genome rearrangement where the state space consists of all permutations of length *N* (corresponding to the order of genes 1, ..., *N*) and an elementary move consists of an inversion of the order of the elements of the permutation between positions *i* and *j*, where this pair is chosen uniformly at random from all such pairs between $\{1, \ldots, N\}$. In this case the state space has size *N*! and the graph is *d*-regular for $d = \binom{N}{2}$.

Both of the graphs we have just described have more structure than mere *d*-regularity. To describe this we recall the concept of a Cayley graph. Suppose we have a (non-abelian or abelian) group $\mathcal{G}$ together with a subset $\mathcal{S}$ of elements of $\mathcal{G}$, with the properties that $1_{\mathcal{G}} \notin \mathcal{S}$ and $s \in \mathcal{S} \Rightarrow s^{-1} \in \mathcal{S}$. Then the *Cayley graph* associated with the pair $(\mathcal{G}, \mathcal{S})$ has vertex set $\mathcal{G}$ and an edge connecting *g* and *g'* whenever there exists some element $s \in \mathcal{S}$ for which $g = g' \cdot s$. To recover the above graph on aligned sequences of length *N* over an *r*-letter alphabet, we may take $\mathcal{G}$ as the (abelian) group $(\mathbb{Z}_r)^N$ and the set $\mathcal{S}$ of all *N*-tuples that are the identity element of $\mathbb{Z}_r$ except on one coordinate. To recover the graph described above for unsigned genome rearrangements we may take $\mathcal{G}$ to be the (non-abelian) symmetric group on N letters and $\mathcal{S}$ to be the elements corresponding to inversions.

The demonstration that such graphs are Cayley graphs has an important consequence: it implies that they also have the following property. A graph $\mathcal{G}$ is said to be *vertex-transitive* if, for any two vertices u and v there is an automorphism of $\mathcal{G}$ that maps u to v. Informally, a graph is vertex-transitive if it "looks the same, regardless of which vertex one is standing at." Clearly a (finite) vertex-transitive graph must be d-regular for some d, and it is an easy and standard exercise to show that every Cayley graph is vertex-transitive (however not every vertex-transitive graph is a Cayley graph, and not every regular graph is vertex-transitive). Thus, there are three properly nested classes of graphs:

$$\text{Cayley graphs} \subset \text{vertex-transitive graphs}$$
$$\subset \text{regular graphs}$$

Given a connected graph $\mathcal{G}$ a (simple) *random walk on a graph* is a walk on the vertices of $\mathcal{G}$ that, from any given position, selects as its next state one of the neighboring vertices (selected uniformly at random). This random process forms a reversible Markov chain. The proof of the following result is given in Appendix 9.1.

Lemma 9.7.1. Suppose W_0, W_1, ... is a random walk on a d-regular graph G. Then, for any two distinct vertices u, v, and any $n \geq 0$,

$$\Pr(W_n = v | W_0 = u) \leq \frac{1}{d} \tag{10}$$

Furthermore, if G is vertex-transitive then

$$\Pr(W_n = u | W_0 = u) = \Pr(W_n = v | W_0 = v) \tag{11}$$

Consider now a continuous-time Markov process $(X_t ; t \geq 0)$ on a finite state space R, and with rate matrix Q. Thus, for any two distinct states α, β, $Q_{\alpha\beta}$ is the instantaneous rate at which state α changes to state β. Suppose that for some fixed positive integer d and some fixed positive real number q we have the following property: for each state $\alpha \in R$ there is some neighborhood $N(\alpha) \subseteq R - \{\alpha\}$ of size d for which, for all $\beta \neq \alpha$ we have

$$Q_{\alpha\beta} = \begin{cases} q, & \text{if } \beta \in N(\alpha) \\ 0, & \text{otherwise} \end{cases} \tag{12}$$

Associated with any such process there is a corresponding graph with vertex set R and where the edge set E is defined by $E = \{\{\alpha, \beta\}: Q_{\alpha\beta} \neq 0, \alpha \neq \beta\}$. Note that this graph is d-regular, and substitution events under a model satisfying (12) corresponds to a random walk on the associated graph. Accordingly we will call any continuous-time Markov process that satisfies (12) a *d-regular walk process*. The equilibrium distribution of any such process is uniform.

Lemma 9.7.2. Let $(X_t; t \geq 0)$ be a d-regular walk process. Then, for any two distinct states α, β, and any values s, $t \geq 0$,

$$\Pr(X_{t+s} = \beta | X_t = \alpha) \leq \frac{1}{d}$$

Proof. For this Markov process, consider the associated graph (R, E). Let M denote the random number of transitions between states, during the interval between time t and $t + s$. Then $\Pr(X_{t+s} = \beta | X_t = \alpha)$ can be written as

$$\sum_{m \geq 0} \Pr(X_{t+s} = \beta | M = m, X_t = \alpha)$$
$$\times \Pr(M = m | X_t = \alpha). \tag{13}$$

Now, $\Pr(X_{t+s} = \beta | M = m, X_t = \alpha)$ is precisely the probability that for a random walk W_n on the graph (R, E) we have $W_m = \beta$ conditional on $W_0 = \alpha$, and by Lemma 9.7.1 this is at most $\frac{1}{d}$. Applying this to the expression for $\Pr(X_{t+s} = \beta | X_t = \alpha)$ given by (13) completes the proof.

The following result shows that for such a Markov process if d is much larger than $2n^2$ (the number of species) then any character generated on a tree with n species will almost certainly be homoplasy-free on that tree.

Proposition 9.7.3. Suppose characters evolve on a phylogenetic tree T according to a d-regular walk process. Let $p(T)$ denote the probability that the resulting randomly generated character χ is homoplasy-free on T. Then

$$p(T) \geq 1 - \frac{(2n - 3)(n - 1)}{d}$$

where $n = |X|$.

Proof. Consider a general Markov process on T with state space R. Suppose that for each arc (u, v) of T and each pair α, β of distinct states in R, the conditional probability that state β occurs at v given that α occurs at u is at most p. Then, from Proposition 7.1 of Semple and Steel (2003) we have $p(T) \geq 1 - (2n - 3)(n - 1)p$. By Lemma 9.7.2 we may take $p = \frac{1}{d}$. The result now follows.

As an example to illustrate Propostion 9.7.3 consider the simple model for random inversions of (unsigned) gene orders mentioned above. If we have L genes then $d = \binom{L}{2}$ and so if we have (say) $n = 10$ genomes each consisting of the same set of $L = 100$ (unsigned) genes that have evolved on a phylogenetic tree, the probability that this character is homoplasy-free on that tree is at least 0.97.

9.8 Results for infinite and large state spaces

Finally, we turn to the question of how many characters we need to reconstruct a large tree if the characters evolve under a Markov model on a large state space.

Markov models for genome rearrangement such as the (generalized) Nadeau–Taylor model (Nadeau and Taylor 1984; Moret *et al.* 2002) confer a high probability that any given character generated is homoplasy-free on the underlying tree, provided the number of genes is sufficiently large relative to $|X|$ (Semple and Steel 2002). In this setting the appropriate limiting model is to assume that every time a substitution occurs a completely new and unique state arises: such a model may be viewed as the phylogenetic analogue of what is known in population genetics as the 'infinite alleles model' of Kimura and Crow (1964).

Mossel and Steel (2004a) recently investigated such a 'random cluster' model on a phylogenetic tree T, which operates as follows. For each edge e let us independently either cut this edge—with probability $p(e)$—or leave it intact. The resulting disconnected graph (forest) G partitions the vertex set $V(T)$ of T into non-empty sets according to the equivalence relation that $u \sim v$ if u and v are in the same component of G. This model thus generates random partitions of $V(T)$, and thereby of X by connectivity, and we will refer to these partitions

as $\bar{\chi}$ and χ, respectively. For an element $x \in X$ we will let $\chi(x)$ denote the equivalence class containing x. We call the resulting probability distribution on partitions of X the *random cluster model* with parameters (T, p) where p is the map $e \mapsto p(e)$. A central result from Mossel and Steel (2004d) was that the number of characters required to correctly reconstruct a fully resolved phylogenetic tree with n leaves grows (with n) at the rate $\log(n)$ provided upper and lower limits to p are specified (and the upper limit is less than 0.5). More precisely, let us suppose for the rest of this section that each value $p(e)$ lies between a value $p_{\min}$ and value $p_{\max}$ where $0 < p_{\min} \leq p_{\max} < 0.5$.

For this model Mossel and Steel (2004d) established the following result: if one independently generates at least

$$\frac{2}{\beta} \log\left(\frac{n}{\sqrt{\epsilon}}\right) \tag{14}$$

characters under this model, where

$$\beta = p_{\min}\left(\frac{1 - 2p_{\max}}{1 - p_{\max}}\right)^4 \tag{15}$$

then with probability at least $1 - \epsilon$, T is the only phylogenetic tree on which the characters are homoplasy-free; furthermore T can be reconstructed from the characters in polynomial time (simulations conducted by Dezulian and Steel (2004) show that even fewer characters may suffice for accurate tree reconstruction than (14) requires, although a logarithmic dependence on n is still provably necessary).

We now provide a similar result for certain regular walk processes on a finite state space. We will show that for a subclass of d-regular walk processes, and provided d grows at least as fast as $n^2 \log(n)$ (where n is the number of leaves of T), then we can generate enough homoplasy-free characters to reconstruct T correctly.

First we describe a subclass of regular walk processes. Suppose that R is a group, and for some subset S (closed under inverses and not containing the identity element of R) we have $Q_{\alpha\beta} = q$ if and only if there exists some element $s \in S$ for which $\beta = \alpha \cdot s$, otherwise for any distinct pair α, β we have $Q_{\alpha\beta} = 0$. Such a process we will call a *group walk process (on the generating set S)*. Clearly a group walk process is a regular process, and the graph (R, E) associated with the regular walk process is the Cayley graph for the pair (R, S). Random walk processes have a further useful property on trees: for each arc $e = (u, v)$ of $T = (V, E)$ consider the event $\Delta(e)$ that the state that occurs at v is different from the state that occurs at u (i.e. there has been a net transition across the edge). By Lemma 9.7.1 (and the fact that the Cayley graph for (R, S) is vertex transitive), it follows that the events $(\Delta(e), e \in E)$ are independent. Let $p'_{min} = \min\{\Pr(\Delta(e)): e \in E\}$, $p'_{max} = \max\{\Pr(\Delta(e)): e \in E\}$, and for any $\epsilon > 0$ let

$$c_\epsilon = \frac{1 + \log(\frac{1}{\sqrt{\epsilon}})}{\beta' \epsilon} \tag{16}$$

where $\beta' = p'_{min}\left(\frac{1 - 2p'_{max}}{1 - p'_{max}}\right)^4$.

We are now ready to state a result for certain Markov processes on large (but finite!) state spaces, which brings together several ideas presented above. Informally, Theorem 9.8.1 states that, for a group walk process, a growth of around $n^2\log(n)$ in the size of the generating set is sufficient (with all else held constant) for producing a sequence of homoplasy-free characters that define T.

Theorem 9.8.1. Suppose characters evolve independently on a fully resolved phylogenetic tree T according to a group walk process on a generating set of size d, where

$$d \geq c_\epsilon \cdot n^2 \log(n)$$

with c_ϵ given by (16) and with $p_{max} < \frac{1}{2}$. Then with probability at least $1 - 2\epsilon$ we can correctly reconstruct the topology of T by generating $\lceil\frac{2}{\beta}\log(\frac{n}{\sqrt{\epsilon}})\rceil$ characters and applying a method such as MP or maximum compatibility.

As an example, consider the group walk process for (unsigned) gene-order reversal mentioned earlier. In this case, for L genes, we have $d = \binom{L}{2}$. Theorem 9.8.1 shows that provided L grows at the rate (with n) at least some constant times $n\sqrt{\log(n)}$ then one can hope to recover fully resolved phylogenetic trees with n leaves from a (logarithmic with n) number of such independent gene-order characters.

Outline of the proof of Theorem 9.8.1. A detailed proof of Theorem 9.8.1 can be found in Mossel and Steel (2004b). Here we simply outline the argument and indicate how it depends on earlier results.

Generate $k = \lceil\frac{2}{\beta}\log(\frac{n}{\sqrt{\epsilon}})\rceil$ characters under a group walk process satisfying condition (12) on a rooted phylogenetic tree. Consider the event H that all of these characters are homoplasy-free on T. Since a group walk process is a regular walk process, satisfying (12), using Proposition 9.7.3 it can be shown that $\mathcal{P}[H] \geq 1 - \epsilon$. Furthermore the probability that T will be correctly reconstructed (using MP or maximum compatibility) from k characters produced by a coupled random cluster model (with $\beta = \beta'$) is at least $1 - \epsilon$ by (14) (recalling that $p_{max} < \frac{1}{2}$). Now, the original k characters induce the same partitions as the coupled random cluster characters whenever event H holds, and $\mathcal{P}[H] \geq 1 - \epsilon$. Consequently, by the Bonferroni inequality, the joint probability that event H holds and that the k characters produced by the coupled process recover T is at least $1 - 2\epsilon$. Thus the probability that the original k characters recover T is at least this joint probability, and so at least $1 - 2\epsilon$, as claimed.

We end this section by noting that a related result—namely the statistical consistency of MP for certain Markov processes on a sufficiently large state space—was established in Steel and Penny (2000). The main difference between that result and Theorem 9.8.1 is that statistical consistency is a limiting statement; it says that as the number of characters becomes large, the probability of recovering the correct tree converges to 1. Theorem 9.8.1 meanwhile provides an explicit bound on the probability of correctly reconstructing the correct tree from a certain given number of characters.

9.9 Concluding comments

MP has continued to provide mathematicians with a rich variety of problems for study. Often these problems have led to elegant and surprising solutions, including the bichromatic binary tree theorem (Carter *et al.* 1990; Erdős and Székely 1993; Steel 1993), the min-max theorem of Erdős and Székely (1992), and the guaranteed embedding of MP trees in median networks due to Bandelt *et al.* (1995). In this chapter we have considered further problems, particularly those concerning the statistical aspects of applying MP to character data on a large state space, and for which some solutions have been proposed. However the reader would be wrong to conclude that MP for even two-state character data is completely understood. Indeed the following problem is still open: under the two-state Poisson process is there a value $p > 0$ so that MP is statistically consistent for all fully resolved trees (having any number of leaves) under the constraint $p(e) = p$ for all edges of the tree? The fact that such a basic question is still open suggests there still await challenges for investigators in future.

9.10 Acknowledgments

We thank the New Zealand Marsden Fund and the New Zealand Institute for Mathematics and its Applications (NZIMA) for supporting this research. We also thank Andrew Hugall for posing a question that led to Theorem 9.6.1, and Joseph Felsenstein, Michael Sanderson, and Cécile Ané for helpful comments on an earlier version of this chapter.

Appendix 9.1 Proof of Proposition 9.4.4, and Lemma 9.7.1

Proof of Proposition 9.4.4. Let $X = \{0, 1\}^k$ and let $\mathcal{C} = \{A_i | B_i, i = 1, \ldots k\}$ where $A_i := \{x \in X : x_i = 1\}$ and $B_i = X - A_i$. We claim that $\mathcal{C}$ has the property described. To this end, suppose that $A | B$ is an X-split that is compatible with every character in $\mathcal{C}$. Let $\mathbf{1} = (1, 1, \ldots, 1) \in X$. Without loss of generality (by interchanging A and B, as well as A_i and B_i if necessary) we may suppose that $\mathbf{1} \in A$ and, for each i, $\mathbf{1} \in A_i$. Note that, by definition, $|A_i| = 2^{k-1}$; also we have $A_i \cap A \neq \emptyset$ for all i. Thus the compatibility of $A | B$ with $A_i | B_i$ ensures that

$$\text{for each } i \text{ either} A_i \subseteq A \text{ or } A \subseteq A_i \text{ or } B_i \subseteq A \quad (17)$$

We then consider two cases:

$(i) \quad |A| < 2^{k-1}$

$(ii) \quad |A| \geq 2^{k-1}$

In case (i) condition (17) and the equality $|A_i| = 2^{k-1}$ ensures $A \subseteq A_i$ for all i. But this means that $A = \{(1, 1, \ldots, 1)\}$ and so $A | B$ is a trivial character. In case (ii) condition (17) and the equality $|A_i| = 2^{k-1}$ ensures that for each i either $A_i \subseteq A$ or $B_i \subseteq A$; in the first case we will let $y_i = 0$ and in the second case we will let $y_i = 1$. Let $y = (y_1, \ldots, y_k)$. Then $A = X - \{y\}$ and so again $A | B$ is a trivial character.

Proof of Lemma 9.7.1. We prove the first claim by induction on n. The result trivially holds for $n = 0$, and for $n = 1$ we have $\Pr(W_1 = v | W_0 = u) \in \{0, \frac{1}{d}\}$ since the graph is d-regular, and so (10) holds. Suppose (10) holds for $n = k$. Then by the elementary theory of Markov chains,

$$\Pr(W_{k+1} = v | W_0 = u)$$
$$= \sum_w \Pr(W_1 = w | W_0 = u) \Pr(W_k = v | W_0 = w)$$
$$(18)$$

Letting $N(u)$ denote the set of vertices that neighbor u the right-hand term in (18) is

$$\frac{1}{d} \sum_{w \in N(u)} \Pr(W_k = v | W_0 = w)$$
$$= \frac{1}{d} \sum_{w \in N(u)} \Pr(W_k = w | W_0 = v)$$
$$(19)$$

where the equality in (19) arises since the chain-transition matrix is symmetric and so $\Pr(W_k = v | W_0 = w) = \Pr(W_k = w | W_0 = v)$. Combining (18) and (19) we have

$$\Pr(W_{k+1} = v | W_0 = u)$$
$$= \frac{1}{d} \sum_{w \in N(u)} \Pr(W_k = w | W_0 = u) \leq \frac{1}{d}$$

so that (10) holds for $n = k + 1$, establishing the induction step and thereby the lemma.

The proof of (11) in Lemma 9.7.1 is similar but easier.

V

Parsimony and genomics

CHAPTER 10

Using phylogeny to understand genomic evolution

David A. Liberles

10.1 Introduction

As genome-sequencing projects have propagated, comparative genomics has emerged as a method of choice for understanding protein function. Simple approaches for comparing sequences, like relative entropy (Shenkin *et al.* 1991) or binary transformations of gene-content comparisons (Gaasterland and Ragan 1998; Pellegrini *et al.* 1999) have been presented. However, phylogenetic methods that explicitly consider evolutionary history are not only more powerful, but enable additional types of analysis drawing on knowledge in parallel fields, such as ecology, anthropology, and geology. This chapter will focus both on methodological issues and on their application to real genomic-scale problems.

Parsimony and maximum likelihood are two phylogenetic approaches that are used and often compared side by side. While the choice between them has been contentious at times, they frequently give similar results and, where they don't, they can complement each other. Maximum likelihood works well when a good model is available. Parsimony works well when a good model does not or cannot exist, as for very complex processes, and also along very short branches where multiple events per position (as in a sequence) are extremely infrequent.

Both methods can be used to estimate ancestral states in a phylogenetic tree. Fitch (1971) famously provided an algorithm for parsimony reconstruction of ancestral character states in a rooted phylogenetic tree. This approach is depicted in Fig. 10.1. Variations on this approach, including branch length weighting, have been implemented

more recently (Liberles 2001). Increasingly sophisticated maximum-likelihood approaches for determining ancestral sequences have also been developed (Yang *et al.* 1995b; Koshi and Goldstein 1996; Pupko *et al.* 2000, 2002). Parsimony-based ancestral character reconstruction is fast and can be performed easily in large-scale genomic applications.

Both explicit ancestral sequence reconstruction (from either parsimony or maximum likelihood) and maximum likelihood methods can be used to estimate the evolution that occurred along any given branch of a phylogenetic tree. Using explicitly reconstructed ancestral sequences, one can examine the difference between nodes connected by a branch (see Fig. 10.2). This gives a reconstructed picture of evolution that is predicted to have occurred along any branch of interest in a phylogenetic tree.

10.2 Gene sequence evolution

In a phylogenetic tree based upon gene sequences, branches correspond to periods of evolution following speciation events or to periods of evolution following gene duplication or gene transfer events. Genes related most recently by a node representing a speciation event are called orthologs, while genes related most recently by a node representing a gene duplication event are called paralogs. All such genes related by common ancestry are called homologs. A species tree is frequently derived from either the fossil record or from sets of genes that are believed to be

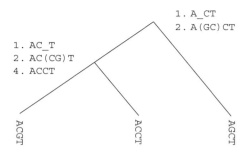

Ordered set of actions for Fitch parsimony on a rooted tree

1. Going up the tree from the extent species, at each node take the intersection of possible characters from the descendants.
2. If the intersection is a null set, then take the union. Do this for all nodes until you reach the root. You now have the preliminary nodal set. Now work back down the tree.
3. At a node, if the preliminary nodal set contains all of the characters present in the final set of the ancestor, go to 4, otherwise go to 5.
4. Eliminate all characters from the set that are not in the final set of the immediate ancestor. Continue with the next node.
5. If the set was formed by a union, go to 6, otherwise go to 7.
6. Add to the set any characters not present that appear in the final set of the immediate ancestor. Continue with the next node.
7. Add any characters not present that are present in both the final set of the immediate ancestor and the current set in at least one of the two descendants. Continue with the next node.
8. Finally, eliminate possible links involving mutations to characters added in steps 3–7.

Figure 10.1 Ancestral sequences are calculated over a rooted phylogenetic tree according to the approach of Fitch (1971). At each of the two nodes, the sequence obtained after each step is indicated.

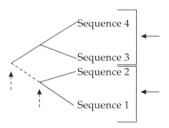

Figure 10.2 In looking for significant events that have occurred between sequence 1 or 2 and sequence 3 or 4, pairwise comparison or phylogenetic analysis to determine evolution along a branch are possible. The pairwise comparisons will average over four branches, while phylogenetic methods allow individual analysis of each of the four branches (for example, the differences between reconstructed ancestral sequences at nodes connected by a branch). If a function-changing event occurred somewhere along the dashed branch, analysis considering only that branch will have a lower signal-to-noise ratio than the pairwise comparison, increasing the chance of detection.

orthologous. Understanding the evolution of genes in a genome in the context of the species tree requires mapping of the gene tree (and the events it represents) on to a species tree representing the history of life on Earth. These concepts are displayed in Fig. 10.3.

10.3 Mapping gene trees onto species trees

Several approaches are available for doing this mapping. Goodman and coworkers (1979) introduced a rigid parsimony approach to mapping gene trees on to species trees. More recently, a Bayesian approach has been developed (Arvestad *et al.* 2003). In these approaches, fixed binary gene and species trees are used. However, not all sections of a genome necessarily show the same ancestral history, especially in periods where rapid successive speciation events may have led to differential fixation of shared ancestral polymorphisms, or where lateral gene transfer has been common. An alternative soft parsimony-based approach that allows for non-binary species trees and uncertainty in gene trees has recently also been developed (Steffansson 2004). Specifically,

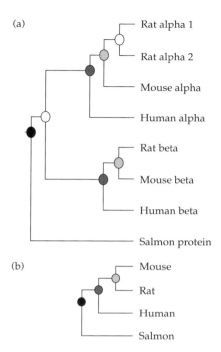

Figure 10.3 (a) A gene tree is indicated for an idealized gene family. Gene duplication events are shown with white circles, while speciation events are shown in various shades of grey. Orthologs are proteins related by a speciation event at the last common ancestor in a phylogenetic tree, while paralogs are related by a gene duplication event. With respect to each other, rat alpha 1 and rat alpha 2 are paralogs, as are the alpha and beta proteins with respect to each other. All mammalian proteins are co-orthologs of the salmon protein, as are both rat alpha proteins with respect to the mouse and human alpha proteins. (b) The species tree for rat, mouse, human, and salmon is shown. Speciation nodes that correspond to each other are indicated in the same shades of grey in a mapping of the gene tree from (a) on to this species tree.

this approach considers gene trees that map on to non-binary nodes in species trees with different resolutions as equally parsimonious to those that resolve non-binary nodes consistently throughout the gene tree. Further work will continue to improve methods for mapping of real data on to relevant species trees, as well as improve the species trees themselves.

With the framework that has been established above, it is now possible to analyze various genomic data in the context of species trees. Koonin and coworkers (Koonin *et al.* 2004; see also Chapter 11 in this book) have done this with gene content, with quite interesting results. The analysis from

Koonin's work is based upon complete genomes. This allows a definitive statement about presence and copy number, or absence of genes from a genome. Gene families, like those found in HOVERGEN (Duret *et al.* 1994), the Master Catalog (Benner *et al.* 2000), and The Adaptive Evolution Database (TAED; Liberles *et al.* 2001), are based upon gene or domain families, or independently evolving units in the case of the Master Catalog. Independently evolving units are pieces of a gene that are found as a self-contained gene in at least one organism or secondarily are found in conjunction with gene segments that are self-contained in another species. Families based upon genes or independently evolving units permit an assessment of more species (including those without complete genomes), but only allow a statement of presence and minimum copy number, not of absence. This can all be combined to give an increasingly comprehensive picture of the genes common to various last common ancestral points in the tree of life, which is presented in Chapter 11 (see also Koonin *et al.* 2004) using a Dollo parsimony approach (allowing gene loss but not *de novo* gene gain) to gene content from various completed genomes.

Such approaches contrast with the non-phylogenetic analyses done using a binary transformation of gene-content data from various complete genomes, called, ironically enough, phylogenetic profiling (Pellegrini *et al.* 1999). This is used as a method to identify functions in bacterial genes without known functions. The principle behind phylogenetic profiling is that proteins performing basic interacting functions for an organism will be conserved together. Nonidentical profiles can actually be analyzed using a parsimony analysis over a species tree (Liberles *et al.* 2002). Incorporating phylogeny into the analysis improves its performance, but the method is still not trustworthy for blind prediction of gene functions without additional information from databases or experiments (Marcotte *et al.* 1999; Liberles *et al.* 2002).

10.4 Understanding gene function

Phylogeny can also be used to understand gene function in other ways. Within-sequence evolution in gene families can be understood in a phylogenetic context. This of course includes evolutionary divergence of both paralogs and orthologs.

To begin with paralogs, Ohno (1970) saw gene duplication as the driving force for innovation. Gene duplication, under a purely neutral mechanism, led to relaxation of selective constraint on both duplicate copies. Both were then free to explore sequence space until one copy no longer achieved the basic function necessary in the genome. That copy remained free to evolve, while the other copy became constrained to uphold the ancestral function. Possible fates for the freely evolving copy were neofunctionalization (the evolution of a new function) and pseudogenization (the loss of gene function).

This theory has been extended by Lynch and coworkers (Force *et al.* 1999), who have proposed a third fate: subfunctionalization. Subfunctionalization occurs when part of the sequence or its regulatory regions becomes modified or inactivated in one copy while another region becomes modified or inactivated in the other copy. Both copies are then required in the genome to perform the ancestral function. Subfunctionalization can also be viewed as a transition state to neofunctionalization, where the sequence freed from constraint in each copy can evolve to either optimize the original activity or develop a new activity.

Neofunctionalization, however, is not limited to paralogs. Both paralogs and orthologs can evolve new functions under neutral or positive selection pressures. An innovative combination of phylogeny, ancestral sequence reconstruction, and evolutionary theory was presented by Messier and Stewart (1997) in examining the evolution of primate lysozyme orthologs.

Under a neutral evolutionary model, the rate of substitution at nucleotide positions that can change the encoded amino acid (called the non-synonymous nucleotide substitution rate, Ka, or dN) should be equal to the rate of substitution at nucleotide positions where substitution does not change the encoded amino acid (called the synonymous nucleotide substitution rate, Ks, or dS). Most protein-encoding genes in a comparison of closely related species show a Ka/Ks ratio significantly less than 1. This is not surprising and is indicative of negative selection or conservation. Most proteins have been optimized over millions of years of evolution for a given function. Therefore, any given mutation is more likely to decrease fitness and be selected against, reducing the rate of substitution at such sites. In rarer cases, in which adaptation involving modification of protein function is one of several possible causes, mutations increase fitness and are selected for, resulting in $Ka/Ks \gg 1$. This is called positive selection.

10.5 Case studies of gene-family evolution

We now turn back to the innovation of Messier and Stewart (1997). Lysozyme is a bacteriolytic enzyme that is widespread among species. Colobine monkeys are the only primates with a foregut, where bacteria ferment edible plant material before passing digested food to a true stomach with high levels of lysozyme. Other primates have a 'simple stomach' with a different anatomy. Instead of just comparing the Ka/Ks ratios between extant species, Messier and Stewart also calculated ancestral sequences at various nodes in the phylogenetic tree and calculated Ka/Ks ratios over the tree along these branches. This implicated not just the branch leading to colobine monkeys as being under positive selective pressure, but also, unexpectedly, the branch leading to hominids (see Fig. 10.4). The events driving this positive selection during the emergence of hominids are not clear, but may correspond to dietary changes. Leptin, another dietarily important protein, also appears to have been under positive selective pressure at the same time (Benner *et al.* 1998).

Coupling ancestral sequence reconstruction to phylogeny in searching for periods of positive selection pressure allows more-precise dating of such selective regimes and increases the power in detecting them, when compared with pairwise calculation involving extant sequences (again, see Fig. 10.2). Of course, it is also possible to evaluate such scenarios using a likelihood-based approach involving nested models. This was done by Yang (1998) for the lysozyme data set of Messier and Stewart (1997), largely confirming the original results.

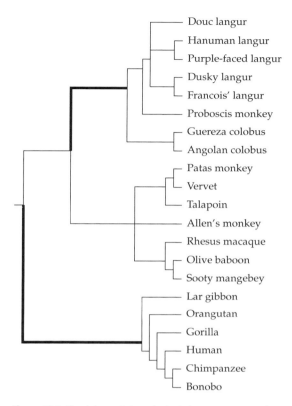

Figure 10.4 The phylogenetic tree of primate lysozyme sequences from Messier and Stewart (1997) is shown. After reconstruction of ancestral sequences and calculation of *Ka/Ks* ratios along each branch, the two branches shown with thick lines showed evidence of positive selection pressures for adaptation of the encoded lysozyme protein.

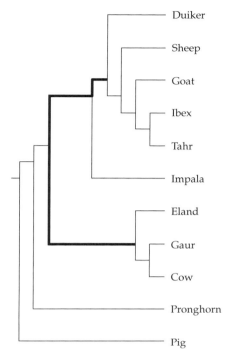

Figure 10.5 The phylogenetic tree of ruminant myostatin sequences from Tellgren *et al.* (2004) is shown. After reconstruction of ancestral sequences and calculation of *Ka/Ks* ratios along each branch, the three branches shown with thick lines showed evidence of positive selective pressures for adaptation of the encoded myostatin protein.

Another interesting case study is that of myostatin. Myostatin has been implicated in the double-muscling phenotype in cattle and other mammals, where some breeds of cattle have twice the number of a specific type of muscle fiber due to mutation in myostatin. From comparative sequencing of various ruminant species (relatives of cattle), myostatin was found to be under positive selective pressure during the divergence of bovids and Antilopinae (sheep, goats, and close relatives; Tellgren *et al.* 2004). This was demonstrated using the approach of Messier and Stewart (1997) as well as that of Yang (1998), as seen in Fig. 10.5. From this analysis, a key protein regulating skeletal muscle appears to have changed function as ruminants diverged, possibly enabling phenotypic divergence. Interestingly, this protein

has also been under positive selective pressure following gene duplication in teleost fish (Liberles *et al.* 2001) and may be a more general selectable regulator for modulating muscle mass. The myostatin gene itself encodes a signal peptide, a regulatory propeptide, and the mature protein. Positive selective pressures in ruminants have acted on both the regulatory propeptide and the mature protein. The more detailed molecular and structural basis of this positive selection is currently under further investigation.

In another interesting case study, perhaps inspired by *Jurassic Park* (Crichton 1990), Chang *et al.* (2002) sought to reconstruct a visual pigment of the last common ancestor of alligators and birds, which also is the last common ancestor with carnivorous dinosaurs (see Fig. 10.6). Through analysis of the ancestral protein's function, insight in to the visual capabilities of this long-extinct

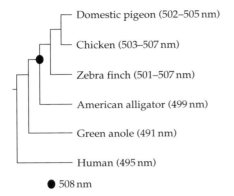

Domestic pigeon (502–505 nm)

Chicken (503–507 nm)

Zebra finch (501–507 nm)

American alligator (499 nm)

Green anole (491 nm)

Human (495 nm)

● 508 nm

Figure 10.6 From Chang *et al.* (2002), the ancestral sequence for an archosaur visual pigment was determined, indicated by the circled node. In their study, additional outgroup sequences besides green anole and human were included. The optimal absorption spectra of proteins from the extant species are indicated. The archosaur sequence was synthesized experimentally and its optimal absorption was determined to be slightly red-shifted compared to the extant sequences.

organism are possible. Using only four ingroup species (American alligator plus three birds: domestic pigeon, chicken, and zebra finch) plus a collection of outgroup species, the ancestral sequence was reconstructed for the ancient archosaur using a phylogenetic approach. Only three positions were found to be ambiguous by method and several variants of the ancestral protein were considered. Ancestral proteins were ultimately synthetically reconstructed in the laboratory and the ambiguous positions proved not to be functionally important. The visible absorption maxima were measured, and the maximum of the protein from the ancient archosaur was red-shifted compared with values reported for most extant birds and reptiles. The further implication was that the ancient archosaur had dim-light vision and may have been nocturnal rather than diurnal.

The pioneering approach of Jermann *et al.* (1995) to study the evolution of function of ribonucleases, which was adopted by Chang *et al.* (2002), is an increasingly important combination of computational phylogenetics and experimental molecular biology/functional genomics used to study protein function. This can be coupled to periods of high *Ka/Ks* ratio or rapid sequence evolution, where any functional changes are assessed before and after periods of positive selective pressure. Ultimately, such approaches may be powerful, not just for understanding nature, but also for identifying key residues that can then be used in protein engineering.

The case studies presented above represent a small number of the growing set of examples studied individually in detail (see Yang and Bielawski 2000). As genome and individual gene-sequencing data have amassed, it has also been possible to apply *Ka/Ks* phylogenetic methods systematically.

10.6 Large-scale analysis

The first approaches to search for positive selection in large data sets did not use phylogeny. An early pairwise comparison of 363 mouse and rat homologs yielded only interleukin-3 as being under positive selection pressure (Wolfe and Sharp 1993). A subsequent systematic comparison by Gojobori and coworkers (Endo *et al.* 1996) examined 3 595 gene families from GenBank. Positive selection in at least half of the pairwise comparisons was seen as evidence for a gene family under positive selective pressure. Using these criteria, only 17 gene families were identified. CDC6, snake neurotoxin, and prostatic steroid-binding protein were the only eukaryotic examples, the latter two in chordates.

A systematic approach using methodology similar to that of Messier and Stewart (1997) was applied to Master Catalog families (Benner *et al.* 2000) in chordates and higher plants. Following a parsimony mapping of gene trees on to species trees, this was collected in a phylogenetic context in the original version of The Adaptive Evolution Database (TAED) (Liberles *et al.* 2001). On the chordate side, 5305 gene families were analyzed with 280 families containing 643 positively selected branches spanning over 63 branches of the National Center for Biotechnology Information (NCBI) taxonomy (Benson *et al.* 2004). A picture for a greater role for positive selection was beginning to emerge. The approach used in the original calculation of this database was approximate, but it was still likely to be conservative, given that it averaged over all sites in a protein.

10.7 Positive selection, protein structure, and coevolution

From many protein structure-function studies, we know that some residues play key scaffolding roles, other residues are involved in surface interactions with solvent, and additional residues perform catalysis, binding, and other functions. In a protein where function is being modified, probably only a subset of these residues corresponding to the nature of the modification is likely to be under positive selective pressure, while the remainder are not.

Further, structure itself can drive positive selective pressure. This can represent a real change, such as coevolving sites modifying a binding affinity. This can also be driven by compensation in interacting residues for slightly deleterious substitutions. As has been seen before for other cases, using phylogeny seems to be the best way to detect intramolecular covariation, from the co-occurence of such sites along such branches separating parsimony-reconstructed ancestors (as in Fukami-Kobayashi *et al.* 2002). This information can actually be used in both structure prediction and phylogenetic reconstruction.

The intermolecular coevolution of proteins can also be studied phylogenetically. Both adaptation and compensatory covariation can explain the correlated evolution of residues in a protein. One interesting case where this has been detected is the evolution of the interaction between leptin and the extracellular domain of its receptor in higher primates. Both show high $\frac{Ka}{Ks}$ periods in several branches during the diversification of primates (Benner *et al.* 1998). Interestingly, both proteins appear to have evolved new gene expression patterns during this period due to the action of transposition in enhancer and splicing regulatory regions of the respective genes (Bi *et al.* 1997; Kapitinov and Jurka 1999).

This type of analysis can be extended to whole pathways, combining genes under positive selection with information on either metabolism or protein interaction. A list of 19 proteins showing evidence for positive selection in *Helicobacter pylori* was published recently (Davids *et al.* 2002). Around the same time, a protein–protein interaction map for *H. pylori* based upon the two-hybrid

system was published (Rain *et al.* 2001). From mapping one data set on to the other, two pathways with multiple positively selected hits were identified (H. Ardawatia and D. A. Liberles, unpublished observations). While the significance of this is still under investigation, this type of approach can increasingly link sequence evolution in the context of phylogeny with the growing field of systems biology.

10.8 Continuous-character ancestral-state reconstruction

Beyond sequence evolution, parsimony and phylogeny have other applications in genomics. Both gene expression and mRNA splicing are important processes regulating how the genome is converted to the proteome. Their regulation, evolution, and ultimately the species-specific effects caused by this combination are, so far, less well understood than sequence evolution.

Data from large-scale gene expression studies and also the relative abundance of alternative mRNA transcripts are continuous rather than discrete characters (as in individual sequence positions). The reconstruction of continuous characters over a phylogeny using the principle of parsimony can be turned in to a minimum-evolution-type distance method (Rossnes 2004). The methodology here is similar (but not identical) to continuous-parsimony approaches like Wagner parsimony (see Kluge and Farris 1969) extended to a rooted tree, without any assumptions of ordered numerical transitions that may not be appropriate for gene expression data. The implementation by Rossnes (2004) is shown by example in Fig. 10.7. A range of values consistent with parsimony or minimum evolution is obtained. The midpoint of this range can be selected if conservatism is desired or an unchanging model of regulatory evolution is anticipated.

Subsequently, looking for branches with significantly different changes in value can be coupled to a traditional reconstruction and branch analysis of the regulatory sequences (upstream regions in the case of gene expression). This can be used to reduce the signal-to-noise ratio in identifying sites with important functional roles

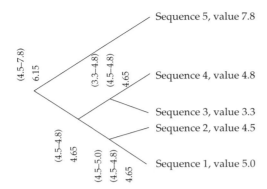

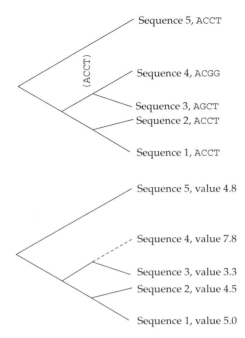

Figure 10.7 For reconstruction of continuous character traits, like relative expression or splice-site usage values, a minimum-evolution distance method has similar properties to a parsimony reconstruction. After going up and then down the tree, the ranges that minimize the total branch length of the tree are indicated next to the nodes. Ultimately midpoint values can be selected if conservatism is desired or if a homogeneous or non-episodic model of evolution is believed. Several measures of along-branch change and significance are possible.

in regulating gene expression or mRNA splicing. Additionally, the reconstructed nodes can be used to address gene expression or relative mRNA abundances of differently spliced variants in ancient organisms. This is shown in Figs 10.7 and 10.8. If one evaluates the absorption maxima of Chang *et al.* (2002), using this approach one obtains a value that is less red-shifted than the sequence evolution gives. However, a fuller characterization of the statistical properties (including variance estimates) of this approach is required to assess any possible incompatibilities of the two results.

On the side of gene expression, Pääbo and coworkers (Enard *et al.* 2002) have produced an innovative data set with comparative expression levels of the same genes in human, chimpanzee, orangutan, and rhesus macaque. This data set has been utilized generally to detect significant changes in expression of genes expressed in human brain relative to the other primates, compared to the changes in liver-expressed genes (Enard *et al.* 2002; Gu and Gu 2003). Interspecific data sets like this, motivated by phylogenetic knowledge, will become increasingly common, paving the way for analyses like that described above.

On the mRNA-splicing side, ASD, the Alternative Splicing Database, has been established for comparative genomics purposes (Thanaraj *et al.*

Figure 10.8 For reconstructed ancestral states of a continuous character, an idealized branch with significant change is shown with a dashed line. The discrete molecular characters that are candidates for regulating the process can be reconstructed simultaneously (for example, regions upstream of a gene for gene expression, or intron/exon boundary sequences for alternative mRNA splicing). In this case, one might point to a CT → GG substitution as a candidate for driving the change seen along the dashed branch. This method benefits from an improvement in signal-to-noise ratio, as previously indicated in Fig.10.2.

2004). This database contains alternative-splicing events available from alignments of expressed sequence tag (EST)/cDNA data sets from human, mouse, rat, cow, chicken, zebrafish, fruit fly, nematode, and mustard weed. As data become available from more closely related species, this data base will become increasingly amenable to phylogenetic analysis.

Starting with a set of the most closely related species, human and the rodents mouse and rat, Modrek and Lee (2003) showed that the most common splice forms were much more conserved than the alternative-splice forms. The alternative forms were viewed as an opportunity for evolutionary innovation. Extending this work, it was shown that the rare splice variants that were conserved were under strong conservative selective pressure (Resch *et al.* 2004). This implies that these splicing patterns have been present in the last common ancestor of the species studied. Extending this analysis phylogenetically as databases like ASD grow in their species coverage will enable a more systematic evolutionary analysis of the role of alternative splicing in cellular biology and regulation. This will ultimately allow a quantifiable view of the role of new splice variants in the generation of evolutionary molecular innovation leading to adaptation.

10.9 Conclusion

Ultimately, evolution of various biological processes including gene content, gene-sequence evolution, gene expression, and mRNA splicing can be collected in a phylogenetic context after mapping onto a species tree. This will enable an understanding of the molecular genomic events underpinning phenotypic evolution. Parsimony will remain a valuable method for pursuing this research goal.

10.10 Acknowledgments

I am grateful to Matthew Betts, Roald Rossnes, Himanshu Ardawatia, and Jessica Liberles for careful reading of this chapter. I also thank Victor Albert for his input and for reading a draft version of the chapter as well.

CHAPTER 11

Dollo parsimony and the reconstruction of genome evolution

Igor B. Rogozin, Yuri I. Wolf, Vladimir N. Babenko and Eugene V. Koonin

11.1 Introduction

The Dollo parsimony method, which was first formalized by J. S. Farris in 1977 (Farris, 1977a, b), is based on the assumption that a complex character that has been lost during evolution of a particular lineage cannot be regained. When applicable, this principle leads to a substantial simplification of evolutionary analysis and provides for unambiguous reconstruction of evolutionary scenarios, which may not be attainable with other methods. In this chapter, we describe applications of Dollo parsimony to the quantitative analysis of the dynamics of genome evolution. Dollo parsimony is the method of choice for reconstructing evolution of the gene repertoire of eukaryotic organisms because, although multiple, independent losses of a gene in different lineages are common, multiple gains of the same gene are improbable. This contrasts with the situation in prokaryotes where the widespread occurrence of horizontal gene transfer makes multiple gains possible, thereby invalidating the Dollo principle. We apply Dollo parsimony to reconstruct the scenario of evolution for the genomes of crown-group eukaryotes by assigning the loss of genes and emergence of new genes to the branches of the phylogenetic tree, and delineate the minimal gene sets for various ancestral forms. A similar analysis, with rather unexpected results, was performed to infer gain vs. loss of introns in conserved eukaryotic genes. We discuss the applicability of the Dollo principle for these and other problems in evolutionary genomics.

11.2 Dollo parsimony for molecular data in the pre-genomic era

Dollo's Law, also known as the Law of Phylogenetic Irreversibility or the Law of Irreversible Evolution, is an important tenet of evolutionary theory, formulated by the Belgian biologist Louis Dollo (1893). It basically states that organisms cannot re-evolve along lost pathways, but must find alternate routes because the same fortuitous combination of mutational events, being completely random, will never repeat. Dolphins, in other words, will never again walk on land with re-evolved pelvic appendages that derive from their current remnant structures that correspond to legs of land animals. They might, however, evolve walking appendages that derive from other biological provenance, especially if there were some selective advantage to do so, say, if the oceans began to dry up. While some non-Darwinian theorists have attempted to use Dollo's Law to promote their cause, Dollo was simply seeking to explain convergence of form in diverse species (e.g. sharks, ichthyosaurs, and dolphins): ichthyosaurs and dolphins look so similar because they have converged to the same (hydrodynamically favorable) shape through independent, parallel paths of degradative evolution.

The Dollo parsimony method was first formalized by Farris (1977a). In its simplest form, the algorithm explains the presence of the complex, derived state 1 by allowing only one forward change $0 \rightarrow 1$ (where 0 is the primitive ancestral state) and as many reversions $1 \rightarrow 0$ as are necessary to explain the observed pattern of states. The method attempts to minimize the required number of $1 \rightarrow 0$ reversions. In molecular studies, Dollo parsimony analysis was often applied for analysis of restriction sites because the loss of an existing restriction site is more probable than a parallel gain of the same site at any particular location (DeBry and Slade 1985).

11.3 Genome evolution

Comparative genomics has already changed our understanding of genome evolution. In what might amount to a paradigm shift in evolutionary biology, genome comparisons have shown that lineage-specific gene loss and horizontal gene transfer (HGT) are not inconsequential freak incidents of evolution but rather extremely common phenomena. To a large degree, these processes have shaped extant genomes, at least those of prokaryotes (Doolittle 1999; Koonin *et al.* 2000; Gogarten *et al.* 2002; Snel *et al.* 2002; Mirkin *et al.* 2003). The extent of gene loss occurring in certain lineages of prokaryotes, particularly parasites, is astonishing: in some cases, >80% of genes in the genome have been lost over approximately 200 million years of evolution (Moran 2002). HGT is harder to document, but a strong case has been made for its extensive contribution to the evolution of prokaryotes (Ochman *et al.* 2000; Koonin *et al.* 2001; Gogarten *et al.* 2002; Mirkin *et al.* 2003). Gene exchange between phylogenetically distant eukaryotes does not appear to be an important evolutionary phenomenon. In contrast, the contribution of gene loss to the evolution of eukaryotic genomes was probably substantial, although the level of genome fluidity observed in prokaryotes is unlikely to have been attained in eukaryotic evolution. A comparison of the genomes of two yeasts, *Saccharomyces cerevisiae* and *Schizosaccharomyces pombe*, showed that, in the *S. cerevisiae* lineage, up to 10% genes have been lost since the divergence of the two species (Aravind *et al.* 2000). In eukaryotic parasites with small genomes, such as the microsporidia, much more extensive gene elimination appears to have occurred (Katinka *et al.* 2001). In contrast, the extent of gene loss in complex, multicellular eukaryotes remains unclear, although

the small number of unique genes in the human genome when compared to the mouse genome (and vice versa) suggests considerable stability of the gene repertoire in mammals (Waterston *et al.* 2002).

11.4 Orthologous and paralogous genes

Sequencing of multiple genomes from diverse taxa provides the data required for quantitative analysis of the dynamics of genome evolution. A prerequisite for such studies is a classification of the genes from the sequenced genomes based on homologous relationships. The two principal categories of homologs are orthologs and paralogs (Fitch 1970; Sonnhammer and Koonin 2002; Storm and Sonnhammer 2003). Orthologs are homologous genes that evolved via vertical descent from a single ancestral gene in the last common ancestor of the compared species. Paralogs are homologous genes, which, at some stage of evolution, have evolved by duplication of an ancestral gene. Orthology and paralogy are two sides of the same coin because, when a duplication (or a series of duplications) occurs after the speciation event that separated the compared species, orthology becomes a relationship between sets of paralogs, rather than between individual genes; genes that belong to such orthologous sets are sometimes termed co-orthologs (Sonnhammer and Koonin 2002).

Robust identification of orthologs and paralogs is critical for the construction of evolutionary scenarios, which include, along with vertical inheritance, lineage-specific gene loss and, possibly, HGT (Snel *et al.* 2002; Mirkin *et al.* 2003). The algorithms for the construction of these scenarios involve, in one form or another, tracing the fates of individual genes, which is feasible only when orthologous (including co-orthologous) relationships are known. In principle, orthologs, including co-orthologs, should be identified by phylogenetic analysis of entire families of homologous proteins, which is expected to define orthologous protein sets as clades (e.g. Sicheritz-Ponten and Andersson 2001). However, for genome-wide protein sets, such analysis remains labor-intensive and error-prone. Thus, procedures have been developed for identification of sets of probable orthologs without explicit use of phylogenetic methods. Generally, these approaches are based on the notion of a genome-specific best hit (BeT), i.e. the protein from a target genome that has the greatest sequence similarity to a given protein from the query genome (Tatusov *et al.* 1997; Huynen and Bork 1998). The central assumption is that orthologs have a greater similarity to each other than to any other protein from the respective genomes due to the conservation of functional constraints. Of course, evolutionary history and sequence similarity can be at odds in some cases, which would invalidate this model (Storm and Sonnhammer 2003). The extent to which this occurs in practice is in fact not known, again due to the enormity of proteome-scale phylogenetic surveys.

When multiple genomes are analyzed using the BeT approach, pairs of probable orthologs detected on the basis of BeTs are combined into orthologous clusters represented in all or a subset of the analyzed genomes (Tatusov *et al.* 1997; Montague and Hutchison 2000). This approach, amended with procedures for detecting co-orthologous protein sets and for treating multidomain proteins, was implemented in the database of Clusters of Orthologous Groups (COGs) of proteins (Tatusov *et al.* 1997, 2003). The current COG set includes approximately 70% of the proteins encoded in 69 genomes of prokaryotes and unicellular eukaryotes (Tatusov *et al.* 2003). The COGs have been extensively employed for genome-wide evolutionary studies, functional annotation of new genomes, and target selection in structural genomics (Koonin and Galperin 2002 and references therein). Recently, we extended the system of orthologous protein clusters to complex, multicellular eukaryotes by constructing clusters of euKaryotic Orthologous Groups (KOGs) for seven sequenced genomes of animals, fungi, microsporidia, and plants; namely humans (Hs), the nematode *Caenorhabditis elegans* (Ce), the fruit fly *Drosophila melanogaster* (Dm), two yeasts, *S. cerevisiae* (Sc) and *S. pombe* (Sp), and the green plant *Arabidopsis thaliana* (At) (Tatusov *et al.* 2003; Koonin *et al.* 2004).

11.5 Matrices of character presence/absence and Dollo parsimony

A simple but critically important concept that was introduced in the context of the COG analysis is a phyletic (phylogenetic) pattern, which is the pattern of representation (presence/absence) of the analyzed species in each COG (Tatusov *et al.* 1997; Koonin and Galperin 2002). Similar notions have been independently developed and applied by others (Gaasterland and Ragan 1998; Pellegrini *et al.* 1999). The COGs show a wide scatter of phyletic patterns, with only a small minority (approximately 1%) represented in all included genomes. Similarity and complementarity among the phyletic patterns of COGs have been successfully employed for prediction of gene functions (Galperin and Koonin 2000; Koonin and Galperin 2002; Myllykallio *et al.* 2002; Levesque *et al.* 2003). Phyletic patterns can be formally represented as strings of ones for presence of a species and zeros for absence of a species; Table 11.1, which can be easily input to a variety of algorithms. The evolutionary parsimony methods are among those that naturally apply to this type of data. Pairs of neighboring genes and intron positions also can be represented as a character matrix and used for parsimony analysis (see discussion below).

A Dollo parsimony tree can be constructed using a matrix of gene (pair of genes/intron) presence/absence and the data-dependent reliability of the tree topology can be assessed in the standard manner using the bootstrap method. The presence vs. absence of a gene in a genome can be naturally treated in terms of character states. The Dollo model is based on the assumption that each derived character state (in this case, the presence of a gene) originates only once, and homoplasies exist only in the form of reversals to the ancestral condition (absence of a gene) in accord with the Dollo principle as formalized by Farris (1977a). In other words, parallel or convergent gains of the derived condition are assumed to be highly unlikely (or impossible, for practical purposes).

The Dollo parsimony principle also can be applied in the opposite direction: assuming a particular species tree topology, a parsimonious scenario of evolution can be constructed. Such a

Table 11.1 Matrix of presence/absence of genes in eukaryotic genomes. For each the number of the KOG and the (predicted) protein function are shown. 1 indicates that the given gene (KOG) is represented in the given species and 0 indicates absence of the gene in the given species. Species abbreviations: At, *Arabidopsis thaliana;* Ce, *Caenorhabditis elegans;* Dm, *Drosophila melanogaster;* Ec, *Encephalitozoon cuniculi;* Hs, *Homo sapiens;* Sc, *Saccharomyces cerevisiae;* Sp, *Schizosaccharomyces pombe*

Species	KOG2207, 3′-5′ exonuclease	KOG4125, acid trehalase	KOG0006, E3 ubiqutin-protein ligase	KOG0090, signal-recognition particle receptor β-subunit	KOG0050, mRNA splicing factor CDC5
At	1	0	0	1	1
Ec	0	0	0	0	1
Sc	0	1	0	1	1
Sp	0	0	0	1	1
Ce	1	0	1	1	1
Dm	1	1	1	1	1
Hs	0	1	1	1	1

scenario is, essentially, a mapping of different types of evolutionary events on to the branches of the phylogenetic tree. Obviously these scenarios, which naturally also include the reconstruction of the character states in all internal nodes of the tree and in the root (when the root position is known), can be of major value for understanding the evolution of the analyzed taxa.

However, in the context of evolutionary genomics, the Dollo principle cannot be assumed to be valid by default. The specific biological context must be examined in order to ascertain whether or not the elementary evolutionary events involved could violate phylogenetic irreversibility by producing homoplastic character gains. Extensive HGT shown to occur during evolution of prokaryotes obviously has the potential to violate the Dollo principle: the 'same' gene (more precisely, a member of the same set of orthologous genes; a COG) can be readily regained via HGT after being lost from the given lineage. In this case, it does not matter whether or not the gene regained via HGT comes from the same lineage as the original (lost) gene because, in this type of analysis, a COG is treated as the basic character. Thus, Dollo parsimony cannot be used for reconstructing evolution of prokaryotic genomes; some form of weighted parsimony should be applied instead (Snel *et al.* 2002; Kunin and Ouzounis 2003; Mirkin *et al.* 2003). Since the relative weights of different elementary events, i.e. lineage-specific gene loss and HGT, are unknown and probably

differ between genes as well as between lineages, evolutionary scenarios produced using these parameters are ambiguous and have to be assessed using external criteria (Mirkin *et al.* 2003).

With eukaryotes, however, the situation is quite different. Although acquisition of bacterial genes via HGT had been an important contribution to the evolution of eukaryotic genomes, at least prior to the emergence of multicellular forms (Doolittle *et al.* 2003), gene exchange between eukaryotic lineages themselves does not appear to occur at an appreciable rate. Therefore, as far as evolution of eukaryotic genomes is concerned, Dollo parsimony yields unambiguous parsimonious scenarios that include only two types of elementary event: loss of genes (or other characters, such as introns or pairs of genes) and emergence of new characters.

In the studies summarized below, we used the PAUP* (Swofford 2002) and DOLLOP (Felsenstein 1996) programs. An important difference between the implementations of Dollo parsimony in these programs should be noted: PAUP* produces unrooted trees while DOLLOP reconstructs rooted trees (Swofford *et al.* 1996). Where root inferences are reported, the latter program was used.

11.6 Dollo parsimony tree based on gene content: application to a crucial problem in animal evolution

The relative positions of nematodes, arthropods and chordates in the phylogeny of animals remain uncertain (for review see Hedges 2002). The traditional tree topology joins arthropods with chordates in a coelomate clade, whereas nematodes, which lack a coelome (a true body cavity), occupy a basal position (e.g. Raff 1996). However, the current leading hypothesis, which is based on phylogenetic trees for 18S rRNA and some additional comparisons of protein-coding genes, joins nematodes with arthropods in a clade of molting animals, Ecdysozoa (Aguinaldo *et al.* 1997; Giribet *et al.* 2000; Peterson and Eernisse 2001; Mallatt and Winchell 2002). The complete genome sequences of the nematodes, insects, and vertebrates provide for the possibility to extend phylogenetic studies to the genomic scale, in order to address this major issue in animal evolution (Mushegian *et al.* 1998; Blair *et al.* 2002; Wolf et al. 2004).

We addressed this problem both by traditional phylogenetic analysis of numerous sets of orthologous genes and by using patterns of gene presence/absence in orthologous sets for tree construction, which is the most straightforward technique in the category of so-called genome-tree methods (Fitz-Gibbon and House 1999; Snel *et al.* 1999; Wolf *et al.* 2002). Presence/absence of a gene in a set of species can be naturally treated as a binary character and the table of such characters can be subjected to either parsimony or distance phylogenetic analysis. After constructing such a character matrix for the complete set of KOGs, we applied the Dollo parsimony method. The rooted tree produced using Dollo parsimony confidently supported the coelomate topology (Fig. 11.1).

Otherwise, however, this tree was at odds with the prevalent taxonomic view (Hedges 2002) in that an animal–plant clade, as opposed to an animal–fungus clade with plants as an outgroup, was observed. This deviation of the gene-content trees from the currently accepted phylogeny is probably due to the varying amount of gene loss in different eukaryotic lineages; in particular, massive gene-loss in yeasts. As discussed previously in the context of prokaryotic genome analysis, the topology of gene-content trees seems to reflect a combination of the phylogenetic signal and other trends in genome evolution that are not necessarily linked to phylogeny, such as parallel gene loss associated with life-style similarities (Wolf *et al.* 2002). The clustering of humans with flies in the gene-content tree points to the congruence in gene repertoires of these animals. In this particular case, the majority of phylogenetic trees for orthologous gene sets point in the same direction, suggesting that the coelomate clade could, after all, reflect the actual course of animal evolution (Wolf *et al.* 2004).

11.7 A parsimonious scenario of gene gain and loss in eukaryotic evolution

As discussed in the previous section, the Dollo parsimony tree based on gene presence/absence shows some conflicts with the accepted phylogenetic tree of the eukaryotic crown group, the principal clades of which have been established with considerable confidence. In particular, some conflicting observations notwithstanding, the consensus of many phylogenetic analyses points to an animal–fungus clade, grouping of microsporidia with the fungi, and a coelomate (chordate–arthropod) clade among the animals (Blair *et al.* 2002; Hedges 2002; Wolf *et al.* 2004). Assuming this tree topology and treating the phyletic pattern of each KOG as a string of binary characters (1 for the presence of the given species and 0 for its absence in the given KOG), the parsimonious scenario of gene loss and emergence during the evolution of the eukaryotic crown group was constructed.

In the resulting scenario, each branch was associated with both gene loss and gain of new genes, with the exception of the plant branch and the branch leading to the common ancestor of

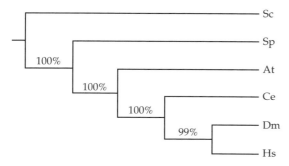

Figure 11.1 Dollo parsimony tree of the eukaryotic crown group based on gene presence/absence. The bootstrap values are indicated for each internal branch. The species abbreviations are as in Table 11.1.

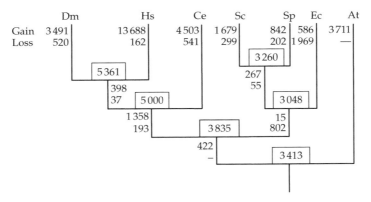

Figure 11.2 Scenario of gene gain and loss during evolution of the eukaryotic crown group derived using Dollo parsimony under a fixed tree topology. The numbers in boxes indicate the inferred number of KOGs in the respective ancestral forms. The numbers next to branches indicate the number of gene gains (emergence of KOGs; top) and gene (KOG) losses (bottom) associated with the respective branches; a dash indicates that the number of losses for a given branch could not be determined. Proteins from each genome that did not belong to KOGs as well as LSEs (lineage-specific gene family expansions) were counted as gains on the terminal branches. The species abbreviations are as in Table 11.1.

fungi and animals, to which gene losses could not be assigned without an additional outgroup (Fig. 11.2). There is little doubt that, once genomes of early-branching eukaryotes are included, gene loss associated with these branches will become apparent. The principal features of the reconstructed scenario include massive gene loss in the fungal clade, additional elimination of numerous genes in the microsporidia, emergence of a large set of new genes at the onset of the animal clade, and subsequent substantial gene loss in each of the animal lineages, particularly in the nematodes and arthropods (Fig. 11.2). The estimated number of genes lost in *S. cerevisiae* after its divergence from the common ancestor with the other yeast species, *S. pombe*, closely agreed with a previous estimate produced using a different approach (Aravind *et al.* 2000).

The parsimony analysis described above involves explicit reconstruction of the gene sets of ancestral eukaryotic genomes. Under the Dollo parsimony model, an ancestral gene (KOG) set is the union of the KOGs that are shared by the respective outgroup and each of the remaining species. Thus, the gene set for the common ancestor of the crown group includes all the KOGs in which *Arabidopsis* co-occurs with any of the other analyzed species. Similarly, the reconstructed gene set for the common ancestor of fungi and animals consists of all KOGs in which at least

one fungal species co-occurs with at least one animal species. Clearly, these are conservative reconstructions of ancestral gene sets because, as mentioned above, gene losses in the lineages branching off the deepest bifurcation could not be detected. Under this conservative approach, 3413 genes (KOGs) were assigned to the last common ancestor of the crown group (Fig. 11.2). More realistically, it appears likely that a certain number of ancestral genes have been lost in all or all but one of the analyzed lineages during subsequent evolution, such that the gene set of the eukaryotic crown group ancestor might have been close in size to those of modern yeasts, or even larger (i.e. 6000–7000 genes).

11.8 Dollo parsimony applied to evolution of eukaryotic gene structure

Most of the eukaryotic protein-coding genes contain multiple introns that are spliced out of the pre-mRNA by a distinct, large RNA–protein complex, the spliceosome, which is conserved in all eukaryotes (Dacks and Doolittle 2001). The positions of some spliceosomal introns are conserved in orthologous genes from plants and animals (Marchionni and Gilbert 1986; Logsdon *et al.* 1995; Boudet *et al.* 2001). A recent systematic analysis of pairwise alignments of homologous proteins from animals, fungi, and plants suggested that 10–25% of introns are ancient (Fedorov *et al.* 2002; Rogozin *et al.* 2003). However, intron densities in different eukaryotic species differ widely, and the location of introns in orthologous genes does not always coincide, even in closely related species (Logsdon *et al.* 1998). Likely cases of intron insertion and loss have been described (e.g. Rzhetsky *et al.* 1997; Logsdon *et al.* 1998), and indications of a high intron-turnover rate have been obtained (Lynch and Richardson 2002). It has been suggested that the proportion of shared intron positions decreased with increasing evolutionary distance and, accordingly, intron conservation could be a useful phylogenetic marker (Nei and Kumar 2001). However, the evolutionary history of introns and, the selective forces that shape intron evolution remain mysterious. Although recent comparisons have revealed the existence of many ancient introns shared by animals, plants, and fungi (Fedorov *et al.* 2002), the point(s) of origin of these introns in eukaryotic evolution and the relative contributions of intron loss and intron insertion in the evolution of eukaryotic genes remain unknown.

We used the KOG data set for evolutionary analysis of intron–exon structure in eukaryotic genes on a genomic scale. For the purpose of this analysis, orthologs from two additional eukaryotic species, the mosquito *Anopheles gambiae* (Ag) and the apicomplexan malarial parasite *Plasmodium falciparum* (Pf), were included in the KOGs using the COGNITOR method (Tatusov *et al.* 1997). Many of the KOGs contain multiple paralogs from one or more of the constituent species due to lineage-specific duplications; among these paralogs, the one showing the greatest evolutionary

conservation (defined as the mean similarity to KOG members from other species) was selected for evolutionary analysis. For a pair of introns to be considered orthologous, they were required to occur in exactly the same position in the aligned sequences of KOG members. Altogether, 684 KOGs were examined for intron conservation; these comprised the great majority, if not the entirety, of highly conserved eukaryotic genes that are amenable for an analysis of the exon–intron structure over the entire span of crown group evolution. The analyzed KOGs contained 21 434 introns in 16 577 unique positions; 5 981 introns were conserved among two or more genomes. Most of the conserved introns were present in only two species, but a considerable number was found in three genomes, and several introns were shared by four to seven species (Table 11.2). A simulation of the intron distribution in the analyzed sample of orthologous gene sets by random shuffling of the intron positions showed that approximately 1% of the observed number of introns shared by two species was expected to occur by chance, whereas none were expected to be shared by three or more species (Table 11.2). It has been proposed that introns insert into coding sequences non-randomly but primarily into "proto-splice sites" (Dibb and Newman 1989). Although the proto-splice model has been questioned as inconsistent with the observed distribution of intron phases (Long and Rosenberg 2000), we considered the potential effect of non-random intron insertion on the apparent evolutionary conservation of intron positions. For this purpose, random simulation was repeated with intron insertion allowed in 10% of the positions in the analyzed genes. Obviously, this led to an increase in the expected number of

Table 11.2 Conservation of intron positions in orthologous gene sets from eight eukaryotic species

Number of species ...	Number of introns – total							
	1	2	3	4	5	6	7	8
Observed[a]	13 406	2 047	719	275	104	25	1	0
Expected	21 368	33	0	0	0	0	0	0
Expected – 10%	20 083	662	8	0	0	0	0	0

[a] The probability that intron sharing in different species was due to chance, P(Monte Carlo) < 0.0001 (this applies both to the analysis of all alignment positions and to the test with 10% of the positions allowed for intron insertion).

shared introns in two or more species, but the excess of introns found in the same position remained substantial and highly statistically significant (Table 11.2). These observations suggest that a substantial majority of introns located in the same position in orthologous genes from different eukaryotic lineages are indeed orthologous, i.e. originate from an ancestral intron in the same position in the respective gene of the last common ancestor of the compared species. Nevertheless, the simulation results under the proto-splice site assumption show that the applicability of Dollo parsimony in this case could be limited. We should note that the magnitude of error introduced by the assumption of irreversible character gain depends directly on the nature of the biological phenomenon involved; in this case, how specific the proto-splice signal is.

The matrix of shared introns in all pairs of analyzed eukaryotic genomes revealed an unexpected pattern (Table 11.3). The number of conserved introns did not drop monotonically with increasing evolutionary distance among the compared organisms. On the contrary, human genes shared the greatest number of introns with the plant *Arabidopsis* instead of with any of the other three animals included in the comparison. In the conserved regions (which give more accurate results given the alignment uncertainties in other parts of genes), 24% of the intron positions in the analyzed human genes were shared with *Arabidopsis* (these comprised approximately 27% of the *Arabidopsis* introns) compared to approximately 12–17% of

introns positions shared by humans with the fly, mosquito, or worm (Table 11.3). The difference becomes even more dramatic when the numbers of introns conserved in *Arabidopsis* and each of the three animal species are compared: approximately three times more plant introns have a counterpart at the same position in orthologous human genes than in the fly or worm orthologs (Table 11.3). Although yeast *S. pombe* and the apicomplexan protist *Plasmodium* have few introns compared to plants or animals, the same asymmetry was observed for these organisms: the numbers of introns shared with *Arabidopsis* and humans are close and are two or three times greater than the number of introns shared with the insects or the worm (Table 11.3).

We then examined the evolutionary dynamics of introns in greater detail by using phylogenetic analysis. It should be noted that the comparative data on intron positions are as conducive to the representation as a character matrix as the phyletic pattern data. Specifically, intron positions were represented as a data matrix of intron presence/absence (encoded, as usual, as 1/0). An example of such a matrix for intron positions in one of the conserved gene clusters (KOGs) is shown in Fig. 11.3. To reconstruct the genome-wide scenario of gene-structure evolution, the matrices for all analyzed KOGs were concatenated to produce one alignment, which consisted of 16 577 columns of ones and zeros. The Dollo parsimony tree that we reconstructed from the matrix of intron presence/absence obviously did not mimic the species tree,

Table 11.3 Conservation of intron positions in eukaryotic orthologous gene sets: the matrix of pairwise interspeicies comparisions

	Pf	Sc	Sp	At	Ce	Dm	Ag	Hs
Pf	971	2	48	137	50	46	54	145
Sc		46	7	3	3	3	4	6
Sp			839	209	98	114	111	308
At				5 589	353	255	254	1 148
Ce					3 465	315	312	948
Dm						1 826	787	802
Ag							1 768	771
Hs								6 930

The diagonal shows the total number of introns in the 684 analyzed genes from the given species. Species abbreviations: At, *Arabidopsis thaliana;* Ce, *Caenorhabditis elegans;* Dm, *Drosophila melanogaster;* Hs, *Homo sapiens;* Ag, *Anopheles gambiae;* Pf, *Plasmodium falciparum;* Sc, *Saccharomyces cerevisiae;* Sp, *Schizosaccharomyces pombe.*

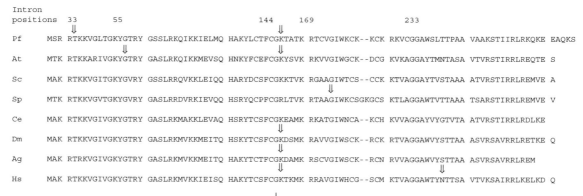

Figure 11.3 Examples of conservation and variability of intron positions in orthologous eukaryotic genes. The data are for KOG0473, ribosomal protein L37. The intron positions are shown directly on the alignment and the conversion of the intron-alignment mapping into a presence/absence matrix is illustrated. 1 indicates the presence of an intron and 0 indicates the absence of an intron in the given alignment position (shown at the top). The species abbreviations are as in Table 11.3.

with humans and *Arabidopsis* forming a strongly supported lineage embedded within the metazoan clade, and another anomalous group formed by yeast *S. pombe* and *Plasmodium* (Fig. 11.4). The topology of these trees supported the notion, already suggested by the pairwise comparisons summarized in Table 11.3, that ancestral introns have been, to a large extent, conserved in plants and vertebrates, but have been extensively lost in fungi, nematodes, and arthropods. Clustering of *Plasmodium* with *S. pombe*, to the exclusion of the other yeast species, *S. cerevisiae*, seems to be due to the fact that *Plasmodium* shared approximately as many introns with *S. pombe* as with worm and insects, but that the total number of introns in *S. pombe* was substantially lower than in animals (Table 11.3). Thus, conservation of intron positions does not seem to be a good source of information for inferring phylogenetic relationships across long evolutionary distances.

Having shown that evolution of introns in eukaryotic genes did not follow the species tree,

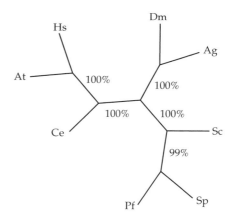

Figure 11.4 Dollo parsimony tree of the eukaryotic crown group based on comparison of intron positions. The bootstrap values are indicated for each internal branch. The species abbreviations are as in Table 11.3.

we applied Dollo parsimony in the opposite direction: given a species tree topology, construct the most-parsimonious scenario for the evolution of gene structure, i.e. the distribution of intron-gain

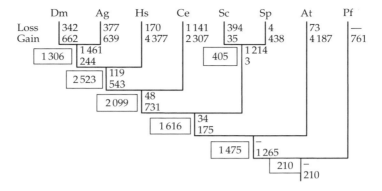

Figure 11.5 The parsimonious evolutionary scenario of intron gain/loss for the most likely topology of the eukaryotic phylogenetic tree. Intron gains and losses are mapped to each species and each internal branch. The numbers next to branches indicate the number of inton losses (top) and gains (bottom) associated with the respective branches; dashes show branches for which losses could not be inferred from the available data. The (minimal) number of introns inferred to have existed in the analyzed set of genes in the respective ancestral forms is indicated in a box next to each internal node of the tree. The species abbreviations are as in Table 11.3.

and -loss events among the tree branches. This approach is completely analogous to the construction of the scenario for gene gain and loss described above. The resulting schema suggests an intron-rich ancestor for the crown group, with limited intron loss in the animal ancestor, but massive losses in yeasts (particularly *S. cerevisiae*), worm, and insects (Fig. 11.5). The differences in the relative rates of intron gain and loss in the terminal branches are remarkable; there is a huge excess of gains over losses in humans and *S. pombe*, and an equally obvious excess of losses in insects and *S. cerevisiae*, whereas *C. elegans* shows nearly identical numbers of gains and losses. All introns shared by *Plasmodium* and any of the crown-group species (at least 210) are assigned to the last common ancestor of alveolates and the crown group, which lived some 1.5–2.0 billion years ago (Hedges 2002). Thus, a substantial fraction of the introns in extant eukaryotic genomes appear to be inherited from a common ancestor of the crown group and the alveolates, i.e. almost from the very onset of eukaryotic evolution. At present, loss of ancestral introns in *Plasmodium* cannot be documented because *Plasmodium* is the outgroup to the crown-group species; neither can losses be assigned to the internal branch that leads to the ancestor of the crown group. Hence we produced a conservative estimate of the number of the most ancient introns in the analyzed gene set, which is likely to be

a substantial underestimate given that *Plasmodium* is a parasite with a highly degraded genome and low intron density. Sequencing and analysis of genomes of other early-branching eukaryotes is expected to substantially increase the number of introns that have survived since the dawn of eukaryotic evolution.

11.9 Dollo parsimony analysis of prokaryotic gene order

As discussed above, Dollo parsimony is hardly applicable to the analysis of evolution of prokaryotic gene repertoires because extensive HGT leads to gross violations of the irreversibility principle. However, it might be possible to come up with (nearly) irreversible, Dollo-compatible characters even in the case of prokaryotic genome evolution. Elements of gene order are, perhaps, the most obvious candidates for the role of such characters in this category. Genome colinearity is preserved only in closely related bacteria and archaea because rearrangements continuously shuffle prokaryotic genomes, gradually breaking ancestral gene strings. Nevertheless, many operons—groups of co-expressed, functionally linked prokaryotic genes (usually, three or four; Jacob *et al.* 1960)—are highly conserved (Mushegian and Koonin 1996; Watanabe *et al.* 1997; Dandekar *et al.* 1998). The elementary unit of gene-order conservation is a contiguous (or tightly linked) gene pair. Independent formation of the same gene pair in several genomes of distantly related prokaryotes is extremely unlikely. The possibility of HGT of operons should be considered more seriously in light of the so-called 'selfish-operon' concept, which posits that operons are often transferred between species as single units (Lawrence and Roth 1996; Lawrence 1999). Depending on the strength of the selfish-operon trend, this could lead to significant regaining of lost characters (gene pairs) and, accordingly, to violations of the Dollo assumption. Nevertheless, phylogenetic analysis using conserved gene pairs as characters for genome comparison appears to be an attractive possibility. Due to the (relatively) high rate of intragenomic rearrangements, the gene-order trees are, at least in theory, especially suitable for resolving the phylogeny of closely related species.

We identified pairs of genes (COGs) whose physical proximity is conserved in several genomes. The presence/absence matrices of these pairs were analyzed using Dollo parsimony and neighbor joining, which produced essentially the same topology (Wolf *et al.* 2001). The results were also very similar to the results of distance-tree analysis of prokaryotic gene orders reported by Korbel and coworkers (Korbel *et al.* 2002). The resulting tree topology showed a good separation between archaea and bacteria and also reproduced well-established, tight bacterial clades, but had a poor resolution at deep branching points. Furthermore, some of the groups seemed to result from preferential HGT between certain prokaryotic lineages (Wolf *et al.* 2001). Thus, the effect of HGT could be too significant for Dollo parsimony to be an appropriate method for tree construction in this case.

11.10 Genomics and Dollo parsimony: validity of the Dollo principle for different types of genomic data

Dollo parsimony assumes that each derived character state originates only once, and homoplasies exist only in the form of reversals to the primitive condition. Obviously, this is not an absolute but a probabilistic notion. It is not physically impossible

for dolphins to re-evolve feet or for yeast to re-evolve the lost system for post-transcriptional gene silencing (Aravind *et al.* 2000) but it appears exceedingly unlikely that these features could reappear in the same form as the lost ones, at least to be considered the same character. A loose thermodynamic analogy seems to fit: movement of a single molecule is completely time-reversible, but the Second Law of Thermodynamics virtually guarantees that any regular configuration of molecules would be irreversibly destroyed by thermal motion. The probabilistic nature of the Dollo law suggests that there would be a continuum of characters differing in their degree of (ir)reversibility. Indeed, regain of a lost gene seems to be impossible, for all practical purposes, barring HGT. By contrast, regain of an ancestral amino acid at a particular site is quite likely, especially given sufficient evolutionary distance separating the compared sequences. The specific problems discussed here cover this entire range. In the case of the eukaryotic gene repertoires, gene loss clearly is irreversible and, accordingly, Dollo parsimony, which simplifies the analysis and allows for more reliable conclusions than other methods of evolutionary reconstruction, is the approach of choice. In the case of intron positions, the Dollo approach can be applied also, but the probability of multiple gains might not be negligible (depending on the biology of the process, which is not yet sufficiently understood), and caution is due. By contrast, Dollo parsimony is not applicable to the study of evolution of prokaryotic genomes due to massive HGT. For the same reason, the attempt to use conserved gene pairs as Dollo characters was not particularly successful either.

With the further growth of genomics and systems biology, the numbers of characters potentially suitable for phylogenetic analysis will continue to grow. Examples of new types of information that are becoming amenable to this type of analysis include various characteristics of gene expression and protein–protein interaction networks. The case studies described here indicate that Dollo parsimony is a useful and potentially powerful methodology of evolutionary genomics, but that its applicability always needs to be gauged against the biology of the specific systems under analysis.

References

Addario-Berry, L., Chor, B., Hallett, M., Lagergren, J., Panconesi, A. and Wareham, T. (2004). Ancestral maximum likelihood of phylogenetic trees is hard. *J. Bioinf. Comp. Biol.* **2**: 257–271.

Aguinaldo, A.M., Turbeville, J.M., Linford, L.S., Rivera, M.C., Garey, J.R., Raff, R.A. and Lake, J.A. (1997). Evidence for a clade of nematodes, arthropods and other moulting animals. *Nature* **387**: 489–493.

Akaike, H. (1973). Information theory and an extension of the maximum likelihood principle. In *Second International Symposium on Information Theory*, Tsahkadsor, Armenia, USSR, September 2–8, 1971 (eds B.N. Petrov and F. Csaaki), pp. 267–281. Budapest, Akademiai Kiado.

Albert, V.A. and Mishler, B.D. (1992). On the rationale and utility of weighting nucleotide sequence data. *Cladistics* **8**: 73–83.

Albert, V.A., Mishler, B.D. and Chase, M.W. (1992). Character-state weighting for restriction site data in phylogenetic reconstruction, with an example from chloroplast DNA. In *Molecular Systematics of Plants*. (eds P.S. Soltis, D.E. Soltis, and J.J. Doyle), pp. 369–403. New York, Chapman and Hall.

Albert, V.A., Chase, M.W. and Mishler, B.D. (1993). Character-state weighting for cladistic analysis of protein-coding DNA sequences. *Ann. Missouri Bot. Gard.* **80**: 752–766.

Albert, V.A., Backlund, A., Bremer, K, Chase, M.W., Manhart, J.R., Mishler, B.D. and K.C. Nixon. (1994). Functional constraints and *rbc*L evidence for land plant phylogeny. *Ann. Missouri Bot. Gard.* **81**: 534–567.

Alfaro, M., Zoller, S. and Lutzoni, F. (2003). Bayes or bootstrap? A simulation study comparing the performance of Bayesian Markov chain Monte Carlo sampling and bootstrapping in assessing phylogenetic confidence. *Mol. Biol. Evol.* **20**: 255–266.

Altschul, S.F. (1989). Gap costs for multiple alignment. *J. Theor. Biol.* **138**: 297–309.

Aravind, L., Watanabe, H., Lipman, D.J. and Koonin, E.V. (2000). Lineage-specific loss and divergence of functionally linked genes in eukaryotes. *Proc. Natl. Acad. Sci. USA* **97**: 11319–11324.

Ariew, A. (1998). Are probabilities necessary for evolutionary explanations? *Biol. Philos.* **13**: 245–253.

Arvestad, L., Berglund, A.C., Lagergren, J. and Sennblad, B. (2003). Bayesian gene/species tree reconciliation and orthology analysis using MCMC. *Bioinformatics* **19** Suppl. 1: i7–i15.

Avise, J.C. (1989). Gene trees and organismal histories: a phylogenetic approach to population biology. *Evolution* **43**:1192–1208.

Avise, J.C. (2000). *Phylogeography: The History and Formation of Species*. Cambridge, MA, Harvard University Press.

Bach, E. (1981). On time, tense and aspect: An essay in English metaphysics. In *Radical Pragmatics* (ed. P. Cole), pp. 63–81. New York, Academic Press.

Baker, A. (2003). Quantitative parsimony and explanatory power. *Br. J. Phil. Sci.* **54**: 245–259.

Bandelt, H.J., Forster, P., Sykes, B.C. and Richards, M.B. (1995). Mitochondrial portraits of human populations using median networks. *Genetics* **141**: 743–753.

Barnes, E.C. (2000). Ockham's Razor and the anti-superfluity principle. *Erkenntnis* **53**: 353–374.

Barrett, M., Donoghue M.J. and Sober, E. (1991). Against consensus. *Syst. Zool.* **40**: 486–493.

Barry, D. and Hartigan, J.A. (1987). Statistical analysis of hominoid molecular evolution. *Stat. Sci.* **2**: 191–210.

Benner, S.A., Trabesinger, N. and Scheiber, D. (1998). Post-genomic science: Converting primary sequence into physiological function. *Adv. Enzyme Regul.* **38**: 155–190.

Benner, S.A., Chamberlin, S.G., Liberles, D.A., Govindarajan, S. and Knecht, L. (2000). Functional inferences from reconstructed evolutionary biology involving rectified databases — an evolutionarily grounded approach to functional genomics. *Res. Microbiol.* **151**: 97–106.

Benson, D.A., Karsch-Mizrachi, I., Lipman, D.J., Ostell, J. and Wheeler, D.L. (2004). GenBank: update. *Nucleic Acids Res.* **32**: D23-D26.

Bi, S., Garilova, O., Gong, D.W., Mason, M.M. and Reitman, M. (1997). Identification of a placental enhancer for the human leptin gene. *J. Biol. Chem.* **272**: 30583–30588.

Blackburn, D.G., (1984). From whale toes to snake eyes: Comments on the reversibility of evolution. *Syst. Zool.* **33**: 241–245.

Blair, J.E., Ikeo, K., Gojobori, T. and Hedges, S.B. (2002). The evolutionary position of nematodes. *BMC Evol. Biol.* **2**: 7.

Bock, W.J. (1973). Philosophical foundations of classical evolutionary classification. *Syst. Zool.* **22**: 375–392.

Boudet, N., Aubourg, S., Toffano-Nioche, C., Kreis, M. and Lecharny, A. (2001). Evolution of intron/exon structure of DEAD helicase family genes in *Arabidopsis, Caenorhabditis*, and *Drosophila. Genome Res.* **11**: 2101–2114.

Boyd, R. (1991). Confirmation, semantics, and the interpretation of scientific theories. In *The Philosophy of Science* (eds R. Boyd, P. Gasper and J.D. Trout), pp. 3–35. Cambridge, MA, MIT Press.

Brady, R.H. (1985). On the independence of systematics. *Cladistics* **1**: 113–126.

Brady, R.H. (1994). Explanation, description, and the meaning of transformation in taxonomic evidence. In *Models in Phylogeny Reconstruction* (eds R.W. Scotland, D.J. Siebert and D.M. Williams), pp. 11–29 special vol. 52, Syst. Assoc. Oxford, Clarendon Press.

Bremer, K. (1988). The limits of amino acid sequence data in angiosperm phylogenetic reconstruction. *Evolution* **42**: 795–803.

Bremer, K. (1994). Branch support and tree stability. *Cladistics* **10**: 295–304.

Brooks, D.R. (1981). Hennig's parasitological method: A proposed solution. *Syst. Zool.* **30**: 229–249.

Brooks, D.R. (1988). Scaling effects in historical biogeography: A new view of space, time, and form. *Syst. Zool.* **37**: 237–244.

Brower, A.V.Z. (2000). Homology and the inference of systematic relationships: Some historical and philosophical perspectives. In *Homology and Systematics: Coding Characters for Phylogenetic Analysis* (eds R. Scotland and R.T. Pennington), pp. 10–21. New York, Taylor and Francis.

Brudno, M., Chapman, M., Göttgens, B., Batzoglou, S. and Morgenstern, B. (2003). Fast and sensitive multiple alignment of large genomic sequences. *BMC Bioinformatics* **4**: 66.

Brudno, M., Poliakov, A., Salamov, A., Cooper, G.M., Sidow, A., Rubin, E.M., Solovyev, V., Batzoglou, S. and Dubchak, I. (2004). Automated whole-genome multiple alignment of rat, mouse, and human. *Genome Res.* **14**: 685–692.

Bryant, D. (2000). A lower bound for the breakpoint phylogeny problem. In *Proceedings of the 11th Annual Symposium on Combinatorial Pattern Matching* (eds R. Giancarlo and D. Sankoff), pp. 235–247. London, Springer Verlag.

Bryant, D. (2003). A classification of consensus methods for phylogenetics. In *Bioconsensus* (eds Janowitz, M.F., Lapointe, F.J., McMorris, F.R., Mirkin, B. and Roberts, F.S.), pp. 163–184. DIMACS Series in Discrete Mathematics and Theoretical Computer Science, vol. 61, Providence, RI, American Mathematical Society.

Buneman, P. (1971). The recovery of trees from measures of dissimilarity. In *Mathematics in the Archaeological and Historical Sciences* (eds Hodson, F.R., Kendall, D.G. and Tautu, P.), pp. 387–395. Edinburgh, Edinburgh University Press.

Burnham, K.P. and Anderson, D.R. (1998). *Model Selection and Inference. A Practical Information-theoretic Approach.* New York, Springer.

Cameron, H.D. (1987). The upside-down cladogram: Problems in manuscript affiliation. In *Biological Metaphor and Cladistic Classification: An Interdisciplinary Perspective* (eds H.M. Hoenigswald and L.F. Wiener), pp. 227–242. Philadelphia, PA, University of Pennsylvania Press.

Camin, J.H. and Sokal, R.R. (1965). A method for deducing branching sequences in phylogeny. *Evolution* **19**: 311–326.

Carillo, H. and Lipman, D. (1988). The multiple sequence alignment problem in biology. *SIAM J. Appl. Math.* **48**: 1073–1082.

Caroll, L. (1872). *Through the Looking Glass.* London, Macmillan.

Carpenter, J.M. (2003). On "Molecular phylogeny of Vespidae (Hymenoptera) and the evolution of sociality in wasps". *Am. Museum Novitates* **3389**: 1–20.

Carter, M., Hendy, M.D., Penny, D., Székely, L.A. and Wormald, N.C. (1990). On the distribution of lengths of evolutionary trees. *SIAM J. Discrete Math.* **3**: 38–47.

Cartmill, M. (1981). Hypothesis testing and phylogenetic reconstruction. *Z. Zool. Syst. Evol.-forsch.* **19**: 73–96.

Chang, B.S.W., Jönsson K., Kazmi, M.A., Donoghue, M.J. and Sakmar T.P. (2002). Recreating a functional ancestral archosaur visual pigment. *Mol. Biol. Evol.* **19**: 1483–1489.

Chang, J. (1996). Full reconstruction of Markov models on evolutionary trees: Identifiability and consistency. *Math. Biosc.* **137**: 51–73.

Chang, J.T. and Kim, J. (1996). The measurement of homoplasy: A stochastic view. In *Homoplasy: the Recurrence of Similarity in Evolution* (eds M.J. Sanderson and L. Hufford), pp. 189–303. Academic Press.

Chase, M.W., Soltis, D.E., Olmstead, R.G., Morgan, D., Les, D.H., Mishler, B.D., Duvall, M.R., Price, R.A., Hills, H.G., Qiu, Y.-L. *et al.* (1993). Phylogenetics of seed plants: An analysis of nucleotide sequences from the plastid gene *rbcL. Ann. Missouri Bot. Gard.* **40**: 528–580.

Crichton, M. (1990). *Jurassic Park*. New York, Ballantine Books.

Crick, F. (1968). The origin of the genetic code. *J. Mol. Biol.* **38**: 367–379.

Crisci, J.V. and Stuessy, T.F. (1980). Determining primitive character states for phylogenetic reconstruction. *Syst. Bot.* **5**: 112–135.

Cummings, M., Handley, S., Myers, D., Reed, D., Rokas, A. and Winka, K. (2003). Comparing bootstrap and posterior probability values in the four-taxon case. *Syst. Biol.* **52**: 477–487.

Dacks, J.B. and Doolittle, W.F. (2001). Reconstructing/deconstructing the earliest eukaryotes: How comparative genomics can help. *Cell* **107**: 419–425.

Dandekar, T., Snel, B., Huynen, M. and Bork, P. (1998). Conservation of gene order: A fingerprint of proteins that physically interact. *Trends Biochem. Sci.* **23**: 324–328.

Darwin, C. (1859). *The Origin of Species by Means of Natural Selection, or the Preservation of Favoured Races in the Struggle for Life*. London, John Murray [1964. Facsimile of 1st edition. Cambridge, MA, Harvard University Press]

Davids, W., Gamieldien, J., Liberles, D.A. and Hide, W. (2002). Positive selection scanning reveals decoupling of enzymatic activities of carbamoyl phosphate synthetase in *H. pylori*. *J. Mol. Evol.* **54**: 458–464.

Davidson, D. (1991). On the individuation of events. *Synthese* **86**: 229–254.

Davis, J.I. and Nixon, K.C. (1992). Populations, genetic variation, and the delimitation of phylogenetic species. *Syst. Biol.* **41**: 421–435.

Davis, J.I., Stevenson, D.W., Petersen, G., Seberg, O., Campbell, L.M., Freudenstein, J.V., Goldman, D.H., Hardy, C.R., Michelangeli, F.A., Simmons, M.P. *et al.* (2004). A phylogeny of the monocots, as inferred from *rbcL* and *atpA* sequence variation, and a comparison of methods for calculating jackknife and bootstrap values. *Syst. Bot.* **29**: 467–510.

Dayhoff, M.O. and Eck, R.V. (1968). *Atlas of Protein Sequence and Structure*. 1967–68. Silver Spring, MD, National Biomed. Res. Foundation.

Dayhoff, M.O. and Park, C.M. (1969). Cytochrome C: Building a phylogenetic hypothesis. In *Atlas of Protein Sequence and Structure. 1969* (ed. M.O. Dayhoff), pp. 7–16 vol. 4. Silver Spring, MD, National. Biomed. Res. Foundation.

DeBry, R.W. and Slade, N.A. (1985). Cladistic analysis of restriction endonuclease cleavage maps within a maximum-likelihood framework. *Syst. Zool.* **34**: 21–34.

De Laet, J. (1997). *A Reconsideration of Three-Item Analysis, the Use of Implied Weights in Cladistics, and a Practical Application in Gentianceae*. PhD Thesis, University of Leuven, Belgium.

De Laet, J. (2003). Parsimony algorithms for characters that are inapplicable in some terminals (Abstract, 21st annual meeting of the Willi Hennig Society, Helsinki 2002). *Cladistics* **19**: 151.

De Laet, J. (2004). When one and one is not two: Parsimony analysis of sequence data (Abstract, 22nd annual meeting of the Willi Hennig Society, New York 2003). *Cladistics* **20**: 81.

De Laet, J. and Smets, E. (1998). On the three-taxon approach to parsimony analysis. *Cladistics* **14**: 363–381.

De Laet, J. and Wheeler, W. (2003). *POY version 3.0.11* (Wheeler, Gladstein and De Laet, May 6 2003). Command line documentation. Available at ftp://ftp.amnh.org/pub/molecular/poy.

de Pinna, M. C. C. (1991). Concepts and tests of homology in the cladistic paradigm. *Cladistics* **7**: 367–394.

de Queiroz, K. (1992). Phylogenetic definitions and taxonomic philosophy. *Biol. Philos.* **7**: 295–313.

de Queiroz, K. (1996). Including the characters of interest during tree reconstruction and the problems of circularity and bias in studies of character evolution. *Am. Nat.* **148**: 700–708.

de Queiroz, K. and Poe, S. (2001). Philosophy and phylogenetic inference: A comparison of likelihood and parsimony methods in the context of Karl Popper's writings on corroboration. *Syst. Biol.* **50**: 305–321.

de Queiroz, K. and Poe, S. (2003). Failed refutations: Further comments on parsimony and likelihood methods and their relationship to Popper's degree of corroboration. *Syst. Biol.* **52**: 352–367.

Dezulian, T. and Steel, M. (2004). Phylogenetic closure operations and homoplasy-free evolution. In *Classification, Clustering, and Data Mining Applications* (Proceedings of the meeting of the International Federation of Classification Societies (IFCS) 2004). (eds D. Banks, L. House, F.R. McMorris, P. Arabie and W. Gaul), pp. 395–416. Springer-Verlag, Berlin.

Dibb, N.J. and Newman, A.J. (1989). Evidence that introns arose at proto-splice sites. *EMBO J.* **8**: 2015–2021.

Dollo, L. (1893). Le lois de l'evolution. *Bull. Soc. Belge Geol. Paleontol. d'Hydrol.* **7**: 164–167.

Donoghue, M.J. and Doyle, J.A. (1989). Phylogenetic analysis of angiosperms and the relationships of Hamamelidae. In *Evolution, Systematics and Fossil History of the Hamamelidae*, vol.1 (eds P. Crane and S. Blackmore), pp. 17–45. Oxford, Clarendon Press.

Donoghue, M.J. and Sanderson, M.J. (1992). The suitability of molecular and morphological evidence in reconstructing plant phylogeny. In *Molecular Systematics of Plants*. (eds P.S. Soltis, D.E. Soltis and J.J. Doyle), pp. 340–368, New York, Chapman & Hall.

Donoghue, M.J., Doyle, J.A., Gauthier, J.A., Kluge, A.G. and Rowe, T. (1989). The importance of fossils in phylogeny reconstruction. *Annu. Rev. Ecol. Syst.* **20**: 431–460.

Doolittle, W.F. (1999). Lateral genomics. *Trends Cell. Biol.* **9**: M5—M8.

Doolittle, W.F., Boucher, Y., Nesbo, C.L., Douady, C.J., Andersson, J.O. and Roger, A.J. (2003). How big is the iceberg of which organellar genes in nuclear genomes are but the tip? *Phil. Trans. R. Soc. Lond. B Biol. Sci.* **358**: 39–57.

Duret, L., Mouchiroud, D. and Gouy, M. (1994). HOVERGEN: A database of homologous vertebrate genes. *Nucleic Acids Res.* **22**: 2360–2365.

Edwards, A.W.F. (1972). *Likelihood*. Cambridge, Cambridge University Press.

Edwards, A.W.F. and Cavalli-Sforza, L.L. (1963). The reconstruction of evolution. *Heredity* **18**: 553.

Edwards, A.W.F. and Cavalli-Sforza, L.L. (1964). Reconstruction of evolutionary trees. In *Phenetic and Phylogenetic Classification* (eds V.H. Heywood and J. McNeill), pp. 67–76 no. 6. London, Syst. Assoc. Publ.

Eisen, M.B., Spellman, P.T., Brown, P.O. and Botstein, D. (1998). Cluster analysis and display of genome-wide expression patterns. *Proc. Natl. Acad. Sci. USA* **95**: 14863–14868.

Enard, W., Khaitovich, P., Klose, J., Zöllner, S., Heissig, F., Giavalisco, P., Nieselt-Struwe, K., Muchmore, E., Varki, A., Ravid, R. *et al.* (2002). Intra and interspecific variation in primate gene expression patterns. *Science* **296**: 340–343.

Endo, T., Ikeo, K. and Gojobori, T. (1996). Large-scale search for genes on which positive selection may operate. *Mol. Biol. Evol.* **13**: 685–690.

Endress, P.K. (1994). *Diversity and Evolutionary Biology of Tropical Flowers*. Cambridge, Cambridge University Press.

Erdős, P.L. and Székely, L.A. (1992). Evolutionary trees: An integer multicommodity maxflow–min-cut theorem. *Adv. Appl. Math.* **13**: 375–389.

Erdős, P.L. and Székely, L.A. (1993). Counting bichromatic evolutionary trees. *Discrete Appl. Math.* **47**: 1–8.

Erdős, P.L., Steel, M.A., Székely, L.A. and Warnow, T. (1999). A few logs suffice to build (almost) all trees (Part 1). *Random Struct Algorithms* **14**: 153–184.

Excoffier, L. and Smouse, P.E. (1994). Using allele frequencies and geographic subdivision to reconstruct gene trees within a species: Molecular variance parsimony. *Genetics* **136**: 343–359.

Farris, J.S. (1966). Estimation of conservatism of characters by constancy within biological populations. *Evolution* **20**: 319–334.

Farris, J.S. (1967). The meaning of relationship and taxonomic procedure. *Syst. Zool.* **16**: 44–51.

Farris, J.S. (1969). A successive approximations approach to character weighting. *Syst. Zool.* **18**: 374–385.

Farris, J.S. (1970). Methods for computing Wagner trees. *Syst. Zool.* **19**: 83–92.

Farris, J.S. (1972). Estimating phylogenetic trees from distance matrices. *Am. Nat.* **106**: 645–668.

Farris, J.S. (1973a). On the use of the parsimony criterion for inferring phylogenetic trees. *Syst. Zool.* **22**: 250–256.

Farris, J.S. (1973b). A probability model for inferring evolutionary trees. *Syst. Zool.* **22**: 250–256.

Farris, J.S. (1977a). Phylogenetic analysis under Dollo's law. *Syst. Zool.* **26**: 77–88.

Farris, J.S. (1977b). Some further comments on Le Quesne's methods. *Syst. Zool.* **26**: 220–223.

Farris, J.S. (1978a). Inferring phylogenetic trees from chromosome inversion data. *Syst. Zool.* **27**: 275–284.

Farris, J.S. (1978b). *Wagner78*. Published by the author.

Farris, J.S. (1979). The information content of the phylogenetic system. *Syst. Zool.* **28**: 483–519.

Farris, J.S. (1982a). Outgroups and parsimony. *Syst. Zool.* **31**: 328–334.

Farris, J.S. (1982b). Simplicity and informativeness in systematics and phylogeny. *Syst. Zool.* **31**: 413–444.

Farris, J.S. (1983). The logical basis of phylogenetic analysis. In *Advances in Cladistics*, volume 2: *Proceedings of the Second Meeting of the Willi Hennig Society*. (eds N. Platnick and V. Funk) pp. 7–36. New York, Columbia University Press.

Farris, J.S. (1983/1994). The logical basis of phylogenetic analysis. In *Advances in Cladistics — Proceedings of the 2nd Annual Meeting of the Willi Hennig Society* (eds N. Platnick and V. Funk. New York, Columbia University Press. Abridged and reprinted in E. Sober (ed.) pp. 7–36. *Conceptual Issues in Evolutionary Biology*, Cambridge, MA, MIT Press, 1994 (page references to the latter).

Farris, J.S. (1986). On the boundaries of phylogenetic systematics. *Cladistics* **2**: 14–27.

Farris, J.S. (1988). *Hennig86*. Published by the author, Port Jefferson Station, New York.

Farris, J.S. (1989a). The retention index and the rescaled consistency index. *Cladistics* **5**: 417–419.

Farris, J.S. (1989b). Entropy and fruit flies. *Cladistics* **5**: 103–108.

Farris, J.S. (1991). Hennig defined paraphyly. *Cladistics* **7**: 297–304.

Farris, J.S. (1997). Cycles. *Cladistics* **13**: 131–144.

Farris, J.S. (1999). Likelihood and inconsistency. *Cladistics* **15**: 199–204.

Farris, J.S. (2001). Support weighting. *Cladistics* **17**: 389–394.

Farris, J.S. and Kluge, A.G. (1986). Synapomorphy, parsimony, and evidence. *Taxon* **35**: 298–315.

Farris, J.S., Kluge, A.G. and Eckardt, M.J. (1970). A numerical approach to phylogenetic systematics. *Syst. Zool.* **19**: 172–189.

Farris, J.S., Källersjö, M., Albert, V.A., Allard, M., Anderberg, A., Bowditch, B., Bult, C., Carpenter, J.M., Crowe, T.M., De Laet, J. *et al.* (1995). Explanation. *Cladistics* **11**: 211–218.

Farris, J.S., Albert, V.A., Källersjö, M., Lipscomb, D. and Kluge, A.G. (1996). Parsimony jackknifing outperforms neighbor-joining. *Cladistics* **12**: 99–124.

Farris, J.S., Källersjö, M. and De Laet, J.E. (2001a). Branch lengths do not indicate support — even in maximum likelihood. *Cladistics* **17**: 298–299.

Farris, J.S., Kluge, A.G. and De Laet, J.E. (2001b). Taxic revisions. *Cladistics* **17**: 79–103.

Fedorov, A., Merican, A.F. and Gilbert, W. (2002). Large-scale comparison of intron positions among animal, plant, and fungal genes. *Proc. Natl. Acad. Sci. USA* **99**: 16128–16133.

Felsenstein, J. (1968). *Statistical Inference and the Estimation of Phylogenies.* PhD Thesis, University of Chicago, Chicago, IL.

Felsenstein, J. (1973). Maximum likelihood and minimum-steps methods for estimating evolutionary trees from data on discrete characters. *Syst. Zool.* **22**: 240–249.

Felsenstein, J. (1978a). Cases in which parsimony and compatibility methods can be positively misleading. *Syst. Zool.* **27**: 401–410.

Felsenstein, J. (1978b). The number of evolutionary trees. *Syst. Zool.* **27**: 27–33.

Felsenstein, J. (1979). Alternative methods of phylogenetic inference and their interrelationship. *Syst. Zool.* **28**: 49–62.

Felsenstein, J. (1981a). Evolutionary trees from DNA sequences: A maximum likelihood approach. *J. Mol. Evol.* **17**: 368–376.

Felsenstein, J. (1981b). Evolutionary trees from gene frequencies and quantitative characters: Finding maximum likelihood estimates. *Evolution* **35**: 1229–1242.

Felsenstein, J. (1981c). A likelihood approach to character weighting and what it tells us about parsimony and compatibility. *Biol. J. Linnean Soc.* **16**: 183–196.

Felsenstein, J. (1982). Numerical methods for inferring evolutionary trees. *Q. Rev. Biol.* **57**: 379–404.

Felsenstein, J. (1983). Methods for inferring phylogenies: A statistical view. In *Numerical Taxonomy* (ed. J. Felsenstein), pp. 315–334. Berlin, Springer-Verlag.

Felsenstein, J. (1988). Phylogenies from molecular sequences: Inference and reliability. *Annu. Rev. Genet.* **2**: 521–565.

Felsenstein, J. (1996). Inferring phylogenies from protein sequences by parsimony, distance, and likelihood methods. *Methods Enzymol.* **266**: 418–427.

Felsenstein, J. (2004). *Inferring Phylogenies.* Sunderland, MA, Sinauer Associates.

Felsenstein, J. and Sober, E. (1987). Parsimony and likelihood: An exchange. *Syst. Zool.* **35**: 617–626.

Feng, D.F. and Doolittle, R.F. (1987). Progressive sequence alignment as a prerequisite to correct phylogenetic trees. *J. Mol. Evol.* **25**: 351–360.

Fink, W.L. (1982). The conceptual relationship between ontogeny and phylogeny. *Paleobiology* **8**: 254–264.

Fitch, W.M. (1970). Distinguishing homologous from analogous proteins. *Syst. Zool.* **19**: 99–113.

Fitch, W.M. (1971). Toward defining the course of evolution: Minimal change for a specific tree topology. *Syst. Zool.* **20**: 406–416.

Fitz-Gibbon, S.T. and House, C.H. (1999). Whole genome-based phylogenetic analysis of free-living micro-organisms. *Nucleic Acids Res.* **27**: 4218–4222.

Force, A., Lynch, M., Pickett, F.B., Amores, A., Yan, Y.L. and Postlethwait, J. (1999). Preservation of duplicate genes by complementary, degenerative mutations. *Genetics* **151**: 1531–1545.

Forster, M.R. (2000). Key concepts in model selection: Performance and generalizability. *J. Math. Psych.* **44**: 205–231.

Forster, M.R. and Sober, E. (1994). How to tell when simpler, more unified, or less *ad hoc* theories will provide more accuracte predictions. *Br. J. Phil. Sci.* **45**: 1–35.

Foulds, L.R. (1984). Maximum savings in the Steiner problem in phylogeny. *J. Theoret. Biol.* **107**: 471–474.

Foulds, L.R. and Graham, R.L. (1982). The Steiner problem in phylogeny is NP-complete. *Adv. Appl. Math.* **3**: 43–49.

Friedman, M. (1983). *Foundations of Space-Time Theories: Relativistic Physics and Philosophy of Science.* Princeton, NJ, Princeton University Press.

Fredman, M.L. (1984). Algorithms for computing evolutionary similarity measures with length independent gap penalties. *Bull. Math. Biol.* **46**: 545–563.

Freudenstein, J.V., Pickett, K.M., Simmons, M.P. and Wenzel, J.W. (2003). From basepairs to birdsongs: phylogenetic data in the age of genomics. *Cladistics* **19**: 333–347.

Frost, D.R. (2000). Species, descriptive efficiency, and progress in systematics. In *The Biology of Plethodontid Salamanders* (eds R.C. Bruce, R.J. Jaeger, and L.D. Houck), pp. 7–29. New York, Kluwer Academic/Plenum Publishing.

Frost, D.R. and Kluge, A.G. (1994). A consideration of epistemology in systematic biology, with special reference to species. *Cladistics* **10**: 259–294.

Frost, D.R., Rodrigues, M.T., Grant, T. and Titus, T.A. (2001). Phylogenetics of the lizard genus *Tropidurus* (Squamata: Tropiduridae: Tropidurinae): Direct optimization, descriptive efficiency, and sensitivity analysis of congruence between molecular data and morphology. *Mol. Phylogenet. Evol.* **21**: 352–371.

Fukami-Kobayashi, K., Schreiber, D.R. and Benner, S.A. (2002). Detecting compensatory covariation signals in protein evolution using reconstructed ancestral sequences. *J. Mol. Biol.* **319**: 729–743.

Funk, V.A. and Brooks, D.R. (1990). *Phylogenetic Systematics as the Basis of Comparative Biology*. Washington, DC, Smithsonian Institution Press.

Gaasterland, T. and Ragan, M.A. (1998). Microbial genescapes: Phyletic and functional patterns of ORF distribution among prokaryotes. *Microb. Comp. Genomics* **3**: 199–217.

Gallut, C. and Barriel, V. (2002). Cladistic coding of genomic maps. *Cladistics* **18**: 526–536.

Galperin, M.Y. and Koonin, E.V. (2000). Who's your neighbor? New computational approaches for functional genomics. *Nat. Biotechnol.* **18**: 609–613.

Gee, H. (2000). *Deep Time: Cladistics, the Revolution in Evolution*. London, Fourth Estate.

Ghiselin, M.T. (1966). On semantic pitfalls of biological adaptation. *Philos. Sci.* **33**: 147–153.

Ghiselin, M.T. (2004). Mayr and Bock versus Darwin on genealogical classification. *J. Zool. Syst. Evol. Res.* **42**: 165–169.

Giribet, G. (2002). Relationships among metazoan phyla as inferred from 18S rRNA sequence data: A methodological approach. In *Molecular Systematics and Evolution: Theory and Practice*, (eds R. DeSalle, G. Giribet, and W. Wheeler), pp. 85–101. Basel, Birkhäuser Verlag.

Giribet, G. and Wheeler, G. (1999). On gaps. *Mol. Phylogenet. Evol.* **13**: 132–143.

Giribet, G., Distel, D.L., Polz, M., Sterrer, W. and Wheeler, W.C. (2000). Triploblastic relationships with emphasis on the acoelomates and the position of Gnathostomulida, Cycliophora, Plathelminthes, and Chaetognatha: A combined approach of 18S rDNA sequences and morphology. *Syst. Biol.* **49**: 539–562.

Giribet, G., Wheeler, W.C. and Muona, J. (2002). DNA multiple sequence alignments. In *Molecular Systematics and Evolution: Theory and Practice* (eds R. Desalle, G. Giribet and W. Wheeler), pp. 107–114. Basel, Birkhäuser Verlag.

Gladstein, D.S. (1997). Efficient incremental character optimization. *Cladistics* **13**: 21–26.

Gogarten, J.P., Doolittle, W.F. and Lawrence, J.G. (2002). Prokaryotic evolution in light of gene transfer. *Mol. Biol. Evol.* **19**: 2226–2238.

Goldman, N. (1990). Maximum likelihood inference of phylogenetic trees, with special reference to a Poisson process model of DNA substitution and to parsimony analyses. *Syst. Zool.* **39**: 345–361.

Goloboff, P.A. (1995). A revision of the south American spiders of the family Nemesiidae (Araneae, Mygalomorphae). Part I: species from Peru, Chile, Argentina, and Uruguay. *Bull. Am. Mus. Nat. Hist.* **224**: 1–189.

Goloboff, P.A. (1993a). Estimating character weights during tree search. *Cladistics* **9**: 83–91.

Goloboff, P.A. (1993b). *Nona*: a tree-searching program. Available at http://www.zmuc.dk/public/phylogeny/Nona-PeeWee/.

Goloboff, P.A. (1993c). *Pee-Wee*: Parsimony and Implied weights. Available at http://www.zmuc.dk/public/phylogeny/Nona-PeeWee/.

Goloboff, P.A. (1994). Character optimization and calculation of tree lengths. *Cladistics* **9**: 433–436.

Goloboff, P.A. (1995). *SPA: Sankoff Parsimony Analysis*, ver. 1.1. Available at http://www.zmuc.dk/public/phylogeny/Nona-PeeWee/.

Goloboff, P.A. (1996a). *PHAST: Phylogenetic Analysis for Sankovian Transformations*, ver. 1.1. Available at http://www.zmuc.dk/public/phylogeny/Nona-PeeWee/.

Goloboff, P.A. (1996b). Methods for faster parsimony analysis. *Cladistics* **12**: 199–220.

Goloboff, P.A. (1998b). Tree searches under Sankoff parsimony. *Cladistics* **14**: 229–238.

Goloboff, P.A. (1999). Analyzing large data sets in reasonable times: Solutions for composite optima. *Cladistics* **15**: 415–428.

Goloboff, P.A. (2003). Parsimony, likelihood, and simplicity. *Cladistics* **19**: 91–103.

Goloboff, P.A. and Farris, J.S. (2001). Methods for quick consensus estimation. *Cladistics* **17**: S26–S34.

Goloboff, P.A., Wheeler, W. and Pol, D. (2003a). Parallel TNT. *Cladistics* **19**: 152 (in Muona, J. (2003). Abstracts of the 21st annual meeting of the Willi Hennig society. *Cladistics* **19**: 148–163.)

Goloboff, P., Farris, J., Källersjö, M., Oxelmann, B., Ramírez, M. and Szumik, C. (2003b). Improvements to resampling measures of group support. *Cladistics* **19**: 324–332.

Goloboff, P., Farris, J. and Nixon, K. (2004). *T.N.T.: Tree Analysis Using New Technology*. Available at www.zmuc.dk/public/phylogeny/tnt.

Goodman, M., Czelusniak, J., Moore, G.W., Romero-Herrera, A.E. and Matsuda, G. (1979). Fitting the gene lineage into its species lineage, a parsimony strategy illustrated by cladograms constructed from globin sequences. *Syst. Zool.* **28**: 132–163.

Goudge, T.A. (1961). *The Ascent of Life. A Philosophical Study of the Theory of Evolution.* Toronto, University of Toronto Press. [1967 reprint]

Grant, T. (2002). Testing methods: The evaluation of discovery operations in evolutionary biology. *Cladistics* **18**: 94–111.

Grant, T. and Kluge, A.G. (2003). Data exploration in phylogenetic inference: Scientific, heuristic, or neither. *Cladistics* **19**: 379–418.

Grant, T. and Kluge, A.G. (2004). Transformation series as an ideographic character concept. *Cladistics* **20**: 23–31.

Greene, B. (2004). *The Fabric of the Cosmos. Space, Time, and the Texture of Reality.* New York, A.A. Knopf.

Greuter, W., McNeill, J., Barrie, F.R., Burdet, H.M., Demoulin, V., Filgueiras, T.S., Nicolson, D.H., Silva, P.C., Skog, J.E., Trehane, P., *et al.* (2000). *International Code of Botanical Nomenclature (St. Louis Code). Regnum Vegetabile* 138. Königstein, Koeltz Scientific Books.

Gu, J. and Gu, X. (2003). Induced gene expression in human brain after the split from chimpanzee. *Trends Genet.* **19**: 63–65.

Gusfield, D. (1997). *Algorithms on Strings, Trees, and Sequences: Computer Science and Computational Biology.* Cambridge, Cambridge University Press.

Hacking, I. (1965). *The Logic of Statistical Inference.* Cambridge, Cambridge University Press.

Hartigan, J.A. (1973). Minimum mutation fits to a given tree. *Biometrics* **29**: 53–65.

Harvey, P.H. and Pagel, M.D. (1991). *The Comparative Method in Evolutionary Biology.* New York, Oxford University Press.

Hasegawa, M. and Kishino, H. (1989). Confidence limits on the maximum-likelihood estimate of the hominoid tree from mitochondrial-DNA sequences. *Evolution* **43**: 672–677.

Hastings, W.K. (1970). Monte Carlo sampling methods using Markov chains and their applications. *Biometrika* **57**: 97–109.

Hedges, S.B. (2002). The origin and evolution of model organisms. *Nat. Rev. Genet.* **3**: 838–849.

Hein, J. (1989a). A new method that simultaneously aligns and reconstructs ancestral sequences for any number of homologous sequences when a phylogeny is given. *Mol. Biol. Evol.* **6**: 649–668.

Hein, J. (1989b). A tree reconstruction method that is economical in the number of pairwise comparisons used. *Mol. Biol. Evol.* **6**: 669–684.

Hein, J.J. (2001). An algorithm for statistical alignment of sequences related by a binary tree. In *Pacific Symposium on Biocomputing 2001.* (eds R.B. Altman, A.K. Dunker, L. Hunter, K. Lauderdale and T.E. Klein), vol. 6, pp. 179–190. Singapore, World Scientific.

Hein, J., Jensen, J.L. and Pedersen, C.N.S. (2003). Recursions for statistical multiple alignment. *Proc. Natl. Acad. Sci. USA* **100**: 14960–14965.

Hendy, M.D. and Penny, D. (1982). Branch and bound algorithms to determine minimal evolutionary trees. *Math. Biosci.* **59**: 277–290.

Hendy, M.D., Foulds, L.R. and Penny, D. (1980). Proving phylogenetic trees minimal with l-clustering and set partitioning. *Math. Biosci.* **51**: 71–88

Hennig, W. (1950). *Grundzüge einer Theorie der phylogenetischen Systematik.* Berlin, Deutscher Zentralverlag.

Hennig, W. (1966). *Phylogenetic Systematics.* Urbana, IL, University of Illinois Press.

Higgins, D.G. and Sharp, P.M. (1988). CLUSTAL: A package for performing multiple sequence alignment on a microcomputer. *Gene* **73**: 237–244.

Huber, K.T., Moulton, V. and Steel, M. (2002). *Four characters suffice to convexly define a phylogenetic tree.* Research Report UCDMA2002/12, Christchurch, New Zealand, Department of Mathematics and Statistics, University of Canterbury.

Huelsenbeck, J.P. and Lander, K.M. (2003). Frequent inconsistency of parsimony under a simple model of cladogenesis. *Syst. Biol.* **52**: 641–648.

Huelsenbeck, J.P. and Ronquist, F. (2001). MrBayes: Bayesian inference of phylogeny. *Bioinformatics* **17**: 754–755.

Huelsenbeck, J.P., Bull, J.J. and Cunningham, C.W. (1996). Combining data in phylogenetic analysis. *Trends Ecol. Evol.* **4**: 152–158.

Huelsenbeck, J., Ronquist, F., Nielsen, R. and Bollback, J. (2001). Bayesian inference of phylogeny and its impact on evolutionary biology. *Science* **294**: 2310–2314.

Huelsenbeck, J.P., Larget, B., Miller, R.E. and Ronquist, F. (2002). Potential applications and pitfalls of Bayesian inference of phylogeny. *Syst. Biol.* **51**: 673–688.

Huelsenbeck, J., Larget, B. and Alfaro, M. (2004). Bayesian phylogenetic model selection using reversible jump Markov chain Monte Carlo. *Mol. Biol. Evol.* **21**: 1123–1133.

Hull, D.L. (1967). Certainty and circularity in evolutionary taxonomy. *Evolution* **21**: 174–189.

Hull, D.L. (1974). *Philosophy of Biological Science.* Englewood Cliffs, NJ, Prentice-Hall.

Hull, D.L. (1975). Central subjects and historical narratives. *History and Theory: Studies Philos. Hist.* **14**: 253–274.

Hull, D.L. (1977). The ontological status of species as evolutionary units. In *Foundational Problems in Special Sciences* (eds R. Butts and J. Hintikka), pp. 91–102. Dordrecht, D. Reidel Pub. Co.

Hull, D.L. (1981). Historical narratives and integrating explanations. In *Pragmatism and Purpose: Essays Presented to Thomas A. Goudge* (eds L.W. Sumner, J.G. Slater and F. Wilson), pp. 172–188, 308–310. Toronto, University of Toronto Press.

Hull, D.L. (1982). Exemplars and scientific change. *Proc. Biennial Mtg. Phil. Sci. Assoc.* **2**: 479–503.

Hull, D.L. (1989). *The Metaphysics of Evolution.* Albany, NY, SUNY Press.

Huson, D.H. and Steel, M. (2004). Phylogenetic trees based on gene content. *Bioinformatics* **20**: 2044–2049.

Huynen, M.A. and Bork, P. (1998). Measuring genome evolution. *Proc. Natl. Acad. Sci. USA* **95**: 5849–5856.

ICZN (1999). *International Code of Zoological Nomenclature, 4th Edn.* London, International Trust for Zoological Nomenclature.

Jacob, F., Perrin, D., Sanchez, C. and Monod, J. (1960). L'Operon: groupe de genes a expression coordonee par un operateur. *C. R. Seance Acad. Sci.* **250**: 1727–1729.

Jenner, R.A. (2004). The scientific status of metazoan cladistics: why current reserach practice must change. *Zool. Scripta* **33**: 293–310.

Jermann, T.M., Opitz, J.G., Stackhouse, J. and Benner, S.A. (1995). Reconstructing the evolutionary history of the artiodactyl ribonuclease superfamily. *Nature* **374**: 57–59.

Jiang, T.L. and Lawler, E.L. (1994). Aligning sequences via an evolutionary tree: Computational complexity and approximation. In *Proceedings of the 26th ACM Symposium on the Theory of Computing*, pp. 760–769. New York, ACM.

Källersjö, M., Farris, J.S., Chase, M.W., Bremer, B., Fay, M.F., Humphries, C.J., Petersen, G., Seberg, O. and Bremer, K. (1998). Simultaneous parsimony jackknife analysis of 2538 *rbcL* DNA sequences reveals support for major clades of green plants, land plants, seed plants and flowering plants. *Plant Syst. Evol.* **213**: 259–287.

Källersjö, M., Albert, V.A. and Farris, J.S. (1999). Homoplasy *increases* phylogenetic structure. *Cladistics* **15**: 91–93.

Kapitonov, V.V. and Jurka, J. (1999). The long terminal repeat of an endogenous retrovirus induces alternative splicing and encodes an additional carboxy-terminal sequence in the human leptin receptor. *J. Mol. Evol.* **48**: 248–251.

Katinka, M.D., Duprat, S., Cornillot, E., Metenier, G., Thomarat, F., Prensier, G., Barbe, V., Peyretaillade, E., Brottier, P., Wincker, P. *et al.* (2001). Genome sequence and gene compaction of the eukaryote parasite *Encephalitozoon cuniculi. Nature* **414**: 450–453.

Kidd, K.K. and Sgaramella-Zonta, L.A. (1971). Phylogenetic analysis: Concepts and methods. *Am. J. Hum. Genet.* **23**: 235–252.

Kim, J. (1996). General inconsistency conditions for maximum parsimony: Effects of branch lengths and increasing numbers of taxa. *Syst. Biol.* **45**: 363–374.

Kimura, M. and Crow, J. (1964). The number of alleles that can be maintained in a finite population. *Genetics* **49**: 725–738.

Kjer, K.M. (1995). Use of rRNA secondary structure in phylogenetic studies to identify homologous positions: An example of alignment and data presentation from the frogs. *Mol. Phylogenet. Evol.* **4**: 314–330.

Kluge, A.G. (1988). Parsimony in vicariance biogeography: A quantitative method and a Greater Antillean example. *Syst. Zool.* **37**: 315–328.

Kluge, A.G. (1989). A concern for evidence and a phylogenetic hypothesis of relationships among *Epicrates* (Boidae, Serpentes). *Syst. Zool.* **38**: 7–25.

Kluge, A.G. (1997a). Testability and the refutation and corroboration of cladistic hypotheses. *Cladistics* **13**: 81–96.

Kluge, A.G. (1997b). Sophisticated falsification and research cycles: Consequences for differential character weighting in phylogenetic systematics. *Zool. Scripta* **26**: 349–360.

Kluge, A.G. (1999). The science of phylogenetic systematics: Explanation, prediction, and test. *Cladistics* **15**: 429–436.

Kluge, A.G. (2001a). Parsimony with and without scientific justification. *Cladistics* **17**: 199–210.

Kluge, A.G. (2001b). Philosophical conjectures and their refutation. *Syst. Biol.* **50**: 322–330.

Kluge, A.G. (2002). Distinguishing "or" from "and" and the case for historical identification. *Cladistics* **18**: 585–593.

Kluge, A.G. (2003a). On the deduction of species relationships: A précis. *Cladistics* **19**: 233–239.

Kluge, A.G. (2003b). The repugnant and the mature in phylogenetic inference: A temporal similarity and historical identity. *Cladistics* **19**: 356–368.

Kluge, A.G. (2004). On total evidence: For the record. *Cladistics* **20**: 205–207.

Kluge, A.G. (2005). Taxonomy in theory and practice, with arguments for a new phylogenetic system of taxonomy. In *Ecology and Evolution in the Tropics: a Herpetological Perspective* (eds M.A. Donnelly, B.I. Crother, C. Guyer, M.H. Wake and M.E. White), pp. 7–47. Chicago, University of Chicago Press.

Kluge, A.G. and Farris, J.S. (1969). Quantitative phyletics and the evolution of Anurans. *Syst. Zool.* **18**: 1–32.

Kluge, A.G. and Farris, J.S. (1999). Taxic homology = overall similarity. *Cladistics* **15**: 205–212.

Koonin, E.V., Aravind, L. and Kondrashov, A.S. (2000). The impact of comparative genomics on our understanding of evolution. *Cell* **101**: 573–576.

Koonin, E.V., Makarova, K.S. and Aravind, L. (2001). Horizontal gene transfer in prokaryotes: Quantification and classification. *Annu. Rev. Microbiol.* **55**: 709–742.

Koonin, E.V. and Galperin, M.Y. (2002). *Sequence-Evolution-Function. Computational Approaches in Comparative Genomics.* Kluwer Academic Publishers, New York.

Koonin, E.V., Fedorova, N.D., Jackson, J.D., Jacobs, A.R., Krylov, D.M., Makarova, K.S., Mazumder, R., Mekhedov, S.L., Nikolskaya, A.N., Rao, B.S., *et al.* (2004). A comprehensive evolutionary classification of proteins encoded in complete eukaryotic genomes. *Genome Biol.* **5**: R7.

Korbel, J.O., Snel, B., Huynen, M.A. and Bork, P. (2002). SHOT: A web server for the construction of genome phylogenies. *Trends Genet.* **18**: 158–162.

Koshi, J.M. and Goldstein, R.A. (1996). Probabilistic reconstruction of ancestral protein sequences. *J. Mol. Evol.* **42**: 313–320.

Kruskal, J. (1983). An overview of sequence comparison. In *Time Warps, String Edits, and Macromolecules: The Theory and Practice of Sequence Comparison* (eds D. Sankoff and J. Kruskal), pp. 1–44. Stanford, CA, CSLI Publications (1999 reprint).

Kumar, S., Tamura, K. and Nei, M. (1993). MEGA: *Molecular Evolutionary Genetics Analysis, vers.* 1.01. University Park, PA, Pennsylvania State University.

Kunin, V. and Ouzounis, C.A. (2003). The balance of driving forces during genome evolution in prokaryotes. *Genome Res.* **13**: 1589–1594.

Lande, R. (1976). Natural selection and random genetic drift in phenotypic evolution. *Evolution* **30**: 314–334.

Larget, B. and Simon, D. (1999). Markov chain Monte Carlo algorithms for the Bayesian analysis of phylogenetic trees. *Mol. Biol. Evol.* **16**: 750–759.

Larson, A. and Losos, J.B. (1996). Phylogenetic systematics of adaptation. In *Adaptation* (eds M.R. Rose and G.V. Lauder), pp. 187–220. San Diego, CA, Academic Press.

Laudan, L. (1990). *Science and Relativism: Some Key Controversies in the Philosophy of Science.* Chicago, University of Chicago Press.

Laudan, R. (1990). What's so special about the past? In *Evolutionary Innovations* (ed. M. Nitechi), pp. 55–67. Chicago, University of Chicago Press.

Lauder, G.V., Leroi, A.M. and Rose, M.R. (1993). Adaptations and History. *Trends Ecol. Evol.* **8**: 294–297.

Lawrence, J. (1999). Selfish operons: The evolutionary impact of gene clustering in prokaryotes andeukaryotes. *Curr. Opin. Genet. Dev.* **9**: 642–648.

Lawrence, J.G. and Roth, J.R. (1996). Selfish operons: Horizontal transfer may drive the evolution of gene clusters. *Genetics* **143**: 1843–1860.

Le Cam, L. (1960). An approximation theorem for the Poisson binomial distribution. *Pacific J. Math.* **10**: 1181–1197.

Le Quesne, W. (1969). A method of selection of characters in numerical taxonomy. *Syst. Zool.* **18**: 201–205.

Lee, D.C. and Bryant, H.N. (1999). A reconsideration of the coding of inapplicable characters: Assumptions and problems. *Cladistics* **15**: 373–378.

Levesque, M., Shasha, D., Kim, W., Surette, M.G. and Benfey, P.N. (2003). Trait-to-gene. a computational method for predicting the function of uncharacterized genes. *Curr. Biol.* **13**: 129–133.

Lewis, P.O. (2001). A likelihood approach to estimating phylogeny from discrete morphological character data. *Syst. Biol.* **50**: 913–925.

Li, S. (1996). *Phylogenetic Tree Construction using Markov Chain Monte Carlo.* PhD Dissertation, Ohio State University, Columbus, OH.

Li, S., Pearl, D.K. and Doss, H. (2000). Phylogenetic tree construction using Markov chain Monte Carlo. *J. Am. Stat. Assoc.* **2000**: 493–508.

Liberles, D.A. (2001). Evaluation of methods for determination of a reconstructed history of gene sequence evolution. *Mol. Biol. Evol.* **18**: 2040–2047.

Liberles, D.A., Schreiber, D.R., Govindarajan, S., Chamberlin, S.G. and Benner, S.A. (2001). The Adaptive Evolution Database (TAED). *Genome Biol.* **2**(8): research0028.1– research0028.6.

Liberles, D.A., Thoren, A., von Heijne, G. and Elofsson, A. (2002). The use of phylogenetic profiles for gene predictions. *Curr. Genomics* **3**: 131–137.

Lidén, M. (1990). Replicators, hierarchy, and the species problem. *Cladistics* **6**: 183–186.

Lipscomb, D.L. (1992). Parsimony, homology, and the analysis of multistate characters. *Cladistics* **8**: 45–65.

Logsdon, Jr., J.M., Stoltzfus, A. and Doolittle, W.F. (1998). Molecular evolution: Recent cases of spliceosomal intron gain? *Curr. Biol.* **8**: R560–R63.

Logsdon, Jr., J.M., Tyshenko, M.G., Dixon, C., J, D.J., Walker, V.K. and Palmer, J.D. (1995). Seven newly discovered intron positions in the triose-phosphate isomerase gene: Evidence for the introns-late theory. *Proc. Natl. Acad. Sci. USA* **92**: 8507–8511.

Long, M. and Rosenberg, C. (2000). Testing the "proto-splice sites" model of intron origin: Evidence from analysis of intron phase correlations. *Mol. Biol. Evol.* **17**: 1789–1796.

Lutzoni, F., Wagner, P., Reeb, V. and Zoller, S. (2000). Integrating ambiguously aligned regions of DNA sequences in phylogenetic analyses without violating positional homology. *Syst. Biol.* **49**: 628–651.

Lynch, M. and Richardson, A.O. (2002). The evolution of spliceosomal introns. *Curr. Opin. Genet. Dev.* **12**: 701–710.

Maddison, D.R. (1991). The discovery and importance of multiple islands of most-parsimonious trees. *Syst. Zool.* **40**: 315–328.

Maddison, D.R. and Maddison, W.P. (2001). *MacClade 4: Analysis of Phylogeny and Character Evolution* (incl. vers. 4.03). Sunderland, MA, Sinauer Associates.

Maddison, D.R., Swofford, D.L. and Maddison, W.P. (1997). NEXUS: An extensible file format for systematic information. *Syst. Biol.* **46**: 590–621.

Maddison, W.P. (1991). Squared-change parsimony reconstructions of ancestral states for continuous-valued characters on a phylogenetic tree. *Syst. Zool.* **40**: 304–314.

Maddison, W.P. (1993). Missing data versus missing characters in phylogenetic analysis. *Syst. Biol.* **42**: 576–581.

Maddison, W.P. and Maddison, D.R. (1992). *MacClade 3: Analysis of Phylogeny and Character Evolution* (incl. vers. 3.04). Sunderland, MA, Sinauer Associates.

Mallatt, J. and Winchell, C.J. (2002). Testing the new animal phylogeny: First use of combined large-subunit and small-subunit rRNA gene sequences to classify the protostomes. *Mol. Biol. Evol.* **19**: 289–301.

Marchionni, M. and Gilbert, W. (1986). The triosephosphate isomerase gene from maize: Introns antedate the plant-animal divergence. *Cell* **46**: 133–141.

Marcotte, E.M., Pellegrini, M., Thompson, M.J., Yeates, T.O. and Eisenberg, D.A. (1999). A combined algorithm for genome-wide prediction of protein function. *Nature* **402**: 83-86.

Martin C. and Paz-Ares J. (1997). MYB transcription factors in plants. *Trends Plant Sci.* **13**: 67–73.

Mau, B. (1996). *Bayesian Phylogenetic Inference via Markov Chain Monte Carlo Methods.* PhD Dissertation, University of Wisconsin, Madison, WI.

Mau, B. and Newton, M. (1997). Phylogenetic inference for binary data on dendrograms using Markov chain Monte Carlo. *J. Comput. Graph. Stat.* **6**: 122–131.

Mau, B., Newton, M. and Larget, B. (1999). Bayesian phylogenetic inference via Markov chain Monte Carlo methods. *Biometrics* **55**: 1–12.

Mayr, E. and Bock, W.J. (2002). Classification and other ordering systems. *J. Zool. Syst. Evol. Res.* **40**: 169–194.

McAllister, J.W. (1996). *Beauty and Revolution in Science.* Ithaca, NY, Cornell University Press.

McAllister, J.W. (2000). Unification of theories. In *A Companion to the Philosophy of Science* (ed. W.H. Newton-Smith), pp. 537–539. Oxford, Blackwell Publishing.

McDade, L.A. (1992). Hybrids and phylogenetic systematics II. The impact of hybrids on cladistic analysis. *Evolution* **46**: 1329–1346.

Messier, W. and Stewart, C.B. (1997). Episodic adaptive evolution of primate lysozymes. *Nature* **385**: 151–154.

Metropolis, N., Rosenbluth, A.W., Rosenbluth, M.N., Teller, A.H. and Teller, E. (1953). Equations of state calculations by fast computing machines. *J. Chem. Phys.* **21**: 1087–1091.

Mickevich, M.F. and Farris, J.S. (1981). *PHYSYS: Phylogenetic Analysis System.* Published by the authors.

Miklós, I., Lunter, A. and Holmes, I. (2004). A "long indel" model for evolutionary sequence alignment. *Mol. Biol. Evol.* **21**: 529–540.

Mindell, D.P. and Thacker, C.E. (1996). Rates of molecular evolution: Phylogenetic issues and applications. *Annu. Rev. Ecol. Syst.* **27**: 279–303.

Mirkin, B.G., Fenner, T.I., Galperin, M.Y. and Koonin, E.V. (2003). Algorithms for computing parsimonious evolutionary scenarios for genome evolution, the last universal common ancestor and dominance of horizontal gene transfer in the evolution of prokaryotes. *BMC Evol. Biol.* **3**: 2.

Mishler, B.D. (1994). Cladistic analysis of molecular and morphological data. *Am. J. Phys. Anthropol.* **94**: 143–156.

Mishler, B.D. (1999). Getting rid of species? In *Species: New Interdisciplinary Essays* (ed. R. Wilson), pp. 307–315. Cambridge, MA, MIT Press.

Mishler, B.D. (2000). Deep phylogenetic relationships among "plants" and their implications for classification. *Taxon* **49**: 661–683.

Mishler, B.D. and Brandon, R.N. (1987). Individuality, pluralism, and the phylogenetic species concept. *Biol. Phil.* **2**: 397–414.

Mishler, B.D. and De Luna, E. (1991). The use of ontogenetic data in phylogenetic analyses of mosses. *Adv. Bryol.* **4**: 121–167.

Mishler, B.D. and Theriot, E. (2000a). The phylogenetic species concept (*sensu* Mishler and Theriot): Monophyly, apomorphy, and phylogenetic species concepts. In *Species Concepts and Phylogenetic Theory: A Debate.* (eds Q.D. Wheeler and R. Meier), pp. 44–54. New York, Columbia University Press.

Mishler, B.D. and Theriot, E.G. (2000b). A critique from the Mishler and Theriot phylogenetic species concept perspective: Monophyly, apomorphy, and phylogenetic species concepts. In *Species Concepts and Phylogenetic Theory: A Debate* (eds Q.D. Wheeler and R. Meier), pp. 133–145. New York, Columbia University Press.

Mishler, B.D. and Theriot, E.G. (2000c). A defense of phylogenetic species concept (*sensu* Mishler and Theriot): Monophyly, apomorphy, and phylogenetic species concepts. In *Species Concepts and Phylogenetic Theory: A Debate* (eds Q.D. Wheeler and R. Meier), pp. 179–184. New York. Columbia University Press.

Modrek, B. and Lee, C.J. (2003). Alternative splicing in the human, mouse, and rat genomes is associated with an increased frequency of exon creation and/or loss. *Nat. Genet.* **34**: 177–180.

Moilanen, A. (1999). Searching for most parsimonious trees with simulated evolutionary optimization. *Cladistics* **15**: 39–50.

Moilanen, A. (2001). Simulated evolutionary optimization and local search: Introduction and application to tree search. *Cladistics* **17**: S12–S25.

Montague, M.G. and Hutchison, 3rd., C.A. (2000). Gene content phylogeny of herpesviruses. *Proc. Natl. Acad. Sci. USA* **97**: 5334–5339.

Moran, N.A. (2002). Microbial minimalism: Genome reduction in bacterial pathogens. *Cell* **108**: 583–586.

Moret, B.M.E., Wang, L.S., Warnow, T. and Wyman, S. (2001). New approaches for reconstructing phylogenies based on gene order. Proc. 9th Int'l Conf. on Intelligent Systems for Molecular Biology ISMB-2001, *Bioinformatics* **17**: S165-S173.

Moret, B.M.E., Tang, J., Wang, L.S. and Warnow, T. (2002). Steps toward accurate reconstruction of phylogenies from gene-order data. *J. Comput. Syst. Sci.* **65**(3): 508–525.

Morgenstern, B. (2004). DIALIGN: Multiple DNA and protein sequence alignment at BiBiServ. *Nucleic Acids Res.* **32**: W33–W36.

Morgenstern, B., Dress, A. and Werner, T. (1996). Multiple DNA and protein sequence alignment based on segment-to-segment comparison. *Proc. Natl. Acad. Sci. USA* **93**: 12098–12103.

Moritz, C. (2002). Strategies to protect biological diversity and the processes that sustain it. *Syst. Biol.* **51**: 238–254.

Mossel, E. and Steel, M. (2004a). A phase transition for a random cluster model on phylogenetic trees. *Math. Biosci.* **187**: 189–203.

Mossel, E. and Steel, M. (2005). How much can evolved characters tell us about the tree that generated them? In *Mathematics of Evolution and Phylogeny* (ed. O. Gascuel). Oxford, Oxford University Press.

Murata, M., Richardson, J. and Sussman, J. (1985). Simultaneous comparison of three protein sequences. *Proc. Natl. Acad. Sci. USA* **82**: 3073–3077.

Mushegian, A.R. and Koonin, E.V. (1996). Gene order is not conserved in bacterial evolution. *Trends Genet.* **12**: 289–290.

Mushegian, A.R., Garey, J.R., Martin, J. and Liu, L.X. (1998). Large-scale taxonomic profiling of eukaryotic model organisms: A comparison of orthologous proteins encoded by the human, fly, nematode, and yeast genomes. *Genome Res.* **8**: 590–598.

Myllykallio, H., Lipowski, G., Leduc, D., Filee, J., Forterre, P. and Liebl, U. (2002). An alternative flavin-dependent mechanism for thymidylate synthesis. *Science* **297**: 105–107.

Nadeau, J.J. and Taylor, B.A. (1984). Lengths of chromosome segments conserved since divergence of man and mouse. *Proc. Natl. Acad. Sci. USA* **81**: 814–818.

Naylor, G.J.P. and Adams D.C. (2001). Are the fossil data really at odds with the molecular data? Morphological evidence for *Cetartiodactyla* phylogeny reexamined. *Syst. Biol.* **50**: 444–453.

Naylor, G.J.P. and Adams, D.C. (2003). Total evidence versus relevant evidence: A response to O'Leary *et al.* (2003). *Syst. Biol.* **52**: 864–865.

Needleman, S.B. and Wunsch, C.D. (1970). A general method applicable to the search for similarities in the amino acid sequence of two proteins. *J. Mol. Biol.* **48**: 443–453.

Neff, N.A. (1986). A rational basis for *a priori* character weighting. *Syst. Zool.* **35**: 102–109.

Nei, M. and Kumar, S. (2001). *Molecular Evolution and Phylogenetics*. Oxford, Oxford University press.

Nelson, G. and Platnick, N.I. (1981). *Systematics and Biogeography: Cladistics and Vicariance*. New York, Columbia University Press.

Newton, M., Mau, B. and Larget, B. (1999). Markov chain Monte Carlo for the Bayesian analysis of evolutionary trees from aligned molecular sequences. In *Statistics in Molecular Biology and Genetics*, vol. 33 (ed. F. Seillier-Moiseiwitsch), pp. 143–162. Bethesda, MD, Institute of Mathematical Statistics.

Neyman, J. (1971). Molecular studies of evolution: A source of novel statistical problems. In *Statistical Decision Theory and Related Topics* (eds S. Gupta and J. Yackel), pp. 1–27. New York, Academic Press.

Nixon, K.C. (1999). The parsimony ratchet, a new method for rapid parsimony analysis. *Cladistics* **15**: 407–414.

Nixon, K.C. (2002). *WinClada*, vers. 1.00.08. Published by the author, Ithaca, New York (distributed through www.cladistics.org).

Nixon, K.C. and Carpenter, J.M. (1993). On outgroups. *Cladistics* **9**: 413–426.

Nixon, K.C. and Little, D.P. (2004). The use of optimality criteria in DNA sequence data and its application in a new computer program (Abstract, 22nd annual meeting of the Willi Hennig Society, New York 2003). *Cladistics* **20**: 90–91.

Nixon, K.C. and Wheeler, Q.D. (1990). An amplification of the phylogenetic species concept. *Cladistics* **6**: 211–223.

Nolan, D. (1997). Quantitative parsimony. *Br. J. Phil. Sci.* **48**: 329–343.

Notredame, C. (2002). Recent progress in multiple sequence alignment: A survey. *Pharmacogenomics* **3**: 1–14.

Notredame, C., Holm, L. and Higgins, D.G. (1998). COFFEE: An objective function for multiple sequence alignments. *Bioinformatics* **14**: 407–422.

Notredame, C., Higgins, D.G. and Heringa, J. (2000). T-Coffee: A novel method for fast and accurate multiple sequence alignment. *J. Mol. Biol.* **302**: 205–217.

O'Hara, R.J. (1988). Homage to Clio, or, toward an historical philosophy for evolutionary biology. *Syst. Zool.* **37**: 142–155.

Ochman, H., Lawrence, J.G. and Groisman, E.A. (2000). Lateral gene transfer and the nature of bacterial innovation. *Nature* **405**: 299–304.

Ochoterena, H. (2004). Independence of alignment and phylogenetic reconstruction and their optimality criteria (Abstract, 22nd annual meeting of the Willi Hennig Society, New York 2003). *Cladistics* **20**: 91.

Ohno, S. (1970). *Evolution by Gene Duplication*. New York, Springer-Verlag.

Oleksiak, M.F., Churchill, G.A. and Crawford, D.L. (2002). Variation in gene expression within and among natural populations. *Nat. Genet.* **32**: 261–266.

Padian, K. (2004). For Darwin, 'genealogy alone' did give classification. *J. Zool. Syst. Evol. Res.* **42**: 162–164.

Parzen, E. (1962). *Stochastic Processes*. San Francisco, Holden-Day.

Patterson, C. (1982). Morphological characters and homology. In *Problems of Phylogenetic Reconstruction*. (eds K.A. Joysey and A.E. Friday), pp. 21–74. New York, Academic Press.

Patterson, C. (1988). The impact of evolutionary theories on systematics. In *Prospects in Systematics* (ed. D.L. Hawksworth), pp. 59–91. Syst. Assoc. Spec. Vol. 36. Oxford, Clarendon Press.

Pellegrini, M., Marcotte, E.M., Thompson, M.J., Eisenberg, D. and Yeates, T.O. (1999). Assigning protein functions by comparative genome analysis: Protein phylogenetic profiles. *Proc. Natl. Acad. Sci. USA* **96**: 4285–4288.

Penny, D., Lockhart, P.J., Steel, M.A. and Hendy, M.D. (1994). The role of models in reconstructing evolutionary trees. In *Models in Phylogeny Reconstruction* (eds R.W. Scotland, D.J. Siebert and D.M. Williams), pp. 211–230. Systematics Assoc. Special vol. 52. Oxford, Clarendon Press.

Penny, D., Hendy, M.D., Lockhart, P.J. and Steel, M.A. (1996). Corrected parsimony, minimum evolution, and Hadamard conjugations. *Syst. Biol.* **45**: 596–606.

Peterson, K.J. and Eernisse, D.J. (2001). Animal phylogeny and the ancestry of bilaterians: Inferences from morphology and 18S rDNA gene sequences. *Evol. Dev.* **3**: 170–205.

Phillips, A., Janies, D. and Wheeler, W.C. (2000). Multiple sequence alignment in phylogenetic analysis. *Mol. Phylogenet. Evol.* **16**: 317–330.

Planet, P.J., DeSalle, R., Siddall, M., Bael, T., Sarkar, I.N. and Stanley, S.E. (2001). Systematic analysis of DNA microarray data: Ordering and interpreting patterns of gene expression. *Genome Res.* **11**: 1149–1155.

Platnick, N.I. (1979). Philosophy and the transformation of cladistics. *Syst. Zool.* **28**: 537–546.

Platnick, N.I. and Cameron, H.D. (1977). Cladistic methods in textual, linguistic, and phylogenetic analysis. *Syst. Zool.* **26**: 380–385.

Platnick, N.I., Griswold, C.E. and Coddington, J.A. (1991). On missing entries in cladistic analysis. *Cladistics* **7**: 337–343.

Pleijel, F. (1995). On character coding for phylogeny reconstruction. *Cladistics* **11**: 309–315.

Pol, D. and Siddall, M. (2001). Biases in maximum likelihood and parsimony: A simulation approach to a 10-taxon case. *Cladistics* **17**: 266–281.

Popper, K. (1957). *The Poverty of Historicism*. London, Routledge and Kegan Paul.

Popper, K. (1959). *The Logic of Scientific Discovery*. New York, Harper and Row [1968 edition].

Popper, K. (1962a). Some comments on truth and the growth of knowledge. In *Logic, Methodology and Philosophy of Science* (eds E. Nagel, P. Suppes and A. Tarski), pp. 285–292. Proc 1960 Internatl. Congress. Stanford, CA, Stanford University Press.

Popper, K. (1962b). *Conjectures and Refutations: The Growth of Scientific Knowledge*. London, Routledge and Kegan Paul.

Popper, K. (1968). *The Logic of Scientific Discovery*. New York, Harper Torchbooks.

Popper, K. (1979). *Objective Knowledge. An Evolutionary Approach*. New York, Oxford University Press.

Popper, K. (1980). Evolution. *New Scientist* **87**: 611.

Popper, K. (1983). *Realism and the Aim of Science*. London, Routledge.

Posada, D. and Crandall, K. (1998). MODELTEST: Testing the model of DNA substitution. *Bioinformatics* **14**: 817–818.

Posada, D. and Crandall, K. (2001a). Selecting models of nucleotide subsitution: An application to human immunodeficiency virus 1 (HIV-1). *Mol. Biol. Evol.* **18**: 897–906.

Pritchard, P.C.H. (1994). Cladistics: The great delusion. *Herpetol. Rev.* **25**: 103–110.

Posada, D. and Crandall, K. (2001b). Selecting the best-fit model of nucleotide substitution. *Syst. Biol.* **50**: 580–601.

Prömel, H.J. and Steger, A. (2000). A new approximation algorithm for the Steiner tree problem with performance ratio 5/3. *J. Algorithms* **36**: 89–101.

Pupko, T., Pe'er, I., Shamir, R. and Graur, D. (2000). A fast algorithm for joint reconstruction of ancestral amino acid sequences. *Mol. Biol. Evol.* **17**: 890–896.

Pupko, T., Pe'er, I., Hasegawa M., Graur, D. and Friedman, N. (2002). A branch-and-bound algorithm for the inference of ancestral amino-acid sequences when the replacement rate varies among sites: Application to the evolution of five gene families. *Bioinformatics* **18**: 1116–1123.

Quine, W.V. (1963). On simple theories of a complex world. *Synthese* **15**: 103–106.

Raff, R.A. (1996). *The Shape of Life: Genes, Development, and the Evolution of Animal Form.* Chicago, IL, University of Chicago Press.

Rain, J.-C., Selig, L., De Reuse, H., Battaglia, V., Reverdy, C., Simon, S., Lenzen, G., Petel, F., Wojcik, J., Schachter, V. et al. (2001). The protein-protein interaction map of Helicobacter pylori. *Nature* **409**: 211–215.

Rannala B. and Yang, Z. (1996). Probability distribution of molecular evolutionary trees: A new method of phylogenetic inference. *J. Mol. Evol.* **43**: 304–311.

Reichenbach, H. (1956). *The Direction of Time.* Berkeley, CA, University of California Press.

Remane, A. (1952). *Die Grundlagen des natürlichen Systems, der vergleichenden Anatomie und der Phylogenetik.* Leipzig, Akademische Verlagsgesellschaft Geest & Portig.

Resch, A., Xing, Y., Alekseyenko, A., Modrek, B. and Lee, C. (2004). Evidence for a subpopulation of conserved alternative splicing events under selection pressure for protein reading frame conservation. *Nucleic Acids Res.* **32**: 1261–1269.

Rexová, K., Frynta, D. and Zrzavý, J. (2003). Cladistic analysis of languages: Indo-European classification based on lexicostatistical data. *Cladistics* **19**: 120–127.

Rice, K.A., Donoghue, M.J. and Olmstead, R.G. (1997). Analyzing large data sets: *rbcL* 500 revisited. *Syst. Biol.* **46**: 554–563.

Rieppel, O.C. (1988). *Fundamentals of Comparative Biology.* Basel, Birkhäuser Verlag.

Rieppel, O. (2003). Semaphoronts, cladograms and the roots of total evidence. *Biol. J. Linn. Soc.* **80**: 167–186.

Rieppel, O. and Kearney, M. (2002). Similarity. *Biol. J. Linn. Soc.* **75**: 59–82.

Rieseberg, L.H. and Soltis, D.E. (1991). Phylogenetic consequences of cytoplasmic gene flow in plants. *Evol. Trends Plants* **5**: 5–84.

Rogers, J. (1997). On the consistency of the maximum likelihood estimation of phylogenetic trees from nucleotide sequences. *Syst. Biol.* **46**: 354–357.

Rogozin, I.B., Wolf, Y.I., Sorokin, A.V., Mirkin, B.G. and Koonin, E.V. (2003). Remarkable interkingdom conservation of intron positions and massive, lineage-specific intron loss and gain in eukaryotic evolution. *Curr. Biol.* **13**: 1512–1517.

Rokas, A. and Holland, P.W.H. (2000). Rare genomic changes as a tool for phylogenetics. *Trends Ecol. Evol.* **15**: 454–459.

Romero, I., Fuertes, A., Benito, M.J., Malpica, J. M., Leyva, A. and Paz-Ares, J. (1998). More than 80 R2R3-MYB regulatory genes in the genome of *Arabidopsis thaliana*. *Plant J.* **14**: 273–284.

Rossnes, R. (2004). *Ancestral Reconstruction of Continuous Characters and its Potential Application to Gene Expression and Alternative Splicing Analysis.* MSc Thesis, University of Bergen, Norway.

Roth, V.L. (1984). On homology. *Biol. J. Linn. Soc.* **22**: 13–29.

Roth, V.L. (1988). The biological basis of homology. In *Ontogeny and Systematics* (ed. Humpries, C.J.). New York, Columbia University Press.

Royall, R. (1997). *Statistical Evidence — a Likelihood Paradigm.* New York, Chapman and Hall.

Ruse, M. (1971). Narrative explanation and the theory of evolution. *Can. J. Phil.* **1**: 59–74.

Russell, B. (1948). *Human Knowledge, its Scope and Limits.* London, George Allen and Unwin.

Rutishauser, R. and Sattler, R. (1989). Complementary and heuristic value of contrasting models in structural biology. III. Case study on shoot-like "leaves" and leaf-like "shoots" in *Utricularia macrorhiza* and *Utricularia purpurea* (Lentibulariaceae). *Botanische Jahrbücher für Systematik* **111**: 121–137.

Rzhetsky, A. and Nei, M. (1992). A simple method for estimating and testing minimum-evolution trees. *Mol. Biol. Evol.* **9**: 945–967.

Rzhetsky, A., Ayala, F.J., Hsu, L.C., Chang, C. and Yoshida, A. (1997). Exon/intron structure of aldehyde dehydrogenase genes supports the "introns-late" theory. *Proc. Natl. Acad. Sci. USA* **94**: 6820–6825.

Saitou, N. and Nei, M. (1987). The neighbor-joining method: A new method for reconstructing phylogenetic trees. *Mol. Biol. Evol.* **4**: 406–425.

Salem, A.H., Ray, D.A., Xing, J., Callinan, P.A., Myers, J.S., Hedges, D.J., Garber, R.K., Witherspoon, D.J., Jorde, L.B. and Batzer, M.A. (2003). Alu elements and horminid phylogenetics. *Proc. Natl. Acad. Sci. USA* **100**: 12787–12791.

Salisbury, B.A. (1999). Strongest evidence: Maximum apparent phylogenetic signal as a new cladistic optimality criterion. *Cladistics* **15**: 137–149.

Salmon, W.C. (1966). *The Foundations of Scientific Inference.* Pittsburgh, PA, University of Pittsburgh Press.

Sanderson, M. and Hufford, L., eds (1996). *Homoplasy.* San Diego, CA, Academic Press.

Sanderson, M. and Kim, J. (2000). Parametric phylogenetics? *Syst. Biol.* **49**: 817–829.

Sanderson, M.J., Purvis, A. and Henze, C. (1998). Phylogenetic supertrees: Assembling the trees of life. *Trends Ecol. Evol.* **13**: 105–109.

Sankoff, D. (1975). Minimal mutation trees of sequences. *SIAM J. Appl. Math.* **28**: 35–42.

Sankoff, D. and Blanchette, M. (1998). Multiple genome rearrangement and breakpoint phylogeny. *J. Comp. Biol.* **5**: 555–570.

Sankoff, D. and Cedergren, R.J. (1983). Simultaneous comparison of three or more sequences related by a tree. In *Time Warps, String Edits, and Macromolecules. The Theory and Practice of Sequence Comparison* (eds D. Sankoff, and J. Kruskal), pp. 253–263. Stanford, CA, CSLI Publications (1999 reprint).

Sankoff, D. and Nadeau, J.H. (eds) (2000). *Comparative Genomics. Empirical and Analytical Approaches to Gene Order Dynamics, Map Alignment and the Evolution of Gene Families.* Dordrecht, Kluwer Academic Publishers.

Sankoff, D. and Rousseau, P. (1975). Locating the vertices of a Steiner tree in arbitrary space. *Math. Program.* **9**: 240–246.

Sankoff, D., Cedergren, R.J. and Lapalme, G. (1976). Frequency of insertion-deletion, transversion, and transition in the evolution of 5S ribosomal RNA. *J. Mol. Evol.* **7**: 133–149.

Sankoff, D., Morel, C. and Cedergren, R.J. (1973). Evolution of 5S RNA and the non-randomness of base replacement. *Nat. New Biol.* **245**: 232–234.

Sarkar, I.N., Planet, P.J., Bael, T.E., Stanley, S.E., Siddall, M., DeSalle, R. and Figurski, D.H. (2002). Characteristic attributes in cancer microarrays. *J. Biomed. Inform.* **35**: 111–122.

Schwikowski, B. and Vingron, M. (1997). The deferred path heuristic for the generalized tree alignment problem. *J. Comput. Biol.* **4**: 415–431.

Schwikowski, B. and Vingron, M. (2003). Weighted sequence graphs: Boosting iterated dynamic programming using locally suboptimal solutions. *Discr. Appl. Math.* **127**: 95–117.

Scriven, M. (1959). Explanation and prediction in evolutionary theory. *Science* **130**: 477–482.

Seitz, V., Ortiz García, S. and Liston, A. (2000). Alternative coding strategies and the inapplicable data coding problem. *Taxon* **49**: 47–54.

Sellers, P.H. (1974). An algorithm for the distance between two sequences. *J. Comb. Theory* **16**: 253–258.

Semple, C. and Steel, M. (2002). Tree reconstruction from multi-state characters. *Adv. Appl. Math.* **28**: 169–184.

Semple, C. and Steel, M. (2003). *Phylogenetics.* Oxford, Oxford University Press.

Shenkin, P.S., Erman, B. and Mastrandrea, L.D. (1991). Information-theoretical entropy as a measure of sequence variability. *Proteins Struct. Funct. Genet.* **11**: 297–313.

Sicheritz-Ponten, T. and Andersson, S.G. (2001). A phylogenomic approach to microbial evolution. *Nucleic Acids Res.* **29**: 545–552.

Siddall, M. (1998). Success of parsimony in the four-taxon case: Long branch repulsion by likelihood in the Farris zone. *Cladistics* **14**: 209–220.

Siddall, M.E. and Kluge, A.G. (1997). Probabilism and phylogenetic inference. *Cladistics* **13**: 313–336.

Sikes, D.S. and Lewis, P.O. (2001). *PAUPRat: PAUP Implementation of the Parsimony Ratchet.* Published by the authors (distributed through www.ucalgary.ca/~dsikes/sikes_lab.htm).

Simmons, M.P. (2004). Independence of alignment and tree search. *Mol. Phylogenet. Evol.* **31**: 874–879.

Simmons, M.P. and Ochoterena, H. (2000). Gaps as characters in sequence-based phylogenetic analyses. *Syst. Biol.* **49**: 369–381.

Simon, D. and Larget, B. (1998). *Bayesian Analysis in Molecular Biology and Evolution (BAMBE)*, version 1.01 beta. Pittsburgh, PA, Department of Mathematics and Computer Science, Duquesne University.

Simpson, G.G. (1964). *This View of Life: The World of an Evolutionist.* New York, Harcourt, Brace and World.

Sinsheimer, J., Lake, J.A. and Little, R.J.A. (1996). Bayesian hypothesis testing of four-taxon topologies using molecular sequence data. *Biometrics* **52**: 193–210.

Slatkin, M. and Maddison, W. (1989). A cladistic measure of gene flow inferred from the phylogenies of alleles. *Genetics* **123**: 603–613.

Slowinski, J.B. (1998). The number of multiple alignments. *Mol. Phyl. Evol.* **10**: 264–266.

Smith R.L. and Sytsma, K.J. (1990). Evolution of *Populus nigra* (sect. *Aigeiros*): Introgressive hybridization and the chloroplast contribution of *Populus alba* (sect. *Populus*). *Am. J. Bot.* **77**: 1176–1187.

Smith, T.F., Waterman, M.S. and Fitch, W.M. (1981). Comparative biosequence metrics. *J. Mol. Evol.* **18**: 38–46.

Smith, V.S., Page, R.D.M. and Johnson, K.P. (2004). Data incongruence and the problem of avian louse phylogeny. *Zool. Scripta* **33**: 239–259.

Smouse, P.E. and Li, W.-H. (1989). Likelihood analysis of mitochondrial restriction-cleavage patterns for the human-chimpanzee-gorilla trichotomy. *Evolution* **41**: 1162–1176.

Snel, B., Bork, P. and Huynen, M.A. (1999). Genome phylogeny based on gene content. *Nat. Genet.* **21**: 108–110.

Snel, B., Bork, P. and Huynen, M.A. (2002). Genomes in flux: The evolution of archaeal and proteobacterial gene content. *Genome Res.* **12**: 17–25.

Sober, E. (1980). Evolution, population thinking, and essentialism. *Phil. Sci.* **47**: 350–383.

Sober, E. (1981). The principle of parsimony. *Br. J. Phil. Sci.* **32**: 145–156.

Sober, E. (1983). Parsimony methods in systematics — philosophical issues. *Annu. Rev. Ecol. Syst.* **14**: 335–357.

Sober, E. (1985). A likelihood justification for parsimony. *Cladistics* **1**: 209–233.

Sober, E. (1986). Parsimony and character weighting. *Cladistics* **2**: 28–42.

Sober, E. (1988a). *Reconstructing the Past: Parsimony, Evolution and Inference.* Cambridge, MA, MIT Press.

Sober, E. (1988b). The conceptual relationship of cladistic phylogenetics and vicariance biogeography. *Syst. Zool.* **37**: 245–253.

Sober, E. (1993). *Philosophy of Biology.* San Francisco, CA, Westview Press.

Sober, E. (1994). Let's razor Ockham's Razor. In *Explanation and its Limits* (ed. D. Knowles), pp. 73–93. Suppl. Philosophy, Roy. Inst. Phil. **27**.

Sober, E. (1996). Parsimony and predictive equivalence. *Erkenntnis* **44**: 167–197.

Sober, E. (2002). Reconstructing ancestral character states — a likelihood perspective on cladistic parsimony. *The Monist* **85**: 156–176.

Sober, E. (2003). Parsimony. In *The Philosophy of Science: an Encyclopedia* (eds S. Sarkar and J. Pfeifer). London, Routledge.

Sober, E. (2004a). The contest between likelihood and parsimony. *Syst. Biol.* **53**: 644–653.

Sober, E. (in press): Is drift a serious alternative to natural selection as an explanation of complex adaptive traits? In *The Spandrels of San Marco Twenty-Five Years After* (ed. D. Walsh). Oxford, Oxford University Press.

Sober, E. and Steel, M. (2002). Testing the hypothesis of common ancestry. *J. Theor. Biol.* **218**: 395–408.

Sokal, R.R. (1986). Phenetic taxonomy: Theory and methods. *Annu. Rev. Ecol. Syst.* **17**: 423–442.

Sokal, R.R. and Camin, J.H. (1965). The two taxonomies: Areas of agreement and conflict. *Syst. Zool.* **14**: 176–195.

Soltis, D.E., Soltis, P.S., Chase, M.W., Mort, M.E., Albach, D.C., Zanis, M., Savolainen, V., Hahn, W.H., Hoot, S.B., Fay, M.F. *et al.* (2000). Angiosperm phylogeny inferred from 18S rDNA, *rbcL,* and *atpB* sequences. *Bot. J. Linn. Soc.* **133**: 381–461.

Sonnhammer, E.L. and Koonin, E.V. (2002). Orthology, paralogy and proposed classification for paralog subtypes. *Trends Genet.* **18**: 619–620.

Steel, M.A. (1993). Distributions on bicoloured binary trees arising from the principle of parsimony. *Discrete Appl. Math.* **41**: 245–261.

Steel, M. (2002). Some statistical aspects of the maximum parsimony method. In *Molecular Systematics and Evolution: Theory and Practice* (eds R. De Salle, R. Giribet and W. Wheeler), pp. 125–139. Basel, Birkhäuser Verlag.

Steel, M. and Penny, D. (2000). Parsimony, likelihood, and the role of models in molecular phylogenetics. *Mol. Biol. Evol.* **17**: 839–850.

Steel, M. and Penny, D. (2004). Two links between MP and ML under the Poisson model. *Applied Math. Lett.* (in press).

Steel, M., Penny, D. and Hendy, M. (1993). Parsimony can be consistent! *Syst. Biol.* **42**: 581–587.

Steel, M., Szekely, L. and Hendy, M. (1994). Reconstructing trees from sequences whose sites evolve at variable rates. *J. Comp. Biol.* **1**: 153–163.

Steffansson, P. (2004). *Inferring Duplication and Loss Events Using Soft Parsimony.* MSc Thesis, Royal Institute of Technology, Stockholm, Sweden.

Stevens, P.F. (1984). Homology and phylogeny: Morphology and systematics. *Syst. Bot.* **9**: 395–409.

Stevens, P.F. (1991). Character states, Morphological variation, and phylogenetic analysis: A review. *Syst. Bot.* **16**: 553–583.

Storm, C.E. and Sonnhammer, E.L. (2003). Comprehensive analysis of orthologous protein domains using the HOPS database. *Genome Res.* **13**: 2353–2362.

Strong, E. and Lipscomb, D. (1999). Character coding and inapplicable data. *Cladistics* **15**: 363–371.

Stuart, J.M., Segal, E., Koller, D. and Kim, S.K. (2003). A gene-coexpression network for global discovery of conserved genetic modules. *Science* **302**: 249–255.

Suzuki, Y., Glazko, G. and Nei, M. (2002). Overcredibility of molecular phylogenetics obtained by Bayesian phylogenetics. *Proc. Natl. Acad. Sci. USA* **99**: 16138–16143.

Swofford, D.L. (1984). *PAUP: Phylogenetic Analysis Using Parsimony.* Champaign, IL, Illinois Natural History Survey.

Swofford, D.L. (1985). *PAUP: Phylogenetic Analysis Using Parsimony,* vers. 2.4. Champaign, IL, Illinois Natural History Survey.

Swofford, D.L. (1990). *PAUP: Phylogenetic Analysis Using Parsimony,* vers. 3.0. (incl. vers. 3.0s). Champaign, Illinois Natural History Survey.

Swofford, D.L. (1991). When are phylogeny estimates from molecular and morphological data incongruent? In *Phylogenetic Analysis of DNA Sequences.* (eds M.M. Miyamoto and J. Cracraft), pp. 295–333. New York, Oxford University Press.

Swofford, D.L. (1993). *PAUP: Phylogenetic Analysis Using Parsimony,* vers. 3.1 (incl. vers. 3.1.1). Champaign, IL, Illinois Natural History Survey.

Swofford, D.L. (2002). *PAUP*: Phylogenetic Analysis Using Parsimony (*and other methods)*, vers. 4 (incl. vers. 4.0b10). Sunderland, MA, Sinauer Associates.

Swofford, D.L., Olsen, G.J., Waddell, P.J. and Hillis, D.M. (1996). Phylogenetic inference. In *Molecular Systematics*, 2nd edn (eds D.M. Hillis, C. Moritz and B.K. Marble), pp. 407–514. Sunderland, MA Sinauer Associates.

Swofford, D., Waddell, P., Huelsenbeck, J., Foster, P., Lewis, P. and Rogers, J. (2001). Bias in phylogenetic estimation and its relevance to the choice between parsimony and likelihoods methods. *Syst. Biol.* **50**: 525–539.

Tatusov, R.L., Koonin, E.V. and Lipman, D.J. (1997). A genomic perspective on protein families. *Science* **278**: 631–637.

Tatusov, R.L., Fedorova, N.D., Jackson, J.D., Jacobs, A.R., Kiryutin, B., Koonin, E.V., Krylov, D.M., Mazumder, R., Mekhedov, S.L., Nikolskaya, A.N., *et al.* (2003). The COG database: An updated version includes eukaryotes. *BMC Bioinformatics* **4**: 41.

Tehler, A., Little, D.P. and Farris, J.S. (2003). The full-length phylogenetic tree from 1 551 ribosomal sequences of chitinous fungi, *Fungi. Mycol. Res.* **107**: 901–916.

Tellgren, Å., Berglund, A.C., Savolainen, P., Janis, C.M. and Liberles, D.A. (2004). Myostatin rapid sequence evolution in ruminants predates domestication. *Mol. Phylogenet. Evol.* **33**: 782–790.

Thanaraj, T.A., Stamm, S., Clark, F., Riethoven, J.J., Le Texier, V. and Muilu, J. (2004). ASD: The Alternative Splicing Database. *Nucleic Acids Res.* **32**: D64-D69.

Thayer, H.S. (1953). *Newton's Philosophy of Nature: Selections from his Writings*. New York, Hafner Publishing Co.

Thompson, J.D., Higgins, D.G. and Gibson, T.J. (1994). CLUSTAL W: Improving the sensitivity of progresssive multiple alignment through sequence weighting, position-specific gap penalties and weight matrix choice. *Nucleic Acids Res.* **22**: 4673–4680.

Thorne, J.L., Kishino, H. and Felsenstein, J. (1991). An evolutionary model for maximum likelihood alignment of DNA sequences. *J. Mol. Evol.* **33**: 114–124.

Thorne, J.L., Kishino, H. and Felsenstein, J. (1992). Inching toward reality: An improved likelihood model of sequence evolution. *J. Mol. Evol.* **34**: 3–16.

Tierney, L. (1994). Markov chains for exploring posterior distributions. *Ann. Stat.* **22**: 1701–1786.

Tuffley, C. and Steel, M. (1997). Links between maximum likelihood and maximum parsimony under a simple model of site substitution. *Bull. Math. Biol.* **59**: 581–607.

Uddin, M., Wildman, D.E., Liu, G., Xu, W., Johnson, R.M., Hof, P.R., Kapatos, G., Grossman, L.I. and Goodman, M. (2004). Sister grouping of chimpanzees and humans as revealed by genome-wide phylogenetic analysis of brain gene expression profiles. *Proc. Natl. Acad. Sci. USA.* **101**: 2957–2962.

Vander Stappen, J., De Laet, J., Gama-López, S., Van Campenhout, S. and Volckaert, G. (2002). Phylogenetic analysis of *Stylosanthes* (Fabaceae) based on the internal transcribed spacer region (ITS) of nuclear ribosomal DNA. *Plant Syst. Evol.* **234**: 27–51.

Vingron, M. (1999). Sequence alignment and phylogeny construction. In *Mathematical Support for Molecular Biology* (eds M. Farach-Colton, F.S. Roberts, M. Vingron and M. Waterman), pp. 53–64. DIMACS Series in Discrete Mathematics and Theoretical Computer Science, vol. 47. Providence, RI, American Mathematical Society.

Vrana, P. and Wheeler, W. (1992). Individual organisms as terminal entities: Laying the species problem to rest. *Cladistics* **8**: 67–72.

Wagner, Jr., W.H. (1952). The fern genus *Diellia*: structure, affinities, and taxonomy. *Univ. Cal. Publ. Bot.* **26**: 1–212, pl. 1–21.

Wagner, Jr., W.H. (1961). Problems in the classification of ferns. In *Recent Advances in Botany*, vol. 1, pp. 841–844. Toronto, University of Toronto Press.

Walsh, D. (1979). Occam's razor: A principle of intellectual elegance. *Am. Phil. Q.* **16**: 241–244.

Wang L. and Jiang, T. (1994). On the complexity of multiple sequence alignment. *J. Comput. Biol.* **1**: 337–348.

Wang, L., Jiang, T. and Lawler, L. (1996). Approximation algorithms for tree alignment with a given phylogeny. *Algorithmica* **16**: 302–315.

Wang, L.S., Jansen, R., Moret, B., Raubeson, L. and Warnow, T. (2002). Fast phylogenetic methods for the analysis of genome rearrangement data: An empirical study. *Proceedings of the Pacifc Symposium on Biocomputing* (PSB 02), pp. 524–535. Singapore, World Scientific.

Watanabe, H., Mori, H., Itoh, T. and Gojobori, T. (1997). Genome plasticity as a paradigm of eubacteria evolution. *J. Mol. Evol.* **44**: S57–S64.

Waterston, R.H., Lindblad-Toh, K., Birney, E., Rogers, J., Abril, J.F., Agarwal, P., Agarwala, R., Ainscough, R., Alexandersson, M., An, P. *et al.* (2002). Initial sequencing and comparative analysis of the mouse genome. *Nature* **420**: 520–562.

Wheeler, Q.D. (1986). Character weighting and cladistic analysis. *Syst. Zool.* **35**: 102–109.

Wheeler, Q.D. and Meier, R. (eds) (2000). *Species Concepts and Phylogenetic Theory: a Debate*, pp. 179–184. New York, Columbia University Press.

Wheeler, W.C. (1994). Sources of ambiguity in nucleic acid sequence alignment. In *Molecular Ecology and Evolution: Approaches and Applications* (eds B. Schierwater,

B. Streit, G.P. Wagner and R. DeSalle), pp. 323–352. Basel, Birkhäuser Verlag.

Wheeler, W.C. (1996). Optimization alignment: The end of multiple sequence alignment in phylogenetics? *Cladistics* **12**: 1–9.

Wheeler, W.C. (1998). Alignment characters, dynamic programming and heuristic solutions. In *Molecular Approaches to Ecology and Evolution* (eds R. DeSalle and B. Schierwater), pp. 243–251. Basel, Birkhäuser Verlag.

Wheeler, W.C. (1999). Fixed character states and the optimization of molecular sequence data. *Cladistics* **15**: 379–385.

Wheeler, W.C. (2001a). Homology and the optimization of DNA sequence data. *Cladistics* **17**: S3–S11.

Wheeler, W. (2001b). Homology and DNA sequence data. In *The Character Concept in Evolutionary Biology* (ed. G.P. Wagner), pp. 303–317. San Diego, Academic Press.

Wheeler, W.C. (2002). Optimization Alignment: Down, up, error, and improvements. In *Techniques in Molecular Systematics and Evolution* (eds R. Desalle, G. Giribet and W. Wheeler), pp. 55–69. Basel, Birkhäuser Verlag.

Wheeler, W.C. (2003a). Implied alignment: A synapomorphy-based multiple-sequence alignment method and its use in cladogram search. *Cladistics* **19**: 261–268.

Wheeler, W.C. (2003b). Search-based optimization. *Cladistics*, **19**: 348–355.

Wheeler, W.C. (2003c). Iterative pass optimization of sequence data. *Cladistics* **19**: 254–260.

Wheeler, W.C. and Gladstein, D.S. (1994). MALIGN: A multiple sequence alignment program. *J. Hered.* **85**: 417–418.

Wheeler, W.C. and Hayashi, C.Y. (1998). The phylogeny of the extant chelicerate orders. *Cladistics* **14**: 173–192.

Wheeler, W., Gladstein, D. and De Laet, J. (2003). POY, ver. 3.0.11. Available at ftp://ftp.amah.org/pub/molecular/poy.

Wiley, E.O. (1975). Karl R. Popper, systematics, and classification: A reply to Walter Bock and other evolutionary taxonomists. *Syst. Zool.* **24**: 233–243.

Wiley, E.O. (1981). *Phylogenetics: the Theory and Practice of Phylogenetic Systematics.* New York, John Wiley and Sons.

Wilkinson, M. (1995). A comparison of two methods of character construction. *Cladistics* **11**: 297–308.

Wolf, Y.I., Rogozin, I.B., Grishin, N.V., Tatusov, R.L. and Koonin, E.V. (2001). Genome trees constructed using five different approaches suggest new major bacterial clades. *BMC Evol. Biol.* **1**: 8.

Wolf, Y.I., Rogozin, I.B., Grishin, N.V. and Koonin, E.V. (2002). Genome trees and the tree of life. *Trends Genet.* **18**: 472–479.

Wolf, Y.I., Rogozin, I.B. and Koonin, E.V. (2004). Coelomata and not Ecdysozoa: evidence from genome-wide phylogenetic analysis. *Genome Res.* **14**: 29–36.

Wolfe, K.H. and Sharp, P.M. (1993). Mammalian gene evolution: Nucleotide sequence divergence between mouse and rat. *J. Mol. Evol.* **37**: 441–456.

Woodger, J.H. (1929). *Biological principles: A critical study.* New York, Harcourt, Brace and Co.

Wrinch, D. and Jeffreys, H. (1921). On certain fundamental principles of scientific inquiry. *Phil. Mag.* **42**: 369–390.

Yang, Z. (1994). Maximum likelihood phylogenetic estimation from DNA sequences with variable rates over sites: Approximate methods. *J. Mol. Evol.* **39**: 306–314.

Yang, Z. (1996). Phylogenetic analysis using parsimony and likelihood methods. *J. Mol. Evol.* **39**: 294–307.

Yang, Z.H. (1998). Likelihood ratio tests for detecting positive selection and application to primate lysozyme evolution. *Mol. Biol. Evol.* **15**: 568–573.

Yang, Z.H. and Bielawski, B. (2000). Statistical methods for detecting molecular adaptation. *Trends Ecol. Evol.* **15**: 496–503.

Yang, Z. and Rannala, B. (1997). Bayesian phylogenetic inference using DNA sequences: A Markov chain Monte Carlo method. *Mol. Biol. Evol.* **14**: 717–724.

Yang, Z., Goldman, N. and Friday, A. (1995a). Maximum likelihood trees from DNA sequences: A peculiar statistical estimation problem. *Syst. Biol.* **44**: 384–399.

Yang, Z.H., Kumar, S. and Nei, M. (1995b). A new method of inference of ancestral nucleotide and amino acid sequences. *Genetics* **141**: 1641–1650.

Yeates, D. (1992). Why remove autapomorphies? *Cladistics* **8**: 387–389.

Index